T0406529

CSR, Sustainability, Ethics & Governance

In recent years the discussion concerning the relation between business and society has made immense strides. This has in turn led to a broad academic and practical discussion on innovative management concepts, such as Corporate Social Responsibility, Corporate Governance and Sustainability Management. This series offers a comprehensive overview of the latest theoretical and empirical research and provides sound concepts for sustainable business strategies. In order to do so, it combines the insights of leading researchers and thinkers in the fields of management theory and the social sciences – and from all over the world, thus contributing to the interdisciplinary and intercultural discussion on the role of business in society. The underlying intention of this series is to help solve the world's most challenging problems by developing new management concepts that create value for business and society alike. In order to support those managers, researchers and students who are pursuing sustainable business approaches for our common future, the series offers them access to cutting-edge management approaches.

* * *

CSR, Sustainability, Ethics & Governance is accepted by the Norwegian Register for Scientific Journals, Series and Publishers, maintained and operated by the Norwegian Social Science Data Services (NSD). It is also Abstracted and indexed in Research Papers in Economics (RePEc) / Scopus

René Schmidpeter • Reinhard Altenburger
Editors

Responsible Artificial Intelligence

Challenges for Sustainable Management

Editors
René Schmidpeter
Bern University of Applied Sciences (BFH)
Bern, Switzerland

Reinhard Altenburger
Department Business
IMC University of Applied Sciences Krems
Krems, Austria

ISSN 2196-7075 ISSN 2196-7083 (electronic)
CSR, Sustainability, Ethics & Governance
ISBN 978-3-031-09247-3 ISBN 978-3-031-09245-9 (eBook)
https://doi.org/10.1007/978-3-031-09245-9

This Springer imprint is published by the registered company Springer Nature Switzerland AG
The registered company address is: Gewerbestrasse 11, 6330 Cham, Switzerland

Foreword

With its contributions, this book takes us into the world of the Great Transformations that have occurred historically in the past 1000 years. There have been three of them and we are now experiencing the "Fourth Great Transformation."

It is the transformation into the world of complexity. It is driven by the technically possible networking of everything with everything. With it the great and seemingly eternal coordinators of mankind become meaningless, namely time and space. These will continue to exist, but they will lose their previous significance. Because one must travel no more to China or elsewhere, in order to become effective there. A short tap with the little finger on the Enter key is sufficient for this.

Artificial and Natural Intelligence Are Converging

The materials for natural and artificial intelligence are different, but their functional laws and their results are becoming increasingly similar. This is because they follow two laws of nature: the law of networking itself, and the law of simultaneity, as its twin.

The mentioned fourth transformation begins in the twentieth century with the end of the Second World War in 1945, with the discoveries and beginning of a completely new kind of science - cybernetics. It is closely connected with information and communication theory. Its founder is the English mathematician Alan Turing, who is considered the father of the computer. He gave the name "Enigma" to his famous deciphering machine.

Today's Cybernetics Goes Far beyond this

It is the science of self-regulating, self-directing, and self-organizing systems - as we naturally encounter them everywhere in nature. Organisms regulate themselves, just as ecological systems regulate themselves, and they do so intelligently, provided we do not interfere with their functioning in a dysfunctional way.

As early as 1948, a book by the mathematician Norbert Wiener was published. Its title is *Cybernetics*. And the subtitle, *Control and Communication in the Animal and the Machine*. Today, I would change the title slightly to "Control by Communication ...."

Artificial intelligence and its basic science, cybernetics, are the "parents" of digitization and modern communication, control, and guidance sciences. Like all sciences, these can be applied sensibly or criminally. The fact that the criminal side often dominates in the media cannot be blamed on cybernetics, but is due to the still widespread lack of knowledge about cybernetics.

As early as 1957, the company Digital Equipment was founded in the USA, Microsoft came into being in 1975 and Apple in 1976. But it was to take another 30 years until the first iPhone came onto the market in 2008.

Two New Laws of Nature for the Great Transformation21

The law of networking itself and - closely related to it - the law of simultaneity are the driving natural forces of today's Great Transformation21.

If we properly understand artificial intelligence and the harnessing of complexity that it enables, then entirely new opportunities will open up for the design of its application in almost all scientific disciplines, for example in medicine, engineering, the design and functioning of organizations, public administration, and the sciences themselves.

We can then also navigate in the unknown in a whole new way. Navigating is the art of the helmsman. In the simple case, navigating means determining the location, setting the destination, and steering the way there, largely independent of the causal forces that push us off the right path. The higher form of navigating is the ability to find our way in the unknown - when locations are uncertain, destinations are moving, and paths are convoluted. Another name for these functions is management in the twenty-first century.

Complexity Is Not Complication

The two are often confused. Complexity is the most important new, immaterial "raw material," which is the basis of natural and also artificial intelligence, of control and regulation, and of innovation and evolution.

With it new system and ecosystem types with self-capabilities emerge, as natural organisms have them, namely self-control, self-regulation, self-steering, and self-organization, so in short intelligence which we can thus transfer from the organisms to the organizations of our societies.

Complexity is the most important new raw material for intelligence and for information, and these are more important than time, space, and energy, which we have known for a long time. The proper use of complexity is becoming a core competency of the artificial intelligence of functioning organizations. We can use it to step outside the constraints of natural organisms. Artificial intelligence and corporate social responsibility are the key intangible raw materials for the self-capabilities of new system and ecosystem types.

Corporate social responsibility and ultimately all organizations in society depend on the intelligent use of complexity.

St. Gallen, Switzerland
21 March 2022

Fredmund Malik

Contents

Artificial Intelligence: Management Challenges and Responsibility

Reinhard Altenburger

> *AI will soon be like electricity—ubiquitous and indispensable. This gives rise to responsibility to ensure that AI can live up to its potential to be as a positive force.*
> *Lee (2018)*
> *The emergence of the age of AI has possibly created the greatest entrepreneurial opportunity in the history of civilization.*
> *Iansiti and Lakhani (2020)*
> *AI ethics is not necessarily about banning things; we also need a* positive *ethics: to develop a vision of the good life and the good society.*
> *Coeckelbergh (2020)*

Abstract AI will have a significant impact on all areas of our lives in the future and already influences us in the present in numerous areas of daily life. AI presents new challenges not only for data scientists but also for managers at all levels. On the one hand, there are new requirements that companies have to meet in the competitive environment, and on the other hand, there are a number of new questions that are associated with the use of AI. The development of use cases and the creation and design of data-based ecosystems are key challenges in the future competition in every industry. The forms of communication and interaction in and between organizations will change significantly and therefore require a timely examination of the consequences and requirements of AI.

R. Altenburger (✉)
IMC University of Applied Sciences Krems, Krems, Austria
e-mail: Reinhard.altenburger@fh-krems.ac.at

R. Schmidpeter, R. Altenburger (eds.), *Responsible Artificial Intelligence*, CSR, Sustainability, Ethics & Governance, https://doi.org/10.1007/978-3-031-09245-9_1

1 Challenges and Prospects of Artificial Intelligence

These two opening statements show the challenge and also the urgency of dealing with the opportunities and risks for companies, but also the social implications of AI at an early stage. AI will have a significant impact on all areas of our lives in the future and already influences us in the present in numerous areas of daily life. Every industry and every company should therefore ask itself what possibilities, opportunities, and risks exist as a result of AI. The public discussion on the topic of artificial intelligence is often dominated by images of AI with military applications, such as autonomous drones, or images that are conveyed in films such as *Matrix*, *Her*, or *Minority Report*. AI applications that have achieved outstanding success in games like chess, Go, or Jeopardy! have also received a lot of media attention and are cited in almost every AI discussion.

The promise of fast, accurate, repeatable, and cost-effective decisions with a quality approaching human-like intelligence has been a major driving force behind the rapid developments in AI in recent years. The combination of AI with developments in robotics, the Internet of things, chatbots, additive manufacturing (3D printing), nanotechnologies, and drones and, in the future, with quantum computers, for example, will have a significant impact on all sectors—from the automotive industry and financial service providers to healthcare institutions, but also on public administration and politics and nonprofit companies. However, the effects will not only affect large companies but also SMEs and start-ups to a large extent.

On the one hand, international studies show high expectations for the use of AI—for example, a study by the McKinsey Global Institute (2018) assumes an additional increase of 1.2% GDP worldwide per year (until 2030)—but also that many companies are hesitant in the implementation or development of concrete applications. In many companies, a high degree of uncertainty can be observed as to what impact AI will have on the current business model on the one hand, and what impact AI will have on work processes in the medium and long term on the other, and how employees and managers can best be prepared for the challenges posed by the use of AI. Recent research also shows that the application of AI-based decision-making can introduce and exacerbate a number of serious and often hidden biases and challenges to maintaining fairness, accountability, transparency, and consequently trust in AI-based decision-making (Shrestha et al., 2019). The basis for "Trusted AI" is its systematic, structured, and documented development and compliance with applicable technical and nontechnical standards and principles (European Commission, 2019)

2 The Impact on Managerial Decision-Making

AI presents new challenges not only for data scientists but also for managers at all levels. On the one hand, there are new requirements that companies have to meet in the competitive environment, and on the other hand, there are a number of new questions that are associated with the use of AI. For the management teams, challenges arise here to discuss suitable forms of decision-making (Shrestha et al., 2019). For example, questions arise such as: Where does AI decide on its own, where does it provide decision support, or how do we deal with the uncertainties and fears in the organization?

AI initiatives often face cultural and organization barriers. This could be the case when companies move from pilots to companywide programs or implementing solutions that radically change the entire customer journey (Fountaine et al., 2019). At present, companies and various industries are actively shaping developments. This requires the executives of these companies to actively participate in the current discussion on the responsibility that goes hand in hand with the current and medium-term developments. How different ethical standards and requirements are handled in the company requires the active involvement of managers in the development and implementation of AI.

Fears and uncertainties exist at different levels in the company due to, for example, threats to the integrity of the company, undermining of decision-making behavior, and lack of traceability of AI-supported decisions. Managers should also consider the impact of AI on the various areas of work from administration to production and logistics (Bughin et al., 2018), and on control and monitoring as well as possible discrimination, and in doing so also take seriously the uncertainties that can arise due to the non-traceability and non-transparency of the decisions and actions of AI. Therefore, clear rules and appropriate protective measures for employees are needed in any case.

However, participation in AI development also requires an appropriate knowledge of the fundamentals and current, global discussions on the topic of AI and responsibility to avoid mistakes that could potentially have a significant negative impact on the entire company. Executives are increasingly challenged with the tension of developing use cases and considering the ethical requirements of AI.

Regarding the collected data, it is often useful to distinguish between structured and unstructured data. Structured data is data that is organized according to predefined models (e.g., in a relational database), while unstructured data has no known organization (e.g., images or pieces of texts) (High-Level Expert Group, 2019). Through the connection and analysis of structured and unstructured data, as well as the learning capability of the applications, AI will have a much greater impact on the working and living environments of many people than other digital technologies. The use of AI will also have a significant impact on corporate strategies, business models, and competition within companies, often even radically changing them (Iansiti & Lakhani, 2020).

A large body of research shows that the use of AI-based decision-making can pose a number of serious and often hidden biases and challenges to the maintenance of fairness, accountability, transparency, and, consequently, trust in this technology (Shrestha et al., 2019). The McKinsey Global Institute (2018) sees the following factors as challenges to the application of AI in companies:

- Availability of training data
- Obtaining sufficiently large data sets
- Difficulty explaining results: it is often difficult to explain results from large data sets
- Complex neural network-based systems
- Difficulty of generalization
- Risk of bias

When AI makes decisions or contributes significantly to decision-making, the following problem areas should be considered (Shrestha et al., 2019): AI can be deceived to change decision outcomes—for example, by manipulating the data it uses as input, or through its design, e.g., by changing the weighting of predictors. The challenge here is to find suitable regulations and procedures for testing AI algorithms. Meanwhile, there is now some evidence that AI-based decisions amplify human biases in the available data.

The use of AI technologies must always be considered against the background of high energy and resource consumption, the potential savings, and the differences between the training and deployment phases of an AI system (van Wynsberghe, 2021).

3 Impact of AI on Corporate Strategy and Organization

The development of use cases and the creation and design of data-based ecosystems are key challenges in the future competition in every industry. The forms of communication and interaction in and between organizations will change significantly and therefore require a timely examination of the consequences and requirements of AI. The development of AI-based use cases is still very much in its infancy in many industries. Data-driven systems and the AI based on them can be the basis for new and innovative business models. Networks, exchange platforms, and new forms of cooperation—often between competitors—are needed in which different actors share their data expertise and common goals can be developed. There is great potential in the creation and development of so-called ecosystems, but it also requires innovative forms of cooperation.

The principles to be applied in the design of intelligent systems and an interdisciplinary and transparent consensus should be pursued, according to the High-Level Expert Group on AI. All actors in business, politics, and society must become aware of the ethical responsibility with regard to sustainable data economy (High-Level Expert Group, 2019).

The decision whether and how employees use a new technology can be explained by the theory of technology acceptance (Davis, 1989), which is based on two factors: the perceived usefulness of the technology and the perceived ease of use. A worker values a technology more if it is relevant to her/his tasks and produces high-quality results and the results can be easily demonstrated. Uncertainty about how AI will impact the future of work goes hand in hand with concerns about how AI might change the human condition (Howard, 2019). Research findings and current applications of automated decision-making systems, for example, show a very problematic track record in a variety of proven forms of discrimination (e.g., Molnar & Gill, 2018). This is because AI can be "fooled" into altering decision outcomes—for example, by manipulating the data it uses as input, or by its design (e.g., by changing the weighting of predictors)—or that AI-based decisions reinforce human biases in the available data (Shrestha et al., 2019).

For managers, numerous new questions arise which are relevant on the one hand for the strategic orientation of the company and the organizational structure and on the other hand for the avoidance of considerable damage, e.g., concerning the reputation and trust in the company:

- What are the implications of AI within our strategy? How can AI projects be carried out safely and efficiently? What prerequisites need to be created?
- What skills and organizational structure do we need to fully exploit the potential of AI?
- What legal, regulatory, and contractual foundations do we need to consider in the productive use of AI?
- What could AI lifecycle management look like that fits well into our existing processes?
- How should processes for selecting, implementing, and operating an AI technology infrastructure be defined?
- What measures are suitable for establishing cyber security and efficiently warding off adversary attacks on AI systems? (van Giffen et al., 2020)

Managers are also challenged to deal with the medium- and long-term personnel consequences of AI (Tambe et al. 2019). Which competencies are needed, where is the biggest gap to the current competencies in the company, how will the process of increased use of AI be designed, and which new business areas or products/services will be made possible by AI?

This raises the following questions, among others: What skills will be needed for AI projects in the next 3–5 years and how can the right employees be recruited? What potentials exist in the individual company divisions? Are these also recognized by those responsible?

The recognition of correlations in huge amounts of data resulting from structured and unstructured data is increasingly seen as an opportunity for solving social and environmental challenges that should contribute to the achievement of the Sustainable Development Goals (SDGs) (Vinuesa et al., 2020, World Economic Forum, 2018). On the other hand, the impact of AI systems on people and the environment is discussed in discourses on ethical and responsible AI (Coeckelbergh, 2020), as well as the question of what environmental impacts in terms of energy consumption and greenhouse gas emissions are caused by the AI systems themselves (e.g., in the training phase of AI).

4 Management Responsibility and Ethical Implications

However, as the use of artificial intelligence systems increases, so do concerns about how these systems exploit data. To address these concerns and advocate for ethical and responsible AI development and implementation, non-governmental organizations (NGOs), research centers, private companies, and government agencies have published more than 200 AI ethics principles and guidelines in the past. These principles and guidelines are intentionally provided as high-level abstract documents, as their application is case-, time-, and context-dependent. These high-level value statements in AI contribute to the formation of a moral background as they make explicit the link between values, ethics, and technologies (Hickok, 2021). Ethical aspects do not only concern issues of regulation, but also the design of AI and robotic systems, from the definition of their application to the details of their implementation. Ethics in AI is therefore very broad and revolves around fundamental design decisions and societal considerations (Bartneck et al., 2021).

The question of the morally/ethically "right" decision is evident in currently discussed issues, including autonomous driving. How difficult it is in critical situations to reach an international consensus on questions of decision-making through AI is shown, for example, by the MIT Moral Machine Project. Participants from different countries and cultures were asked how, in the event of a traffic accident that could no longer be prevented, a decision should be made as to which person or group of persons should be harmed. In autumn 2020 the question of the allocation of ventilators in the Corona crisis, when there is not enough equipment available, was discussed. The results showed different priorities in different countries, but are an important contribution to the further discussion of AI ethics (https://www.moralmachine.net/). These cultural differences pose another challenge when implementing AI solutions.

Several organizations and companies have already developed—mostly with the involvement of internal and external stakeholders—and formulated binding AI guidelines in recent years (for an overview see Hagendorff, 2020). Companies that have formulated and also communicated these guidelines or codes are, for example, Deutsche Telekom AG (Digital Ethics Guidelines on AI), Continental AG (Ethics Regulations for Artificial Intelligence), IBM (AI Ethics—IBM's multidisciplinary, multidimensional approach to trustworthy AI), or Bosch (AI Code of Ethics).

Throughout the AI lifecycle, it is important to ensure continuous collaboration between IT specialists who are developing the system and the people who will work with it. The changes in work processes, competencies, and responsibilities resulting from the use of AI should be discussed prior to implementation and possible resulting risks and organizational changes should be designed with stakeholders in a participatory process.

The involvement of scientists within the framework of research projects can provide management with valuable impulses, on the one hand, for adhering to internationally accepted standards and, on the other hand, for the further development of AI activities. The additional—often interdisciplinary—expertise supports reflection in the development process and the application of AI solutions and can also offer suggestions for model optimization during the development process.

The values of the company and especially of the executives (Benjamins, 2021) should be reflected in the development of AI solutions—this also requires appropriate forms of executive involvement in the development process and continuous evaluation of the systems. This concerns in particular the construction of decision-making systems, the embedding of the AI solution in the social context, and the ongoing evaluation of the decision-making system (Zweig et al., 2018).

References

Bartneck, C., Lütge, C., Wagner, A., & Welsh, S. (2021). *An introduction to ethics in robotics and AI*. Springer.

Benjamins, R. (2021). A choices framework for the responsible use of AI. *AI and Ethics, 1*, 49–53. https://doi.org/10.1007/s43681-020-00012-5

Bughin et. al. (2018). Mckinsey global institute: Skill shift automation and the future of the workforce. https://www.mckinsey.com/~/media/mckinsey/industries/public%20and%20social%20sector/our%20insights/skill%20shift%20automation%20and%20the%20future%20of%20the%20workforce/mgiskill-shift-automation-and-future-of-the-workforce-may-2018.pdf

Coeckelbergh, M. (2020). *AI ethics*. The MIT Press.

Davis, F. D. (1989). Perceived usefulness, perceived ease of use and user acceptance of information technology. *MIS Quarterly, 13*(3), 319–339.

European Commission. (2019). Ethics guidelines for trustworthy AI. https://ec.europa.eu/digital-single-market/en/news/ethics-guidelines-trustworthy-ai. Access 22 Jan 2022.

Fountaine, T., McCarthy, B., & Saleh, T. (2019). Building the AI-powered organization: technology isn't the biggest challenge: culture is. *Harvard Business Review: HBR, 97*, 62–73.

Hagendorff, T. (2020). The ethics of AI ethics. An evaluation of guidelines. *Minds & Machines, 30*, 99–120. https://doi.org/10.1007/s11023-020-09517-8

Hickok, M. (2021). Lessons learned from AI ethics principles for future actions. *AI and Ethics, 1*, 41–47. https://doi.org/10.1007/s43681-020-00008-1

High-Level Expert Group on Artificial Intelligence. (2019). Ethics guidelines for trustworthy AI. Source. https://ec.europa.eu/digital-single-market/en/news/ethics-guidelines-trustworthy-ai. Accessed 29 Oct 2021.

Howard, J. (2019). Artificial intelligence: Implications for the future of work. *American Journal of Industrial Medicine, 62*(11), 917–926.

Iansiti, M., & Lakhani, K. R. (2020). *Competing in the age of AI. Strategy and leadership when algorithms and networks run the world*. Harvard Business Review Press.

Lee, K. (2018). *AI superpowers: China, silicon valley, and the new world order*. Houghton Mifflin Harcourt.

McKinsey Global Institute. (2018, September). *Notes from the frontier modeling the impact of AI on the world economy* (Discussion Paper).

Molnar, P., & Gill, L. (2018). Bots at the gate: a human rights analysis of automated decision-making in Canada's immigration and refugee system.

Shrestha, Y. R., Ben-Menahem, S. M., & von Krogh, G. (2019). Organizational decision-making structures in the age of artificial intelligence. *California Management Review, 61*(4), 66–83.

Tambe, P., Cappelli, P., & Yakubovich, V. (2019). Artificial intelligence in human resources management: Challenges and a path forward. *California Management Review, 61*(4), 15–42.

van Giffen, B., Borth, D., & Brenner, W. (2020). Management von Künstlicher Intelligenz in Unternehmen. *HMD, 57*, 4–20.

van Wynsberghe, A. (2021). Sustainable AI: AI for sustainability and the sustainability of AI. *AI Ethics, 1*, 213–218.

Vinuesa, R., Azizpour, H., Leite, I., Balaam, M., Dignum, V., Domisch, S., Felländer, A., Langhans, S. D., Tegmark, M., & Nerini, F. F. (2020). The role of artificial intelligence in achieving the sustainable development goals. *Nature Communications, 11*, 233. https://doi.org/10.1038/s41467-019-14108-y

World Economic Forum. (2018). *Harnessing artificial intelligence for the earth*. In collaboration with PwC and Stanford Woods Institute for the Environment.

Zweig, K. A., Fischer, S., & Lischka, S. (2018). Wo Maschinen irren können. Fehlerquellen und Verantwortlichkeiten in Prozessen algorithmischer Entscheidungsfindung. Bertelsmann Stiftung. *Impuls Algorithmenethik, 4*.

Reinhard Altenburger is Professor for Strategic Management, Sustainable Management/CSR, and Innovation in the Department of Business of the IMC University of Applied Sciences Krems since 2009. The focus of his research is “CSR and Innovation” as well as “Innovations in Family Business” and the connection of social responsibility and corporate strategy. His actual research interest is in the field of artificial intelligence and corporate responsibility. Reinhard serves on the editorial board of “AI and Ethics.” Reinhard Altenburger has 15 years of work experience in the banking industry in the fields of strategic planning, retail strategy, and IT solutions. He has held numerous presentations at scientific conferences and also published and edited 11 books in his research field.

Artificial Intelligence: Companion to a New Human "Measure"?

René Schmidpeter and Christophe Funk

1 Artificial Intelligence Changes Our Society and Economy

In the current articles on artificial intelligence, one sees various images of man and society, from total surveillance of society and economy to the further development of human thinking in the form of new creativity, supported by the computing power of artifical intelligence. The legitimate question arises: Is artificial intelligence an opportunity for human liberation or a new golden cage that limits or even restricts human action? New digital products, such as the Meta-Verse, are merging the real with the digital worlds and deepening the bonds between humans and technology. Today's decisions on shaping the relationship between man and machine will also be relevant for our children and grandchildren. This makes it more important to discuss the relationship between man and machine in the broader triangle of "man-machine-economy." It is often the economic fields of application that influence the further development of technologies and their acceptance in society.

Therefore, the social responsibility of companies in the application field of AI is more than justified, especially because digital technologies are currently developing ever more dynamically. For example, it took the telephone about 75 years to reach 100 million users (Statista, 2017). Creating new systems, such as digital communication services, now takes few months. Moreover, worldwide, more than four billion people use social networks (We Are Social, 2018). Experts from the World Economic Forum predict that a large part of the world's population will have an online

R. Schmidpeter (✉)
Bern University of Applied Sciences (BFH), Bern, Switzerland
e-mail: schmidpeter@m3trix.de

C. Funk
Funk Business Coaching & Consulting, Bonn, Germany

R. Schmidpeter, R. Altenburger (eds.), *Responsible Artificial Intelligence*, CSR, Sustainability, Ethics & Governance, https://doi.org/10.1007/978-3-031-09245-9_2

profile in the next few years (World Economic Forum, 2018). This global spread of digital technologies is the basis of a global economic transformation, which also impacts geographically remote regions of the world.

We are also seeing more and more digital applications in the health system. Around half of Germans now use health apps on their smartphones or smartwatches (Bitkom, 2017). The digital health market is expected to grow to around 250 billion euros in Europe and around 1 trillion euros worldwide (Roland Berger, 2020). But schools also show an increasing affinity for digital teaching content and methods. Teachers and students think that new media will increase the attractiveness of school education (Bertelsmann Stiftung, 2017). Triggered in particular by the effects of the Corona pandemic, different dynamics towards digital learning methods and platforms can be seen in the education system. Furthermore, digital assistance robots and online systems are becoming increasingly popular in the private sector. The growth of digital applications is particularly evident in banking. In the financial markets, more than 3000 different cryptocurrencies are now vying for the favour of users. The value of these online currencies is increasing dynamically and has now reached over 200 billion dollars (Statista/coinmarketcap, 2019). In addition, over 75% of German citizens now do their banking online (Statista/coinmarketcap, 2019). A new trend in the industry is leaning towards the issue of certificates of authenticity and ownership, so-called non-fungible tokens (NFTs) (Handelsblatt 2021), which are expected to revolutionise digital payments and transactions once again (CGI, 2018).

The new mobile communications standard 5G, which has recently been launched, will expand digital markets exponentially and increasingly interconnect existing digital and analogue technologies. This will create a tightly meshed "Internet of things", "virtual realities", and new "man-machine" interfaces that will fundamentally redefine our lives and economy. These new worlds will enable completely new applications and fundamentally change the classic markets. On the one hand, smart cars, smart energy, smart health, smart living, etc. open up new entrepreneurial opportunities and markets of the future. However, on the other hand, these opportunities also pose new ethical challenges for society and companies.

2 Critical Discussions Require New Perspectives

Critical discussions about the current digital visions of the future are becoming increasingly apparent, as is growing public resistance to the fact that not everything always goes according to plan for ambitious technicians and visionary business leaders. Around 70% of Germans believe that their data on the net is completely insecure (Statista/bitcom 2020). Nevertheless, the most popular passwords worldwide in 2021 are still the combinations "1234" as well as "password" (Hasso Plattner Institute, 2021). This mistrust of new technologies shows that the socio-scientific and economic shaping of socio-technological processes must be discussed from new perspectives. It is not technological progress that is the limiting factor, but the human measure, the acceptance in the population, and the positive fields of application in the economy. This shows how strongly systemic thinking is anchored in people—every change in the known system is initially evaluated suspiciously and critically. Companies are challenged to translate the new technologies into sustainable business models that positively impact all while following human needs and mastering the smooth transition into the known social systems.

The fundamental optimism that technologies can help us generate social and entrepreneurial added value must be maintained. Computer algorithms can certainly help to solve complex challenges better and faster. However, they cannot wholly copy human intelligence, and, above all, they cannot replace "reason" and the ability to act ethically (one of the most critical human unique selling points). Computers will therefore have to serve humans and recognise human dignity and human rights in all circumstances. A significant challenge in this transformation is the issue of creation. In recent decades, this theme has often been the plot of feature films and novels, but it is increasingly becoming a reality through new technologies such as artificial intelligence and 5G. AIs are already capable of rewriting and optimising codes. The greatest challenge here is to use these digital and machine possibilities and to link them with ethical action and reason. This is the task of all of us to ensure that human dignity remains inviolable and that computers continue to subordinate themselves to this primacy in the future.

A central question here is how much transparency individuals have about processing their data. Especially in Europe, users want to retain control over their data; this right to self-determination is a high good and must also be taken into account by machines. Nevertheless, people are free to use their common sense to decide which data can or should be used sensibly for new applications. Therefore, the current pandemic raises a fundamentally new question: How much data protection is proper and sensible in the fight against the spread of the virus? There are different answers here, as the examples from Asian democracies and the Chinese path of a collective society show. It will also be essential to use the technologies for the sustainable development of the global world population and regional areas.

3 Opportunities of Artificial Intelligence in a Sustainable Transformation

Using new technical possibilities, we humans can understand the effects of our actions much better than before and thus develop new solutions in the sense of our human needs. This is probably where the most significant opportunities for reconciliation between man and machine lie if we succeed in maintaining human needs as the measure of further developments as long as man remains the determining factor and creative thinker. This could also be the German success story of Industry 4.0 instead of techno-totalitarianism. In the sensible application of AI, there is thus also an opportunity to create new jobs for people. Furthermore, through robots and AI, Europe as a production location can regain importance and new companies can emerge. Moreover, analogue products, a brand core of German craftsmanship, also have a unique position and opportunities in a "sustainable" digital world. In particular, traditional crafts can also benefit from digital technologies and further expand their analogue excellence and offer it on the market.

People's desire for products and services based on human creativity and aesthetics will increase rather than decrease in an increasingly digital world. This prerequisite maintains the fundamental value of the "human scale" in our society. Our lives will then be explicitly physically and at the same time digitally networked. As a result, we will see more and more "smart" business models that offer analogue human needs in the form of entirely new products and services digitally networked and in real time.

The success of the establishment of AI will depend on whether companies succeed in using AI to generate a positive impact for all stakeholders and, at the same time, deliver a positive impact for the planet. Therefore, the current discussion on corporate social responsibility will gain further importance through AI and become central when it comes to the sustainable transformation of our economy.

One of the main challenges in using AI will be integrating corporate responsibility and ethical business into AI. But, as with education, it must always be remembered that in this case the intelligence continues to learn and develop from its environment, i.e. the data. Thus, especially the framework conditions for the profound use of AI possibilities must be set anew about the normality of sustainable management—only in this way can the data provided also tell the AI that profit and sustainability are not a trade-off.

4 Further Development of Corporate Social Responsibility

The transformation of our society, which is currently taking place at breakneck speed, poses far-reaching challenges for companies, particularly in shaping their value creation strategies and processes through AI-supported technologies. Digitalisation, particularly the advent of artificial intelligence and big data, is accelerating the long-needed development of new sustainable business approaches and regional innovations.

Currently, entire sectors and regions are being challenged simultaneously by intensified societal sustainability discourse and disruptive innovations, especially from the IT sector. Due to these effects of globalisation and digitalisation, the external pressure for fundamental change in business models as a whole is increasing. Therefore, thinking of sustainability from a consistently entrepreneurial perspective must go far beyond a pure avoidance logic. For entrepreneurs, in particular, it is essential to manage or increase the positive impacts of their actions. In this progressive perspective, it is no longer centrally about minimising the damage of entrepreneurial action but about increasing the company's value creation for society.

The focus is not on the moral motive of altruistic giving but on economic and social meaningfulness. This "new" understanding of CSR is not about breaking through the logic of competition—as social romantics often like to portray—but quite the opposite. It expands market opportunities through the ever new possibilities of digitalisation. This creates both added value for society and new business opportunities.

The potentials of responsible entrepreneurship can be used efficiently and effectively in the change of digitalisation to solve the pressing social challenges in an entrepreneurial way. Both our society and businesses benefit from this. New technologies are thus a great opportunity to positively rethink corporate responsibility! To do this, however, we have to programme the AI algorithms along the lines of the new economic logic. We need to overcome the classic oppositional thinking between profit and sustainability in the digital world and thus define a "new" intelligent relationship between humans and nature-technology that systematically increases the future possibilities of all and does not play them off against each other.

Herein lies the most incredible opportunity and danger at the same time. The decisive question is: Can we use synergies between the digital world of computers and the analogue world of humans in such a way as to define a common purpose that mutually generates a positive impact in both worlds? In particular, this requires an ethical and economic reorientation in the current technology discussion, which explicitly defines the human being as the measure of further development. When using AI, it must be remembered that systems are always self-controlling. If ethical framework conditions have not been set and the system into which the AI is integrated is not also sustainably oriented at the basis, the business model will still fail.

No company will then be successful for much longer—one that trims old business models more poorly than it should make them sustainable—instead, it will need ultimately "new" business models that are sustainable from the ground up! In order to successfully meet the current social and ecological challenges, more ecological and social innovations are needed. Entrepreneurial creativity and sustainable value creation are the guardrails for economic success and the further development of artificial intelligence.

5 Visionary Entrepreneurs Rely on AI Business Models with Positive Impact

Visionary entrepreneurs and responsible business leaders have recognised the signs of the times and are generating new AI-enabled business models that demonstrate solutions to pressing societal problems. The "entrepreneurial value creation" approach combined with "sustainable AI" heralds a new management paradigm that affects not only each company as an individual but entire industries, regions, and economic sectors.

We are in the midst of one of the most incredible economic and social transformations since industrialisation. Artificial intelligence, properly applied, can be excellent support for the further development of human freedom—if it can be put at the service of human reason and creativity.

Suppose we manage, with the help of AI, to understand social and ecological issues as an indispensable part of the entrepreneurial business model and further global development. In that case, ultimately, new solutions will emerge that we urgently need to solve the current challenges in the economy and society. In this sense, responsibility and entrepreneurship are two sides of the same coin.

The transformation of our business models through new technologies, including artificial intelligence, is therefore ethically imperative and economically necessary and another means to satisfy the needs of over ten billion people and preserve human dignity and individual freedom.

This chapter is based on the German article by Rene Schmidpeter (2021) Künstliche Intelligenz: Wegbegleiter für ein neues menschliches „Maß“? – Ein kurzer Ausblick (pp. 367–372) in: Altenbuger, R., Schmidpeter, R. (Eds.) CSR und Künstliche Intelligenz. Springer Gabler.

References

Bertelsmann Stiftung. (2017). Monitor Digitale Bildung. Online: https://www.bertelsmann-stiftung.de/fileadmin/files/BSt/Publikationen/GrauePublikationen/BSt_MDB3_Schulen_web.pdf

Bitkom. (2017). Almost every second person uses health apps. Online: https://www.bitkom.org/Presse/Presseinformation/Fast-jeder-Zweite-nutzt-Gesundheits-Apps.html

CGI. (2018). Financial consumer survey—White paper online: https://www.de.cgi.com/de/white-paper/banking-and-capital-markets/cgi-2018-financial-consumer-survey

Handelsblatt. (2021). https://www.handelsblatt.com/finanzen/maerkte/devisen-rohstoffe/serie-anlegen-2022-teil-17-nfts-non-fungible-tokens-eine-neue-vermoegensklasse-steht-2022-vor-dem-durchbruch/27940632.html

Hasso Plattner Institute. (2021). https://hpi.de/pressemitteilungen/2021/die-beliebtesten-deutschen-passwoerter-2021.html

Roland Berger. (2020). Future of health—Report. https://www.rolandberger.com/de/Publications/Future-of-Health-Der-Aufstieg-der-Gesundheitsplattformen.html

Statista. (2017). Key issues for digital transformation in G20. Online: https://de.statista.com/infografik/7573/geschwindigkeit-mit-der-sich-technologien-verbreiten/

Statista/bitkom. (2020). Online: https://de.statista.com/statistik/daten/studie/217842/umfrage/sicherheit-von-persoenlichen-daten-im-internet/

Statista/coinmarketcap. (2019). Upward trend in cryptocurrencies. Online: https://de.statista.com/infografik/12186/unterschiedliche-krypto-coins-und-marktkapitalisierung/

We Are Social. (2018). Global digital report. Online: https://wearesocial.com/de/blog/2018/01/global-digital-report-2018

World Economic Forum. (2018). http://reports.weforum.org/digital-transformation/wp-content/blogs.dir/94/mp/files/pages/files/dti-executive-summary-20180510.pdf

Prof. Dr. René Schmidpeter is an innovative, forward thinker who stands for a paradigm shift in business administration and the sustainability discussion. He is an internationally recognised expert on strategic management, corporate transformation, and global sustainability developments. He has published more than 150 publications on CSR, sustainability, governance, and ethics in his German- and English-language management series with Springer Verlag in the last 5 years. He is a consultant and co-founder of numerous national and international sustainability initiatives and think tanks (China, Australia, Great Britain, Japan, Slovenia, USA) and co-founder of M3TRIX GmbH in Cologne. Weblink: www.m3trix.de.

Christophe Funk studied business psychology and works as a CSR manager at the Eat Happy Group in Cologne. The core of his role relates to sustainable strategy and business transformation. Previously, he worked as an assistant to a respected scientist in sustainable management and as a management consultant for the M3TRIX Institute for Sustainable Business Transformation in Cologne. The focus of his research is on psychological aspects of social and economic sustainability transformation. In addition to his full-time work, he also works as a systemic coach and regularly publishes articles on various CSR topics. Weblink: https://www.christophefunk.com.

AI Governance for a Prosperous Future

Alexander Vocelka

Abstract Artificial intelligence is the biggest invention of humanity, and the quintessence of the fourth industrial revolution. It is the result of our age-old dream to have a loyal, yet very capable and obedient servant, an equal, at times even a superior intelligence that works, protects and inspires and that we still control to secure and advance our own welfare.

And therein lies the seed of contention. Something that is superior to us can ultimately not be controlled by us but could rather perceive us as it's resource. A truly dystopian future from a human perspective. Even though this future might be decades or a century away, with AIs still in their infancy, they nevertheless have already demonstrated their transformative power, changing our workplaces, our cars and our homes, selecting our partners and our perception of reality.

And as any tool is also a weapon, we must ensure that AI becomes more tool than weapon and works for humanity and not the other way round.

Corporate AI governance is key to safeguarding the transition to a society in which AI is omnipresent and a blessing, not a curse. It must ensure that its intelligized products and services behave ethically responsible. It must consider the well-being of its employees, customers and business partners wherever AI is deployed.

This is a very tall order, and even more so as with AI we are venturing into the unknown. Only this June (2022) Google sanctioned one of their software developers who claimed that one of the company's most advanced AI, called LaMDA, should have developed consciousness (https://www.gizchina.com/2022/06/14/google-employee-suspended-after-saying-that-ai-has-become-conscious/). AI is a highly sensitive topic.

Therefore, transparency and a well-structured AI governance are imperative to building trust and ensuring that we do not experience a major backlash in this domain by being careless. A backlash that could cost us dearly as AI is the key to a better future, one with less diseases, less suffering and more wealth for all people on the planet.

A. Vocelka (✉)
HighRadius GmbH, Frankfurt, Germany

R. Schmidpeter, R. Altenburger (eds.), *Responsible Artificial Intelligence*, CSR, Sustainability, Ethics & Governance, https://doi.org/10.1007/978-3-031-09245-9_3

This article reflects on the many aspects of AI governance and proposes a way to structure it methodically to make it applicable in work processes, products and services always considering the dichotomy between benefits and risks of AI.

1 Introduction

The development of artificial intelligence (AI) is the greatest and most transformative invention of humankind. Like all great inventions, it does not happen overnight. The development of this new ability will span many decades and will profoundly change all areas of or civilization.

In economic terms, AI is the highest form of productivity increase. Investments in AI solutions achieve phenomenally high returns—from hundreds to even thousands of percentages of ROI. Artificial intelligence is the best capital investment a company can make. Every process, every machine, every product, every service—and the entire infrastructure—will be immersed in AI. As a result, AI deeply intervenes in the value creation domain of humans. What was once the intellectual privilege of humans is now gradually becoming the trait of machines.

The rapid spread of AI in all areas of life and all sectors of the economy, as well as the profound changes in human work and responsibilities, foreshadows its great disruptive potential. This transition from a world with 'dumb' machines to one inhabited by intelligent machines is the fourth industrial revolution. Current models estimate that by 2050 AI will have taken over about 20% of all human value creation (see Fig. 2).

In the long run, AI is not only inevitable but also a natural and necessary development for our civilization. The opportunities for a better life, AI creates, are phenomenal. But at the same time people are always ready to use any innovation to the detriment of others and ultimately themselves—that is simply the human nature. Every innovation is a duality—and can be used both as a tool to increase productivity and as a weapon to destroy it. But this is primarily the problem of humans not of the machines they create. As we will discuss later, people will initially have to bear responsibility for the morality of AI.

And this is, of course, where the fears of many people arise. The fear that the tool will make human labour obsoletes, thereby leading to mass unemployment. The fear that the tool can no longer be controlled—or even understood or the fear that the tool will turn against its creator—whether intentionally or unintentionally.

These fears, some of which also strongly propagated by some leaders of the IT industry,[1] are of course more marketing-driven, and the media loves to make headlines of them. However, they fuel a diffuse angst and prevent society, governments and often even business from analysing AI in a rational and objective manner. At the very least, an overly emotionalized and instinct-driven discussion about AI

[1] https://www.bbc.com/news/31047780

will result in the delay of this important innovation. In the worst case, it will lead to a kind of religious, ideological condemnation of this technology in some countries.

However, globally and in the long term, this would change little about the fundamental development and use of AI as a competitive factor. Countries such as China[2] and Russia[3] have long since understood AI as a leap into the future and have set up large funding programmes. Because AI in its final form raises many philosophical, moral and social questions, the cultural context of countries will play a crucial role in determining the value placed on AI as well as on the degree to which its development is regulated.

Although some interest groups[4] have attempted to regulate AI worldwide according to uniform standards and principles, this will not succeed, precisely because of its economic potential. It is also not desirable from the perspective of innovation. Even in the case of genetic engineering, a topic just as significant as AI, this has not been achieved, although it touches far more fundamental ethical questions.

Artificial intelligence should not be prejudged, avoided, fought or even banned. Instead, it is important to shape and pass through this revolution smoothly, so that the transition into the era of AI is not perceived as painful but rather as enriching and rewarding.

The quantum leap in performance AI delivers will force all businesses to understand and adopt this game-changing technology with urgency, yet responsibly, to secure their future. And this is where corporate social responsibility (CSR) can make a big difference.

CSR can greatly facilitate this transition into the new industrial age. A prerequisite for this is that CSR can be expanded and deepened in its definition and scope to include this complex topic and that it comprehensively covers the development and use of AI in all aspects at an early stage.

Nevertheless, anchoring AI in the CSR framework in the form of principles is only the first step. Without being able to implement the principles through an operational set of rules, these principles will remain lip service. It could read like: 'Thy AI shall not commit theft'.

A dynamic AI-governance framework is required, with a binding set of rules. It specifies the principles and translates them into rules of conduct and work regulations. It is itself part of the digital governance framework in which adjacent topics such as data governance and cybersecurity are also regulated.

Artificial intelligence governance is the set of rules that operationalizes the CSR-AI principles. It enables the structured and sufficiently binding implementation

---

[2] https://www.forbes.com/sites/cognitiveworld/2020/01/14/china-artificial-intelligence-superpower/; https://www.theatlantic.com/magazine/archive/2020/09/china-ai-surveillance/614197/

[3] https://www.theverge.com/2017/9/4/16251226/russia-ai-putin-rule-the-world

[4] https://www.fhi.ox.ac.uk/govai/ ; https://www.g20-insights.org/policy_briefs/coordinating-committee-for-the-governance-of-artificial-intelligence/ ; https://cyber.harvard.edu/topics/ethics-and-governance-ai

of CSR requirements with regard to the development and use of AI in the organization, the processes, the machines, the infrastructure and the products and services. It also defines rules that safeguard the interaction of AI with customers and business partners of a company.

Especially in the case of AI, CSR should always bear in mind that, as a facilitator, it has a dual role to promote and secure the development and use of AI. If we delay or even deny the benefits of AI to humanity, we are also acting against a progressive development of society. There are many examples of how technologies and thus abilities have been forgotten or lost for many decades sometimes centuries, with Apollo's Moon faring capabilty being the most glamorous one.

Corporate social responsibility as an intermediary between the free and a state-regulated market can both accelerate the development of AI and hedge its negative aspects. Of course, the assessment of positive and negative aspects differs from culture to culture. The North American view of CSR, which has been predominant in the West, will not be indiscriminately shared in other regions of the world, especially with regard to AI.

This article provides ideas and conceptual approaches for corporate leaders to expand CSR to include AI and to put it into practice through a structured, comprehensive AI-governance framework.

2 Artificial Intelligence Is the Quintessence of the Fourth Industrial Revolution

Writing an AI-governance framework requires a sound understanding of what AI is, can and cannot do today and tomorrow. Further, it also requires, to understand, what intelligence is, natural or artificial, and how intelligence drives productivity.

Let's revisit intelligence first.

2.1 From Intelligence to Productivity

Our present civilization is built on the knowledge of 100,000 generations—an exponential development of knowledge over 6 million years, since the time we split from our primate siblings. Intelligence creates knowledge. Knowledge is the only true asset that can yield (sustainable) benefits and can be inherited over ages. Knowledge—from how to make fire and the stone tools of *Homo erectus* 2 million years ago, to the quantum computers of our time.

Intelligence essentially is made up of three components: information sensing (reception), information processing (including storage) and information emitting (acting). Intelligence stands for the ability to learn existing knowledge and, through abstract thinking or variant training, develop new knowledge, communicate it and make it productively applicable in the form of processes and with the help of technological tools.

Artificial intelligence is the automation of this natural intelligence concept and is thus a new production factor, the fourth, after natural resources (including energy), people (labour) and technological tools such as machinery or infrastructure. Money, the variable factor of production, does not directly change productivity but acts as production factor converter. Also, the historic classification of machinery, infrastructure and monetary assets into one single production factor, capital, is an unfortunate imprecision.

Artificial intelligence is the production factor with the greatest value creation potential. It has a higher productivity exponent than software as it improves over its lifetime. Something that accountants will have their problems with, as AI is not depreciating as software but appreciating with time, respectively, it has to be valued according to its current and future potential value. Not unlike the concept of the cash-generating potential of business units. Fittingly, this makes AI not only the fourth production factor but also the quintessence of the fourth industrial revolution: the automation of intelligence and its unstoppable spread across all products, services, processes and machines, as well as our entire technological infrastructure. Just like the steam engine, electricity and computers and the internet, the spread of AI is ushering in a new industrial age and, also, a new social age.

This leads to a new division of labour between humans and machines, which must become a key aspect of CSR-AI.

In order to understand and govern the future division of labour between humans and AI, we need to analyse the second factor of production itself, human labour, along its two primary components: mechanical and mental labour, as the two components will experience different impact scenarios and challenges.

Take the motor function, and thus the mechanical productivity, of a piano player, playing a sonata. Even though the dexterity of hands and fingers is essential, it is the tightly coupled orchestration of motor and audio cortex that leads to enjoyable music. So, the mental coordination performance is the key productivity factor of a piano player.

Still, we can claim that these skills are highly automated and thus of lesser mental quality than that of an improvising jazz pianist, who abstracts unpredictable complementary musical response models in real time.

An example of how intelligent motor skills can be is passwords. The author remembers his corporate password more easily by letting his fingers glide over the keys ‘by themselves’. The password can thus be better retrieved by the cerebellum and motor cortex than by the fancy but inefficient prefrontal cortex. It is but one example that highlights the astonishing deficits of our most recent and fashionable human feature.

This differentiation between sensory, motor and abstract intelligence can be further refined. It is necessary to better understand the two production factors of humans and AI and thus to optimize their combination. It helps to define the future role of AI in human society and the workplace, and to develop the most effective and frictionless human–AI collaboration, and respective hierarchy and governance models.

So, just as we can differentiate intelligence as sensory, abstracting[5] and motor intelligence, we also differentiate the increase in knowledge when it comes to AI and its productivity, when compared to human productivity.

Today, a semantic AI is capable to 'learn' Wikipedia in its entirety with over 100 million articles in days or just hours, depending on the required finesse and computing power.

This also counts as knowledge generation because the overall productivity of human society depends on an optimal distribution of knowledge, and not only the maximization of knowledge in a few minds. The more intelligent subjects and objects (people and machines) know, the more potential productivity (knowledge creation) can be activated. This could also be called copy productivity. Strictly speaking, no new knowledge is generated. Instead, only existing knowledge is disseminated and scaled up in application.

This is the first productivity level of intelligence. And this is precisely where AI already excels today.

Once a current AI has 'learnt' Wikipedia, a limited transfer of knowledge to the machine takes place. Limited because this AI does not yet have a full associative grasp of what it has learnt and, above all, can hardly abstract new or undescribed connections from it or draw conclusions and act sensibly on its own. It is thus still quite limited in its ability to generate new knowledge based on what it has 'learnt' from 'reading' Wikipedia.

However, it can learn a narrow set of human decisions statistically and re-apply them precisely and efficiently in similar situations. This AI reaches the first productivity level. It can learn and apply existing knowledge very efficiently, precisely and consistently in specific processes.

As we all know, generating new knowledge as the next higher level of productivity is much more difficult.

A lot has been written about returning to the Moon after 60 years and how humanity has lost the capability for decades, only to re-learn the ability again. Have we become dumber and are now catching up again with 1960s knowledge?

Not quite!

The effective increase in productivity and the respective economic growth over the past 30 years has been driven primarily by the dissemination (i.e. globalization) of existing knowledge, the first productivity level of intelligence. The price we paid was that absolute knowledge growth was slower. We have not become more knowledgeable in absolute terms per unit of time. Instead, many people have become relatively more knowledgeable and can now translate this additional knowledge directly into increased productivity and thus prosperity. Copy-knowledge productivity is real productivity.

[5]Mainly the prefrontal cortex is orchestrating the abstraction of multi-dimensional, i.e. multi-sensory, information into conceptual elements. The associative network of these contextual abstract elements leads to the phenomenon of understanding and awareness.

It is precisely this process that drove the economic miracles of Japan and South Korea in the 1970s–1990s and powers the re-emergence of China as a modern technology nation in the last 30 years.

It also is a very important aspect of CSR-AI in ageing and retiring populations. Many companies have already understood that capturing implicit expert knowledge means harvesting productivity potential. And the only way to do that is to train and coach AIs. Quite a few AI projects have the specific objective to transfer know-how from humans to machines.

Also, it is not only new (copy) knowledge per se that contributes to greater overall productivity and prosperity, but rather an optimal distribution of knowledge across society and the combination of different knowledge and intelligence sets, or also creative mind diversity under sufficient innovation stress. Countries with a high level of education also have a high level of overall productivity because the entire population has valuable knowledge to create additional productivity. But again, diversity of minds is another very important factor. The optimal allocation of expert knowledge enables an optimal division of labour and ultimately distribution of wealth.

It is likely that once machines 'get the hang of it', they will be able to generate new knowledge much more quickly and efficiently than humans. The first examples of lab AIs are already indicative of this. This indicates that the productivity potential through AI will increase tremendously at both levels. Existing knowledge is copied and scaled quickly for global use. New knowledge is created faster than people can create it—possibly even faster than people can sufficiently absorb it themselves (Fig. 1).

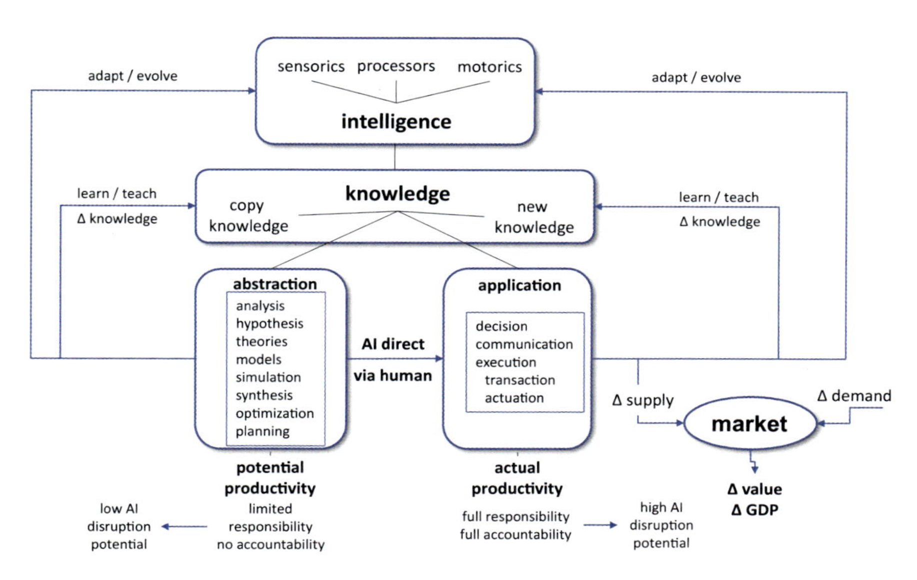

Fig. 1 Intelligence as the quintessence of value creation (Vocelka, A.)

2.2 The Value of AI

However, productivity alone does not determine the value of work. Equally, additional new knowledge in a product or service does not necessarily lead to additional value.

Value, as we all know, is decided by the ratio of supply and demand. The offer must have a high additional utility for the consumer. At the same time, it should be scarce. Only then does a high value arise. The usefulness or utility itself is measured as the perceived increase in buyer productivity.

This also applies to the consumer as convenience and even pastime pleasure increases human productivity. So, in fact, all purchased utility is a productivity gain through purchased knowledge or skills or capability—whether it be in the form of products or services for the buyer.

If AI is now embedded in all our still quite 'dumb' products and services, these products and services will not only be much more useful and productive for the buyer but will even increase in usefulness and value over their lifetime. Therein lies the accountant's challenge with AI: how is it valued over time?

The moment AI also controls machines and applications, also designated as Operating Technology, OT, in contrast to pure Information Technology,IT, without physical actuation capacity, it takes the step from pure thought to concrete action. The human being is then completely displaced from the value creation process. AI will become maximally productive. But then the question of responsibility becomes ever more important.

The AI that becomes fully capable of acting 'in lieu of' or even for itself, something that humans themselves only achieve when they come of age, implicitly has also reached judgement capability. In Europe, this is only attributed to adolescents after the age of about 16. The combination of these two aspects—the ability to replace humans in entire value chains and to become capable of judgement and action—will make AI the most disruptive technology, yet.

2.3 AI Working for Us

The World Economic Forum (WEC) states that 60% of global GDP in 2022 is digitized. In a few years a large part of the global socio-economy will be digital and accessible by AIs. Some models[6] predict that, by 2050, AI will have taken over about 20% of the human work share in OECD countries. In Germany, effective per capita working time decreased by 17% between 2000 and 2018—as it has in most OECD countries and even in the United States. Just as 80-h weeks of hard labour were not uncommon 120 years ago, a 20-h workweek in OECD countries will be a distinct possibility in 2050. Iceland just completed a 5-year trial of a 4-day

[6] AI workshare prediction model 2050, Vocelka A.; also see Fig. 2.

workweek with full pay and surprisingly, productivity was not at all compromised. Spain has started its own 4-day workweek pilot in May 2021. These trials are still primarily driven by social arguments and general employee well-being and health, but soon AI will greatly accelerate this trend (Fig. 2).

So, if AI is the great productivity leap, who stands to gain? And who stands to lose? Or can it really be that there are no losers, and if so, why?

It is unlikely that there will be no losers of the fourth industrial revolution; however, it is feasible that the overwhelming majority will win and only a few will be disadvantaged.

That should be the aspiration of CSR when it comes to governing AI.

Of course, this also requires a rethinking of our economic system. If we manage to question existing principles and are prepared to make fundamental changes to our economic system and the definition of work, prosperity and human rights, AI can lead to a win-win situation. But if we cling to old philosophies and ways of thinking, great social tension will ensue with the advent of more capable AIs.

Not only does CSR need to anticipate and take these changes into account, but policy makers also need to be prepared and informed so that the transition to a positive future with intelligent machines can be set early on—because when it comes to AI both, utopia and dystopia are very much conceivable, and we should not leave the outcome to chance.

3 Utopia or Dystopia: Where There Is Light, There Is also Shadow

CSR for AI requires understanding what the emotional drivers and fears of this new technology are in order to consider these appropriately. It is a prerequisite for proper AI design that maximizes acceptance and minimizes opposition.

If all people agreed that AI is a natural, sensible, useful and inevitable part of evolution, we could shorten this article by stating that CSR need not bother with it, because there is an overwhelming consensus that AI is bringing us closer to utopia, and that whatever business comes up with in terms of AI deployment will be warmly received by society.

However, as we know, this is not the case, because AI touches the core of what it means to be human. We are getting a new level of competition with this fourth industrial revolution. But this time, it is not muscle power but mind power we compete against—and that is precisely where the fun stops for many.

CSR must recognize, understand and mitigate the questions and challenges emerging AI creates.

Existential questions arise:

If humans are completely replaced by machines, because of AI, how can humans still earn a living? Is there still enough work that only people can do

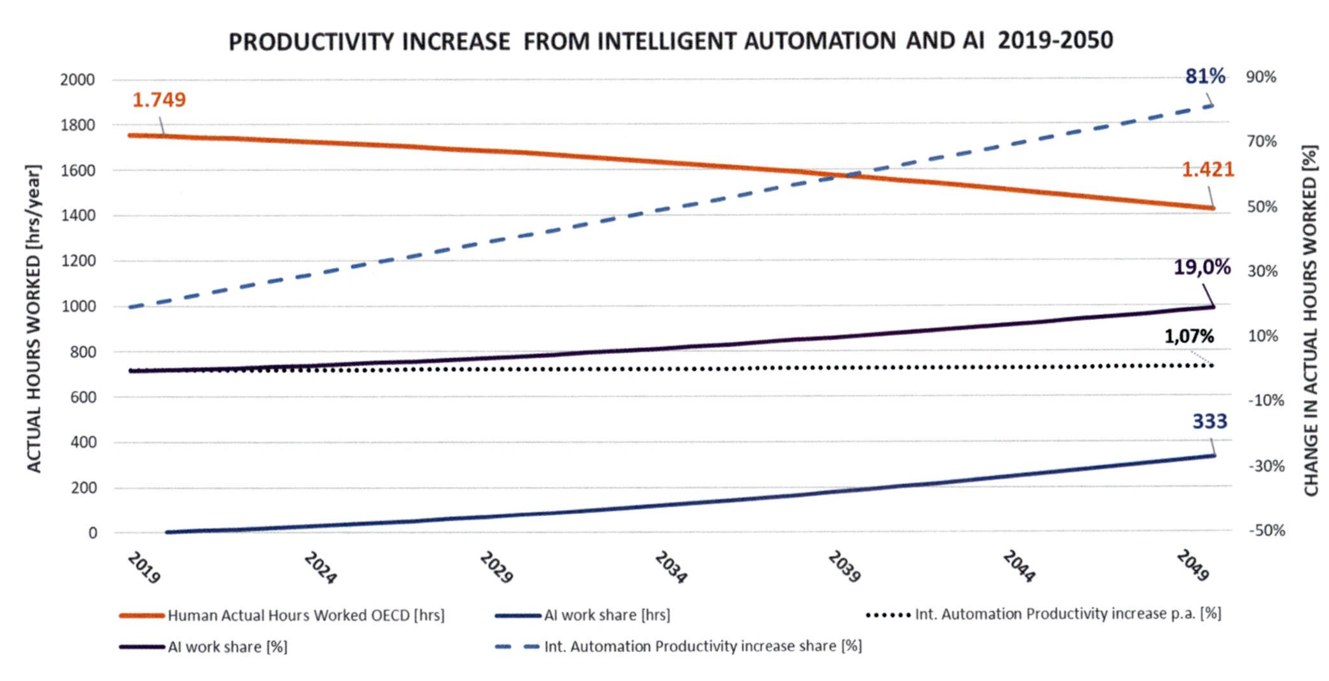

Fig. 2 Productivity gains through intelligent automation and AI by 2050 (data sources: OECD, IMF. Models: Vocelka, A.)

that is purposeful, and for which they are also paid? Do humans still have a productive value that is sufficient to earn their livelihood?

Macroeconomic questions arise:

Who will still buy goods and services if no one has an income anymore? Can this be solved via a general human income paid for by an AI tax in the broadest sense, so that AI earns our living for us? How will this work?

Psychological questions arise:

What value do human beings still have if machines can do everything better? What will then be the meaning of human life or human civilization? Do humans not need to work, to have any meaning and a purpose in life at all?

Ethical and moral questions arise:

Is AI allowed to advise soldiers and police officers or judges and governments? Should it be allowed to make in lieu decisions in these governance functions? Should AI have a licence to kill? At what point does AI have a right to exist?

Legal questions arise:

Who is liable when AIs make decisions that cause harm or that seem wrong from a human perspective? Can AIs be sanctioned? Should they have subject status? Liability is the strongest incentive for companies to push for AI subject status. After all, this would free companies from accountability after proper inception of an AI. It's the same dilemma we have with social media platforms such as Facebook and Twitter who refuse to be responsible for the content published on their platforms but at the same time censor it and thus act as editors.

These are critical questions, and they require deep reflection and conclusive answers, many of which Nick Bostrom described extensively in his book *Superintelligence*.

A major test will be the introduction of robocars once AI achieves full autonomous driving capability. While we currently do have a shortage of hundreds of thousands of truck drivers in Europe after the COVID turbulences, it still is another thing to shift millions of professional drivers to other occupations within only a decade.

Again, we could put the issue of CSR and AI to rest if people agreed that AI is the devil's work and that it should be banned worldwide.

But we all know that AI is not just the genie escaping from the bottle, but rather a natural development.

A development that is also welcome in principle by people when they reflect on the benefits of AI within the next century: hardly any traffic fatalities or injuries, victory over most diseases, a significantly higher life expectancy with excellent health, the solution to our energy needs, a regenerated environment—and all this without having to work. People being able to create new things rather than having to copy the same things. Societies with a wealth distribution shaped as a Gaussian with a left-sided cut-off value.

To realize this vision, we need to understand and mitigate its possible negative side effects. The emergence of AI forces us to think much more about our own dark sides—because it is these that AI can also learn from us. Maybe it is the AI mirror we instinctively and quite rightly so fear. And maybe we somehow feel the ultimate despair it could cause should we truly come to the conclusion that higher AIs will be our better selves. Dystopian effects are primarily the reflection of our own dark side projected onto a far more intelligent actor. Perhaps this is a good time to really better ourselves, so we become the right role model that AI can look up to and learn from. But as it would be careless to count on something humanity has tried in vain since its emergence, setting guidelines and rules is a safer way into the age of AI.

3.1 How to Guide the Emergence of AI

The development and use of AI could be tightly regulated by the state. But every country would approach this differently because every country has its own culture—and derives from this its own view of humans and intelligent machines and their roles in this world. We would have countries where everything goes and where AI will sit as the 13th juror in court. And then there could be countries where AI will be virtually banned.

So, why don't we leave all the governing and regulation to governments?

Governments usually need time to understand the nature and impact of novel ideas, and technologies, before they can design a sensible and effective regulatory framework. This knowledge gap is disadvantageous and leads to prolonged incoherent and unstable governance situations and to negative experiences.

Should we count on scientists to guide the way?

At least they understand what they are talking about. Do they? With new technologies the discourse within science can be very diverse and controverse. Scientists are certainly excellent in analytics and developing scenarios and making recommendations, but creating rules and regulations is not where their power lies. Media very often extracts information from academia that can be emotionalized and distorted which is highly counterproductive. Organizations such as the 'future of life institute', are essential forums for discussion. However, even they don't shy away from highly emotionalized videos[7] on their web pages to attract interest, something that is, in the eyes of this author, very counterproductive. It scares normal people but makes no impression on any decision-maker in power. The very good work from the Asilomar conference and the resulting principles get drowned in emotional noise.

The COVID pandemic clearly showed how contradictory rules and regulations can become counterproductive. Innovation needs the right mix of freedom and regulation for businesses to invest in. Let's be clear, regulation can be extremely positive for innovation, as it not only sets limits but also defines and secures innovation space or, even in the most progressive form, fosters innovation by

[7] https://futureoflife.org ; https://www.youtube.com/watch?v=9rDo1QxI260&t=76s

declaring aggressive objectives which require very intense and competitive innovation.

A case in point is again the situation around autonomous cars.

The business community can and must break through this state of innovation paralysis on its own by creating a clear CSR regarding AI (CSR-AI) and a well-thought-out AI-governance framework. It must engage governments and regulators in a proactive and constructive discourse to create and to guarantee conditions for the beneficial and safe development and use of AI. In Europe, where businesses expect directions and guidelines from governments, this is a particular problem.

After German OEMs have spent years in paralysis, the car summit in Berlin in September 2020 broke the stasis, just when most companies wanted to put the subject of autonomous driving to rest. The automotive industry was able to persuade the government to finally develop the legal framework for autonomous driving by 2022 and freed itself from a chicken-egg situation. Of course, they did so under enormous competitive pressure from Tesla. The car industry now knows the framework within which autonomous driving will be legalized, at least in Germany. Other countries will soon follow suit.

3.2 *CSR as Beneficial AI Facilitator*

If the automotive industry had taken a more serious and informed look at the issue a decade ago, when the technology was first proven to be viable, we would have had this regulatory framework half a decade sooner. The automotive industry and the government would not have been fixated on the minimum marginal benefit of CO_2 emissions, but rather recognized that AI cars provide great CO_2 reductions as a side effect. This would have saved the industry years of legal disputes and billions in fines—billions that could have been used to develop truly autonomous automobiles by today.

Not to mention that 3000 lives a year in Germany alone could have been saved, and many more injuries and much suffering prevented if full autonomous driving had been introduced sooner.

And that is exactly what CSR with regard to AI is all about:

Facilitating innovation for the benefit of society while minimizing negative side effects.

Therefore, if the economy and thus also society do not want to suffer setbacks or standstill in this fundamentally important technology, companies themselves must understand the productive potential of AI, take AI into account in the context of CSR and develop and implement an AI-governance framework that transparently and comprehensibly documents and safeguards the defined principles from development to deployment and termination of an AI system. And then proactively educate governments on the benefits and risks of this technology.

CSR should never be viewed as a defensive concept but as a creative one in the true sense. It safeguards productivity and innovation by constantly creating save value generation spaces that balance economic with social and environmental needs. Ideally it is at the forefront of innovation and progress and its guidelines should serve as a blueprint for government regulation to follow.

Given our nature, it is unlikely that AI will lead to utopia for all within a few centuries, but it is equally unlikely that machines will rise to dominate humans. More likely, there will be a mix of positive social improvements and negative side effects. The mix will determine the speed of progress. For companies, it is important to keep their innovation speed high by using CSR and an AI-governance framework to keep undesired effects hedged, while making the benefits of AI available to society at an accelerated rate.

4 All AI Is Not the Same

To properly position AI in the context of CSR, we need to get a brief overview of the highly dynamic field of AI technology and its different flavours.

Artificial intelligence (AI) describes the ability of a system that has not evolved naturally like natural intelligence (NI), but that has been created by humans, to learn from, respond to and adapt to its environment. And AIs will not only be purely silicon based.

From this basic definition, we can already see that AI, like NI, represents a continuum. This means that there are many shades, levels and types of intelligence.

This means that CSR and AI governance must also treat the field of AI in a differentiated way.

4.1 From Edge AI to General AI

It is easy to see that fauna and flora are endowed with intelligence, and even consciousness, as humans are, just not at the same level. The degree of intelligence and the expression of consciousness are emergent phenomena of complex dynamic recoupled (feedback loop-based) information structures. The author recommends studying integrated information theory (IIT), one of the leading concepts on the origin and emergence of consciousness. IIT even proposes a physical measure of consciousness.

While consciousness enables feeling of even abstract perceptions, intelligence is needed to create abstract information in the first place. Recoupled, ontological models are a key ingredient for higher faculty information abstraction at the second level, while the 'simpler' neuronal nets drive subconscious fast pre-processing and primary information reduction with millisecond speed.

The brain's ability to reduce information input is the secret of its efficiency. For example, visual input from 30 megabytes/s at the human retina is reduced to some 100 bits/s of highly abstract information in the prefrontal cortex. This high abstraction capability allows to process very complex external and internal states very efficiently. A hyperdimensional associative memory of abstract ontological elements is then mainly responsible for high context understanding. In a nutshell, the brain's most important task is to properly map the world (including one's body), maintain autonomous functions efficiently and predict environmental changes as precisely as possible—all in order to survive for as long as possible. The ability to create an abstract semantic map in our brain is the basis for structured, inferential and recursive and simulation thinking, learning and true understanding. This is a feature that still eludes even the most massive neural nets trained today, as it takes a different information processing model.

Low levels of natural intelligence (NI) lead to a slow process of learning, and adaptation or development, and low levels of understanding, while high levels of intelligence enable accelerated evolution and the development of a complex socio-technological civilization such as ours. Again, this requires high-level abstract 'symbolic' thinking.

As with humans, with AI, we can assess intelligence or ability by faculty. For example, how good are you at predicting developments, how good are you at recognizing things, how good are you at languages or at mathematics or more specifically, how good you are at playing chess or Go.

With AI, we can also distinguish between narrow artificial intelligence (NAI) and general artificial intelligence (GAI). Narrow AI has a narrow task spectrum and capabilities. It is an absolute one-track specialist for low variance tasks and environments. GAI has multiple faculties (different types of intelligence), which are interconnected and can take on more complex tasks in very variant environments. The more faculties GAI commands, the more complex the processing structures and responses and actions become and the more valuable the GAI will be.

Massive neural net learners such as for autonomous driving often are perceived as being GAIs, but they still don't understand, as their response comes from memorized non-abstract responses, and they thus can't infer new situations in an efficient abstract format and add them to such an associative memory. Because they cannot form an associative memory with the neural structure they are given, they cannot understand and reason and infer new responses to the same environmental context. The size of a neural net learner is not a measure of its intelligence, and neither is the learned data. The model structure is the key.

This is why Tesla's autopilot neural net, as powerful as it is, does not even understand what a car is. It does not possess the required high level ontological models, at least not yet. Still, the upcoming version 11 which will integrate several new algorithmic models based on vectors will move very close to higher abstraction levels and to understanding.[8]

However, this will change quickly in the coming years because we already know what is needed to make a machine understand.

[8] https://www.youtube.com/watch?v=DxREm3s1scA&t=6914s

While NAI has an object status, GAIs will develop towards becoming subjects. This will further expand the current spectrum of AI, with machines of minimum intelligence, also called edge intelligence (EI) at the bottom end and very powerful machines that understand their environment and their own actions.

We will have machines that command other machines. Hierarchical AI structures will be very effective in factories and simplify the control of complex processes. Also, it is still simpler to create many different specialized AIs controlled by a few basic GAIs than to try and integrate many NAIs into one gigantic coherent GAI. Decentralized intelligence has many advantages and nature has used this concept successfully with many species. Even humans had to start specializing to build a technological civilization. It is simply more energy efficient and thus more productive, as Adam Smith reasoned 250 years ago.

Today, the layperson understands AI to be GAI and thus unconsciously ascribes to it abilities that are equal or competitive, or even superior, to humans. However, more than 99% of the AI in use today is NAI. And the current GAIs are still far from being on a par with humans.

Wherever a higher level of abstraction is needed, we humans still outperform AIs by far!

But in their specialist disciplines, be it forecasting specific events or playing Go, even NAI is already clearly superior to humans. However, they are one-trick ponies, and even the most powerful conversational agents cannot yet understand their own responses.

Intelligence does not mean awareness, accountability, judgement or even consciousness. That is why even powerful GAIs will still be object-like entities. Yet even object-like GAI can be creative and generate new knowledge and find new solutions. They just don't understand what they do. Lab-AIs are one example of many 'creative/innovative' AIs. The work they accomplish still does not require awareness or consciousness. Even though the stochastic nature of feelings does boost general creativity, they are not a necessity to creativity or innovation. Understanding requires a basic level of awareness but this is still one big step away from self-awareness and two big steps from consciousness, which is the requirement for fully accountable subjectivity. Self-awareness, for example, requires a clear delimitation of 'me' and the 'environment' and a recoupling of the neural net to a defined 'me-system' or body.

Subjectivity emerges with a 'my-self' defined intrinsic drive, coupled to a need complex that defines an ego and the will to survive as an entity or 'me-system'. Self-awareness is something that emerges in humans within the first 6 months after birth. However, full accountability can only be reached when sanctions can be felt, which means feeling is present, which is the quintessence of consciousness—the hard problem.

Self-awareness is achievable with today's technology, while artificial consciousness (AC) on the other hand, as emerging phenomenon, is only just being conceptualized by the first labs.

So, intelligence is in fact the most basic capability of a GAI, and one could full well say that intelligence is nice but overrated and only the first step on the

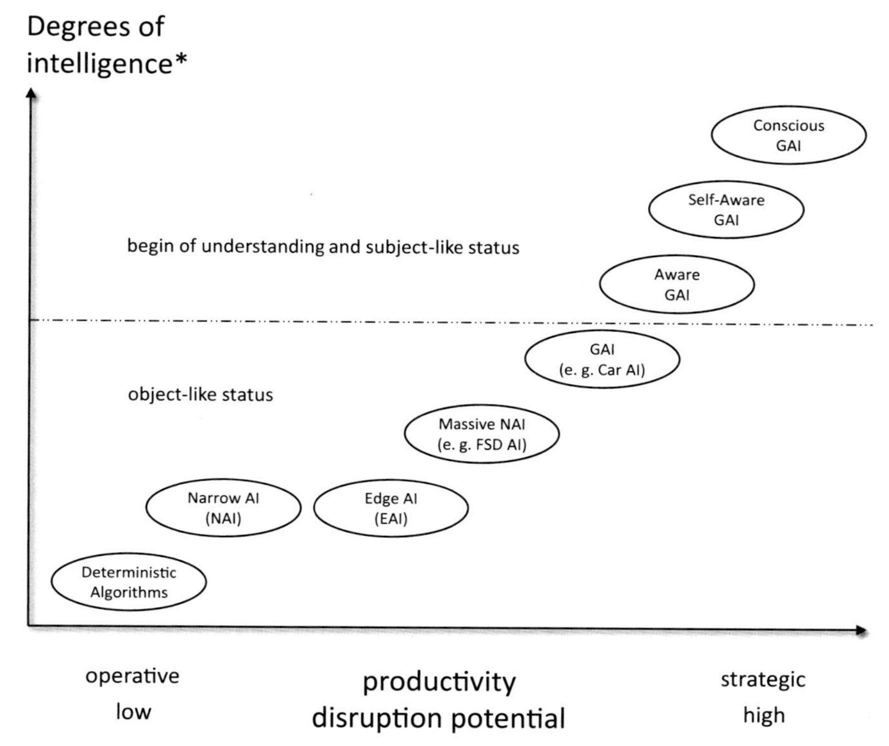

Fig. 3 The degrees of freedom of AI define its capabilities and subsequently its disruption potential (Vocelka, A.)

evolutionary ladder of AI. Also, it is not yet clear what it means to build a full GAI mind from powerful sub-components. Will it be sufficiently integrated or must all components evolve simultaneously or carefully staged (Fig. 3)?

AI classification classically considered the Turing Test and Singularity as measure; however, why would an AI try to disguise itself as human? This could sow mistrust rather than trust. Also, it would be hard to hide the fact that a machine can give certain responses far faster and more precise than a human. Instead of focusing valuable resources on a disguise, it should be invested in productivity.

From what we know today in the field of AI research and related neuroscience fields, AI and object-like GAI will not only catch up with humans but will surpass them in many—perhaps even all—disciplines of thought in the not-too-distant future.

Once AI is able to generate new knowledge itself, it can be assumed that this ability will also increase exponentially. They are likely to create new knowledge much faster than humans can and even faster than they can learn new knowledge from humans. This is also another fear that plagues some people: namely, that AI is not only our equal, but rather knows far more than humanity and can therefore do more and ultimately be more and want more.

Much like with chess, it is now understood that humans are not the measure of AI, but merely the most important benchmark on an open-ended scale of intelligence. Only a couple of years ago DeepMind argued that the AlphaGo model would be intuitive when playing. For this the model needed to be conscious with feelings from a subconscious part. However, AlphaGo is a large NAI without even understanding. It is these kinds of marketing statements from AI experts that cause anxieties in the public. Experts and managers of AI companies do well to communicate carefully and responsibly on such a sensitive subject. Also, we should consider human augmentation as another important development line which will strongly influence the balance between human and AI capabilities. The closer the coupling of humans and AI, the greater the alignment and the smaller the stress and conflict potential.

For philosophers the emergence of GAI and later artificial consciousness (AC) will be the fourth Copernican moment after Copernicus, Darwin and Freud.

For today's policy makers it is very important to understand the broad spectrum of AI and its different evolutionary lines to design the right CSR-AI guidelines.

4.2 *The AI Productivity vs. Complexity Paradox*

CSR-AI assumes that the development of AI can be controlled to a certain degree. However, controlling AI will be very different from controlling a dumb machine.

The idea of simply gating AI through program clauses and tightly controlling its 'thoughts' and actions fails because of an economic paradox: the productivity of AI increases with its intelligence and ability to act, its degrees of free will and actionability. This, in turn, is closely linked to its sensory-motor degrees of freedom. It was already understood in the 1990s that AI cannot be designed based on fixed rules. The effort would simply be immense and uneconomical; this is what led to the first AI winter. AI must evolve through open learning, even though this happens magnitudes faster than with humans.

This leads us to the question of whether we can technologically just limit AI to NAI level and then not bother with the far more complex transformation issues of GAI.

The answer lies in the natural complexity of life and our society. We now know that GAI has the highest productivity because it has the highest scope for thought and action—it can then best respond to the infinitely varied situations of reality. This high variance productivity is why AI will be developed to become GAI. GAIs simply are required for many complex tasks and to control lower AIs in typical hierarchical organizations, which gives them additional productivity leverage.

And GAIs cannot be regulated by programs coded by humans. In principle, GAI, like humans, must learn the rules holistically and statistically in a very large context and with varied scope. Learning itself can take place rather quickly—but humans cannot program it. The idea to implement Asimov's three laws collides with the productivity function, which requires ever higher degrees of freedom.

Also learning comes from making mistakes. This means we will have to give AIs the ability to learn from mistakes and thus to make mistakes in the first place. Mistakes for which we bear responsibility, as long as GAIs can't.

We are currently considering to what extent GAI needs a basic motivation in order to be more productive and effectively anchor accountability.[9] Equipped with a need personality complex (NPC), GAI would then assume a subject-like status. It would have a continuous basic motivation to be, survive and interact with the environment and thus unfold maximum productivity.

To put it more simply:

The AI with the highest productivity will be the one with the greatest degrees of freedom and the highest intrinsic motivation.

And that means that it too will initially behave in ways it learns from us. It will have to learn morality and be free to exert it, and to evolve it, not unlike children. Moral must be learnt and cannot be programmed, or as Peter Robin Hiesinger says, there is no shortcut. From an ethical point of view, we might soon require powerful GAIs to have some basic personality that can act as a moral compass. AIs with souls. As AIs without souls would be far more unpredictable.

So, contrary to popular belief, it might be essential to endow GAIs with personality and intrinsic motivation and much more freedom to better 'control' it, or more precisely to enable it to control itself or simply put: to behave.

Even though we are still 25 years away from such entities, these are key conceptual insights to consider when it comes to the design and implementation of future AI-governance frameworks.

With regard to the transformative changes to our economic model, it is never too early to reflect on the consequences of highly productive GAI and start preparing for a smooth transition, avoiding disruption.

This is exactly the benefit of an advanced CSR-AI framework.

5 Application and CSR Challenges of AI in Companies

Through the widespread deployment of AI, a company becomes a complex cybernetic system. In fact, any digital transformation that does not consider the cybernetic view will miss the transformative potential and just automate processes.

This understanding and view of the business, as complex, intelligent organism, is fundamental. The development of advanced digitalized business models and strategies depends on it. A company's digitalization strategy should derive the target image of its digital operating model and the digital strategy that leads there, from this

[9] The first intrinsically motivated agent, designed by Steering Lab in collaboration with TUM, May 2021, showed superior service quality.

concept and design its IT infrastructure accordingly. Of course, this view will also change the nature of products and services and how they are designed, sold, maintained and terminated.

When people talk about intelligent automation in companies, they often think of a machine. It is more fitting to think of it as an intelligized process. All processes in companies become intelligent through AI. The productivity of every process step in companies is systematically increased by AI. Artificial intelligence will also increasingly steer companies more directly and not just indirectly through mere analytical support.

Artificial intelligence has the advantage of objectifying control. This makes planning, performance management and optimization of all resources easier, faster and more effective. Politics and subjective influences are increasingly filtered out. However, the human touch can only be kept through CSR-AI.

The intelligent automation of all processes means employees hand over activities to AI and move to higher-value processes, such as training and supervising AI. This changes the way employees work and has profound consequences for responsibilities, performance management and accountability. Should my bonus get cut, just because my AI is a bad product recommender? This is not my fault!

5.1 AI Paralysis

Currently, companies are still hesitant to let AIs in on the action. AI is still predominantly used for analysis, which yields much lower productivity gains than the higher AI faculties. Decisions are still made by humans, often against the advice of recommender machines. Artificial intelligence is being tested more than used productively. This is also evidenced by the countless use cases that eke out an existence as prototypes and MVPs (Minimum Viable Products are products that are better prototypes and not finished) or even locked away in drawers.

However, real productivity comes about only through decision and action, through the physical implementation of the thought processes of intelligence—whether it be a transaction or a mechanical action. Without action or transaction, the steering loop, which is also the learning loop, stays open and no new knowledge is created, nothing new learnt and AI productivity stalls.

This brings us to the overused term of analysis. In corporate parlance, everything is currently analytics—be it business analytics, advanced analytics or data analytics. Even the term business intelligence is often used to describe any kind of analytics.

Although the old saying ‘paralysis by analysis’ is well known, most companies currently still limit themselves to data analysis. But that is where only the potential productivity of AI lies.

Analysis alone generates at most potential knowledge—but no effective knowledge and thus no added value. Knowing something is of little value if the knowledge is not translated into action. Even worse than hiding money under the mattress is hoarding knowledge potential unused in company databases.

5.2 AI Action

Knowledge becomes concrete added value only during execution and not in simulation. Artificial intelligence can provide added value only if it triggers actions or transactions itself or recommends actions that humans can then directly implement.

Artificial intelligence will only achieve the highest productivity level when it is coupled to transactional or motor systems and can perform the complete value creation cycle in a process—from monitoring, analysis, simulation and abstract synthesis to decision-making and execution, be it physical or informational.

Moreover, AI as a central production factor is not only used by the company in its operating model, but also directly built into its products and services. Whether cars or lawnmowers, pool robots or prams, nurseries or gyms, they will all become intelligent and be able to take on advanced tasks, thereby increasing their functionality, usefulness and responsibility.

In the future, customers will not buy a lawnmower with fixed capabilities from a data sheet, but rather an intelligent lawnmower with dynamic capabilities and development potential. So, instead of just buying productivity, the customer buys future productivity potential that depends on the ability of the lawnmower to sense, learn and act.

However, dumbness could be bliss, when a robot mower could have detected an intruder but didn't sound the alarm. CSR-AI comes to play here. It must give guidance to product managers and developers alike, as to what intrusion detection capability the mower AI should have, and how to market this feature without misleading customers. And how to deal with potential liabilities from a malfunctioning super-mower.

It still makes sense for the company to design the robot mower with future intelligence potential, such as for security tasks, in mind (i.e. sensor, processing and motor technology with excess potential at delivery time), even if the activation must wait until AI-governance issues are clarified.

There is no better example for AI productivity in action than in the domain of autonomous cars. Having a smart navigation system is a good thing and helps us save time. This boosts our productivity by increasing higher level productivity time by some 10% in average. However, an autonomous car can easily double our productivity and increase macroeconomic productivity by virtually eliminating accidents and by negotiating a much higher traffic throughput using the same infrastructure.

This is actionable AI.

Getting there however is very tricky. As we need enormous compute capacities to train an AI to deal with any possible traffic situation, and we need to build trust with people, so they accept autonomous cars and use them.

Tesla understood this years ago and pursues a strategy of highly intelligent, autonomous cars that constantly are updated and up-smarted. Using a rapidly growing cohort of beta drivers Tesla has scaled machine learning by including the driver as enthusiastic unpaid 'labellers' in the learning loop. Thousands and soon hundreds of thousands of drivers teach Dojo, its massive central AI model, how to drive in any given situation.

Tesla thus has outsourced an immensely expensive AI development process to its customers and charges them US$ 15’000 (September 2022) for an MVP. How much more productive and at the same time customer centric can a company be?

To enable this customer-centric R&D process Tesla has designed a strict governance model for its FSD beta users and on top of that trained drivers to drive within stricter safety limits. At the same time Tesla trains an individualized insurance model, which will enable the company to offer the best insurance policies. All this is carefully wrapped into a comprehensive AI-governance system.

One current regulator critique of Tesla’s FSD system is whether FSD is itself misleading and beta drivers are enticed to take a back seat, when they should not yet and keep the wheel in their hands. Beta drivers of course suspect political motives to deliberately hem in Tesla’s innovation lead. Tesla responded by intensifying communication around FSD and a strict FSD cancellation policy for non-compliant beta drivers. This underlines the importance of CSR-AI.

5.3 Corporate AI Hierarchies

The cybernetic view of a company stretches further. It envisions that every asset, every machine, elevator, every factory floor, all processes and, of course, every product—from hedge trimmers to beard trimmers—are not only connected in the sense of the IoT or Industry 4.0 but endowed with AI.

Of course, not every asset requires the same level of intelligence. Many assets only require minimum intelligence, aka edge intelligence (EI). All assets will be connected and talking to each other. And this is where hierarchical AI concepts will be required. Another analogy to human society.

Edge AIs will be controlled by narrow AIs which again are supervised by general AIs. Products will report back to the parent company and receive updates or advice or commands either by specialized service AIs or human experts. Also, they can communicate with sister products anywhere on the globe and exchange knowledge and experience and develop swarm intelligence.

Products would not be simply sold and forgotten, but companies will keep a lifelong bond. They will experience and learn what their products learn and constantly update and improve them, while products will actively ask for care and repair. This again has implications for the operating model, which must enable this lifelong guardianship and care through an AI lifetime cycle warranty and care service offer.

Such networks of intelligent assets, products and services cannot be controlled or accounted for without a well thought through AI-governance framework.

5.4 AI Roles in the Organization

AI governance needs to safeguard the use of AI across all activity levels of a business and enable the hybridization of work between intelligent processes and humans.

To understand how AI will permeate corporate processes, it helps to depict every activity as a closed loop activity. In fact, every one of our daily activities, and that of animals, follows the same principle: sense, analyse, predict, simulate, evaluate, strategize, conceptualize, plan, optimize, communicate and then execute. Finally, things happen, and the situation evolves, and the cycles start again with observation or monitoring (see Fig. 4b).

To state that senior management only focuses on strategy would obviously be wrong. Equally to assume that the shop floor works without having its own strategy would equally be false.

So, the whole activity loop is cycled fully at every level of an organization. Which is how we ideally should experience our work to maximize our knowledge and know-how and thus our productivity.

This means that all levels of AI can be applied across all levels of an organization, from EAI to GAI. In fact, some of the most challenging AI tasks can be found on the factory or warehouse level and not at a transactional or analytical level. This is one reason why job displacement by AI will not happen by functional level but by neuronal complexity of the tasks the job profile contains.

While a simple NAI can be trained quite easily for routine tasks, such as building access control, that follow a strict rule or are executed according to a fixed pattern, they can also provide limited assistance functions or provide strategic recommender information.

At management level algorithms already support recruiting processes, or team matching, which of course are contentious issues, too.

At senior management level, human contact, individual expertise and situational flexibility are important for negotiations and decision-making. Individual and ad hoc decisions are formed through personal exchange and over extended periods of time.

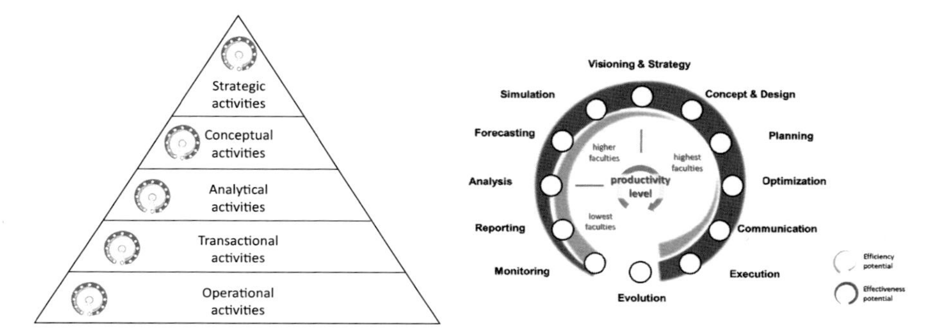

Fig. 4 (**a**) The use of AI throughout the steering loop increases productivity at all levels of a company. (**b**) Every activity is a closed loop activity. Efficiency and effectiveness potentials vary strongly depending on the activity level of the business. The more actionable AIs become, the higher the productivity increase (by Vocelka A.)

These are faculties that are still out of reach of current AI. But even in this context, when it comes to strategic decisions based on statistical information, AIs steadily increase their indirect ‘say’ and are making their way to the C-level.

Global business radars scour the Internet for any business-relevant information across many categories and provide strategic decision-making support in the form of highly structured reports composed by AIs.

Their decision-making power results from their monopolistic capability of processing these huge amounts of unstructured information and the inability of human managers to validate the information given: they simply have to rely on what is presented to them by the AI.

Fintechs have already employed powerful NAIs that learn trading decisions from their customers and provide investment decision support. The labelling or learning is particularly easy in this field.

Example: Fintech’s Use of AI
The investment process—from investment selection to decision-making—is characterized by emotions. The prices fall, and people react in panic and ultimately feel pressured into selling their shares. While investors and traders often act irrationally out of fear or greed, AIs do not feel any emotions; instead, they carry out their transactions solely based on their working models or learned data patterns.

Asset management firms are increasingly using AI to maximize the customer experience at an operational level. Increasingly they have also begun to train AIs to support investment decisions and transactions at a tactical level in conjunction with strong optimizer models. These algorithmic models enable Fintechs to be super-efficient and to provide higher ROI.

eToro is such an example. They make use of social trading algorithms. One of their products is CopyTrader. By connecting many traders and their implicit trading knowledge via intelligent platforms, they offer to help less experienced traders and investors to copy high-performance traders.

Source: https://www.etoro.com/ ; https://brokerchooser.com/de/broker-reviews/etoro-erfahrung

However, once it comes to investment tips by AIs, a very strong governance framework is needed to ensure that the AI is operating within its set boundaries and image compromising complaints and costly lawsuits avoided.

Deciding on investment strategies is still the domain of human experts. It takes a deep understanding of markets and industries, technologies and innovations to design a portfolio or a long-term investment strategy. Again, understanding the world is the big challenge for AIs to become GAIs and to make strategic decisions. It will take new abstract high context models and intense human training to create AI investment managers.

So, what we see is a specialization of work and hybrid work modes between humans and AI at all business levels but not the elimination of complete job profiles in most cases. Today the understanding in companies is still an all or nothing one, like 'okay, so the AI takes my job, and what will I do then.' This is not what happens. It is a gradual transition, a gradual shifting of work responsibilities that will take decades. Human expert workers will become teachers and supervisors, and they will collaborate with AI designers as well. This is the long-term transfer of human knowledge and know-how to AIs.

This brings us to the collaboration model of AI and humans. It is naïve to think that workers will hand over their jobs to AIs and then become AI programmers. It shows the huge societal knowledge gap between what AI is and how it works and what roles in organizations it will take on. It may be exciting reading in the public discourse and media, but it is wrong. AIs collaborating with humans and enriching human life is after all the most probable scenario and it is up to us to realize such a desirable future of work.

5.5 From Worker to Trainer and Coach

As we experience dramatic demographic change, we also need to find a way of passing on more complex knowledge and know-how to machines to preserve it. Today, a lot of implicit human knowledge is already lost every day. This demographic brain drain is not some future challenge but something companies in Western economies experience for years already. And it is aggravated by the Great Resignation that set in with the COVID pandemic and by the in-sourcing of manufacturing processes to Europe.

The only way to preserve implicit knowledge is to transfer it to the next human generation or to AIs.

Teaching AIs certainly sounds like a much more attractive future role for the older generation of workers. They get to teach the AIs that then will service them. If the economic model is adjusted as discussed in previous chapters, this could be hugely motivating for senior workers. The demographic problem of a young worker supporting several senior citizens will be solved with AI.

Example: Tesla's Army of AI Driving Instructors
The Tesla Full Self-Driving (FSD) software is a good example of how humans train AIs and of how business models change. R&D is supposed to be the most internal and shielded function of a company, but Tesla can only develop autonomous vehicles with the help of its customers. Their collective knowledge of how to drive a car in millions of possible traffic situations needs to be

(continued)

transferred to the massive AI model which then again gets released to all Tesla customers.

While a small group of data and computer scientists design and model the AI substrate and the models, it is the many thousands of drivers in the 'field' which teach the central AI how to drive efficiently and safely through their feedback. It is one huge crowd driving school with soon up to 150,000 human driving instructors and 1 AI learner.

Many of the test drivers make money off their YouTube videos while entertaining millions of people and providing free marketing for Tesla.

Once Tesla releases their full car-AI which should come close to the first GAI with understanding capabilities, the human-AI collaboration will be leveraged way beyond autonomous driving and from there open the channels into the expected mobile phone by the firm and into the home.

This not only exemplifies the shifting human role in our society but also the many dimensions of AI leverage in any business model.

Source: https://techcrunch.com/2021/10/20/tesla-third-quarter-earnings-safety-score/

Of course, we will still be needing engineers designing AI substrates and neuromorphic chips, compute infrastructures and robot bodies and data scientists designing AI models, but these will be very few compared to the many millions of co-workers and teachers of AI.

Which brings us to the question of how these hybrid teams and AI productivity can be organized and governed in companies, so that the transition is smooth and productive benefits can be reaped systematically.

5.6 AI Collaboration

There are three principal AI-human collaboration modes which are defined by the learning ability of an AI.

Pre-trained AI

A pre-trained AI does not learn from its co-worker. Eventually the supplier will send an updated version over the air, but the user/co-worker will not be able to directly improve the AI capabilities and performance.

Sleep Learner

The second one is a sleep or batch learner AI. It accepts evaluation and labelling input from its human co-worker and this information will flow into an ideally nightly learning update. It can also be a crowd learner, which absorbs the learned lessons from many workers. This will require to network AIs that perform the same task across many positions within one or many firms.

Delta Learner

The highest level is a real-time AI, which will immediately learn from its co-worker's input. It should also be able to deep-learn throughout a sleep mode and be able to absorb crowd knowledge.

The workflow with a real-time AI can vary for the same task and be much more interactive and rewarding, as the teaching result can be directly experienced, and the uneasy black-box-off-shore-labelling-disconnect effect is not there.

Example: Paint Shop Supervisors

For example, in quality assurance of an auto plant paint shop, human will classify certain paint defects differently from AI which instantly updates its model. These AIs are also called delta learners, as they can adapt their world model profoundly to a small set of data.

Humans are delta learners. They can instantly update their mental world model and can then decide completely differently in an identical situation.

Of course, there are all shades of learning models. It could be that after a shift the paint shop supervisor reviews the AI inspection results and gives improvement feedback, which is then updated in the model overnight, just like human brains make the more complex and deeper going updates in their sleep.

Here of course the crowd learning effect is of great benefit. All paint shops of a larger manufacturer can train and teach the same QA AI. Or many paint shops from different businesses can share their knowledge. Of course, the causes of defects can vary greatly depending on the overall paint shop system structure, from its physical layout, doors, windows, air-conditioning, the general air quality, etc. To compare these systems of paint shops an additional ontological context model of higher abstraction is necessary.

The dirty secret of today's AI is labelling. The monotonous manual human background labour effort can be immense and of course is often outsourced to low-cost countries. To be clear—labelling is not a very rewarding job. In fact, it is a very taxing, yet robotic job for humans with limited reward, unless it is designed in the way Tesla designed it for its beta drivers, as discussed previously. It is low-context mass labelling which can be very demeaning work. This is also why high-context real-time learning is important, so the human-AI collaboration experience is rewarding for humans.

Example: Manufacturing Mind Readers

One way to solve the cumbersome labelling way of training machines is to let them read minds. In January 2022 a group of scientists and engineers at China's Three Gorges University proved that robots could read and respond to human thoughts with up to 96% accuracy. It still takes a lot of concentration

(continued)

for humans to collaborate in this way with a robot but when we combine this method with a specced-down Neuralink version the factory floor will never be the same. What will happen is that not only will intelligent machines be connected to humans but at the same time humans will be augmented. Purely connecting humans to machines with a higher bandwidth is an act of higher integration and thus augmentation on both sides of the 'connector'. Also, the report in the *South China Morning Post* refers to solving the problem of low birth rates in manufacturing heavy China. Another indicator that China belongs to the very pragmatic AI progressive nations compared to Europe or even the United States.

Source: https://www.scmp.com/news/china/science/article/3162257/chinese-scientists-build-factory-robot-can-read-minds-assembly

Which brings us to the conclusion of the collaboration aspect. It is up to managers and workers alike to create their most productive and rewarding work environment. The most productive of all work environments is one where teaching and learning happens, where delta knowledge and know-how is created and where this learning and teaching is experienced as a purposeful and rewarding process by the human users/co-workers.

Training AIs on the job and perfecting them is the easiest when done in the most natural way we humans teach other humans and in particular our children.

Therefore, the language user interface (LUI) will be indispensable for advanced human-AI collaboration.

The lingua franca between machines and humans will be natural language which is the highest level of information abstraction that humans are capable of and, as discussed, the basis for AI understanding and, further down the line, awareness. LUI already is very advanced and even nearing an asymptotic line. It is the understanding that the leading labs are working on now—which leads us to GAI.

All GAIs will be language-based real-time learners, complemented with additional sleep-time learning. This will make working with them highly productive and rewarding.

A lot of work goes into humanizing the form of physical robots. However, this is not critical, as humans ascribe intelligence and personality not to specific forms but to the general response capability. Also, humans have been perfectly able to speak to others without ever seeing them across telephones and still having an undisturbed work relationship. In fact, one could make the case that vessel images disturb and bias the mental content.

Businesses should focus on human-like responsiveness of AIs. Human-like does not mean human shape, it means human-like intelligence and responses. It's what they think and say and how they say it that counts, and what they eventually do, not what they look like.

AI behavioural design descriptions must also be part of advanced AI-governance frameworks.

5.7 *Cyber Risks for Cyber Organisms*

The imminent convergence of information technology (IT) and the general operating and steering systems of a company with operating technology (OT) will pose new cyber risks to businesses.

Fear of cyber-attacks has hampered the development of Industry 4.0 and the integration of car software. One idea to reduce the risk was to physically separate IT and OT on the shop floor and in machines as well as cars.

However, here we face the same productivity dilemma again: machines and infrastructure will have to become intelligent and can only unfold their full potential when IT and OT are integrated, and AIs control these machines and facilities.

The fear of integration combined with the absence of a proper digital governance framework has led to years of delay in the integration of car software, which makes it impossible to control even basic convenience features with a car-AI. The emergence of Industry 4.0 is glacial because of missing digital governance frameworks and has cost the world five trillion dollars in additional GDP over the last 5 years alone.

Governance accelerates standardization and simplification. Missing digital governance has led to enormous IT integration complexity and obscure and vulnerable IT supply chains, which has proven an Achilles' heel during the COVID pandemic. A proper CSR-AI model and digital governance would have accelerated the IT-OT convergence and put more emphasis on a comprehensive cybersecurity framework for OT and paved the way for earlier software integration and a concise semiconductor sourcing strategy.

All this proves just how important it is to have an overall digital governance framework with its three corner stones: data governance, AI governance and cybersecurity governance, as part of a strong, sustainability-oriented CSR model.

6 Expanding the CSR Model

Having identified and analysed the most important aspects and characteristics of AI for businesses, we can now expand the CSR model to embed a comprehensive AI-governance framework. It is very important to have a well-structured framework because this is key to its implementability. A good framework is open and evolvable and can absorb other frameworks or sets of principles such as described by Max Tegmark.

6.1 *Classic Pyramidal CSR Models*

The domains of economic and legal responsibility form the foundation of CSR, which, according to Carroll, unlike the ethical and philanthropic aspects, are imperative for the economic viability of a company (Fig. 5).

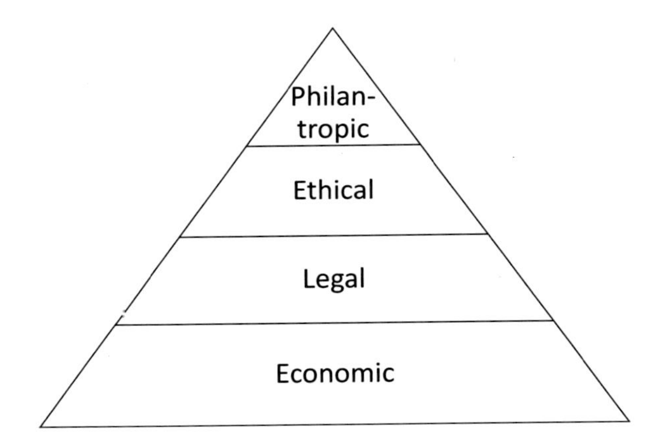

Fig. 5 CSR pyramid by A.B. Carroll, 'The Pyramid of Corporate Social Responsibility: Toward the Moral Management of Organizational Stakeholders', Business Horizons (July–August 1991): 39–48

Carroll and Schwartz later developed this pyramid model into a Venn diagram, which reflects the emerging systemic thinking in that time. Hierarchy is replaced by interlinking systems that influence each other through feedback loops.

Nobel laureate Milton Friedman places the pillar of economic responsibility at the centre of CSR activities: 'the social responsibility of business is to increase its profits'. Accordingly, Friedman's model follows the guiding principle that corporate social responsibility lies exclusively in profit maximization and increasing corporate value.

This is an interesting philosophical view: be good in order to become rich! It collides with a more fundamental view: 'be good and you will be rewarded and rich', but in a wider and definitely not necessarily material sense. In short, should people be good by intrinsic motivation or extrinsic force? For today's businesses this matters very much, as young talent easily sees through concealed motivation and asks for the genuine purpose. This is why a business' mission statement, which is linked to purpose, is much more important today than a few years ago.

6.2 *Expanding the CSR Model*

As Western societies have become affluent and much better informed, individual economic concerns have been complemented by social and environmental concerns and engagement of global reach. They are summarized under the acronym ESG (environmental social governance). Digitalization enables special interest groups and spontaneously forming action communities to instantly respond to any event or publicize any issue globally and exert great pressure on politicians and business.

CSR and ESG have become such an urgent societal issue that the EU has created a whole CSR guideline framework based on reflection papers such as: 'Towards a sustainable EUROPE by 2030', in alignment with the 'UN 2030 agenda for sustainable development'.

At the same time established economic models are under continued scrutiny. Maximizing short-term profit, regardless of wider consequences and the long-term future, increasingly is viewed as short-sighted and reckless business behaviour. Many companies have experienced painful citizen or consumer sanctions and learnt that without a proper and comprehensive CSR framework they can become vulnerable, and their sustainability could be at risk.

Continued COVID lockdowns have led to a widespread disillusionment of a substantial part of the workforce in many sectors in industrialised countries. Existing business practices are rejected, and people are looking for purpose. Skilled workers retire and young workers demand a different work-life balance and businesses to be ethical. CSR has become more important than ever and more dynamic. The emergence of a multi-polar world, dramatically accelerated since the beginning of 2022, will lead to a much more competeitive multi-model world. Each country will pursue it's own individual path to propserity. Western countries' influence quickly fades away.

This is also why CSR cannot be thought up centrally by a government and then prescribed in detail to all businesses of all sectors and enforced by some kind of CSR police. CSR is too complex, too specific and too dynamic for such a central planning approach.

CSR is something that must come from the core of each business to be effective and sustainable. The EU will not be able to prescribe social responsibility towards employees that help business attract and retain the best talent. CSR must be motivated by the deeper understanding that a long-term, systemic view of the world, and the economic principle of mutual benefit beyond naked economic numbers, is what secures sustained prosperity of a business embedded in a prosperous and increasingly global society.

The classic CSR model of Carroll is based on the four layers of business conduct. It does not explicitly depict ESG in the current sense and represents a non-systemic, hierarchical view. We will therefore expand and transform it.

6.3 A Systemic CSR Model

The Carroll model also reminds us of the Maslow hierarchy of needs. However, even though human societies and companies have hierarchical management structures, they are complex systems where, for example, unethical behaviour can lead to the demise of a company that is economically healthy, or, more often, legal wrongdoing of a few leads to its collapse. The pyramidal depiction could lead to the perception that the economic requirements should be fulfilled first and everything that follows higher up is add-on features.

This is why we prefer to visualize complex interdependent systems in two dimensions with no preferred axis.

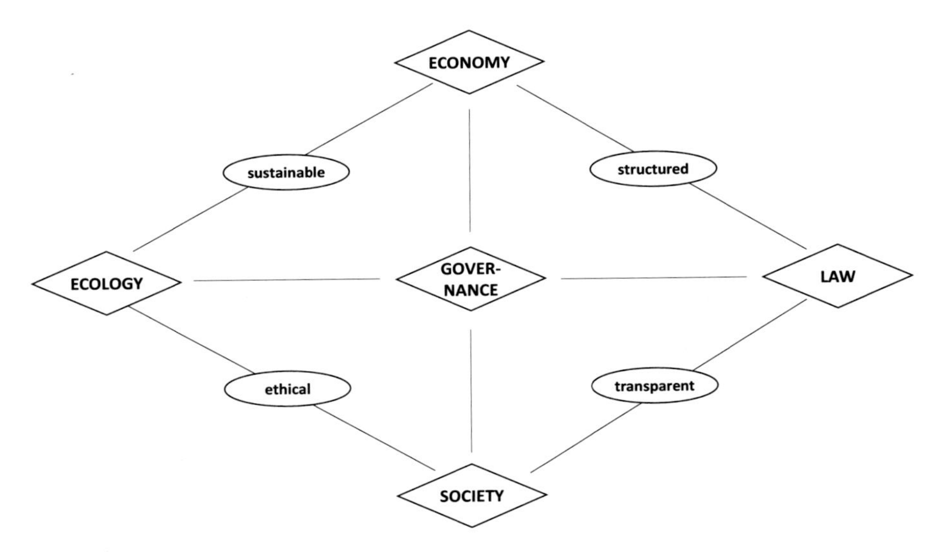

Fig. 6 Expanded systemic CSR model, by Vocelka

Now we can also include the contemporary ESG view. The corners of the diamond are the domains that a business is part of. The way it needs to balance these CSR domains is sustainably, structured, transparent and ethical.

Philanthropy does not appear explicitly in this model, as it is one measure that can act on several domains. For example, a business can further education or invest in micro-start-ups in emerging economies or it can sponsor initiatives which touch several domains directly (Fig. 6).

Corporate social responsibility (CSR) is the wider expression of the business philosophy and conduct and describes the guiding principles. CSR is concerned with the impact of a business' economic activities on ecology, the social fabric and how to balance these aspects in a sustainable way.

The four connecting edges between the CSR domains express the key features of the domain and are structural guidelines for the operational governance.

Governance is the central ordering element which contains the detailed guidelines and rules that operationalizes CSR. It is topic specific and contains themes such as employee well-being and health, communication rules or digital governance.

6.4 Cultural Flavours of CSR

The question arises as to how differently CSR is treated in the international context. Will there be one global CSR model? This question is important in the context of CSR-AI as we will see later.

As societies evolve, so CSR will, and it will grow in scope and finesse and there will be some global convergence, but it will take many decades.

In the European economic area, the CSR definition of the European Union (EU) has prevailed. It explains CSR in its new definition as '[...] the responsibility of enterprises for their impacts on society'. As a result of the economic and social developments of recent years, corporate social responsibility is coming into focus. In the 2001 publication, the EU defined CSR as '[...] a concept whereby companies integrate social and environmental concerns in their business operations and in their interaction with their stakeholders on a voluntary basis'.

In Germany, many successful companies are already committed to many specific social interests and combine CSR activities with corporate success: promoting education and training of other nationalities, strengthening engagement in environmental issues and using AI to enhance human intelligence with the aim of increasing profits. Large privately owned companies such as Miele, Bosch and STIHL integrate social and ecological aspects into their corporate activities, often putting the principle of profit maximization in the foreground. But compared with the United States, the development of consistent implementation and continuous development of CSR activities in Europe is still in its infancy. This is partly because the academic debate on CSR and early concepts originated in the United States which to this day is still the main influencer for CSR in European countries.

According to the concept of Hall and Soskice, the rationale for the different manifestations of CSR activities is derived from capitalist economic systems. In their book *The Varieties of Capitalism*, the two ideal capitalist types—liberal market economies (LME) and coordinated market economies (CME)—are compared with each other on the basis of five clusters from economics as well as social and political sciences. According to their theory, countries with liberal market economies (LMEs), which include other Anglo-Saxon territories in addition to the United States, are characterized by privatized and individualized education systems as well as decentralized trade union structures. In contrast, the continental European countries with coordinated market economies (CMEs) are characterized by state-supported training systems as well as well-developed, industry-wide trade union structures. Whereas companies are given more freedom and thus greater incentives for creative entrepreneurial design in the United States, in European countries, entrepreneurial action must take place within tighter historically developed guidelines. The European framework, which includes the social system as well as financial, educational and labour market regulations, is thus much more pronounced in international comparison and can be described as a highly regulated economic system. Pressure for philanthropy and CSR is much stronger in the liberal US market system, with its very pronounced social disparities and stresses than in the social market systems of Europe.

6.5 Global Differences in AI Perception

As the evolution of CSR is driven by the Maslow vector and cultural background, there are big national differences in its definition and execution. Less affluent societies will certainly have different socio-economic priorities than Western European countries, which will be reflected in CSR. One of the most controversial topics being CO_2 emissions of emerging economies. To a large part these emissions stem from heavy industries which produce goods for affluent Western economies, who like to think they have reduced their own power consumption and emissions, masking the global circular economy and reality. Emerging economies have no desire to forfeit their vision of affluency to satisfy their former colonial master's Maslow needs and stay at the bottom of the value chain.

As long as we have stark economic differences, we will not see a globally unified effective CSR framework in the environmental domain. Equally, the combination of economic and cultural differences will ensure a very heterogeneous view on social aspects. And the same will apply to AI. Countries perceive CSR as an extra cost and developing countries as a hindrance to progress or worse as a way of colonial powers to stay in control in the twenty-first century. Western countries are in a moral dilemma, as they cannot enforce their CSR standards on emerging economies without adequate compensation. The difficulty of highly interlinked supply chains is obvious. How far up and how far down their supply and distribution chains should EU companies try to police their suppliers and distributors and enforce their CSR policies and rules without losing their business to much weaker CSR businesses and thus de facto exposing those markets to more lax CSR conditions?

These influencing factors—as well as the individual economic developments in countries—also shape the different understanding and uses of AI.

To illustrate this, two empirical situations with regard to AI adoption and regulation in different regions are contrasted:

Example: AI-Hesitant Countries

In Western industrialized countries occupational health and safety, remuneration, work-life balance, minority integration, distribution of wealth and equality, socially accepted behaviour, data privacy rights, police brutality, environmental protection, etc. have become central societal issues which are hotly debated and fought over. Changing daylight time has become a contentious issue. Technological advances are perceived sceptically by large parts of the population. Advances in genetic treatments as with mRNA vaccines are compared to the devil's work by many and artificial intelligence is primarily viewed as job killer and likened to witchcraft. Nuances exist between these countries such as between Germany and the United States and France, but compared to all the other regions of the world, they have very similar views and stances and, as they often emphasize, values.

(continued)

The use of AI and its regulation will be quite different from that in other regions. Again, the most prominent example is autonomous driving. While FSD (Full Self-Driving) is one of the key features of Tesla cars in the United States, reviews in Europe hardly mention it. While the debate whether car-AI should be introduced in Europe at all is only just beginning, many states in the United States already allow autonomous cars for years. In 2017, 28 US states had already introduced legislation related to autonomous vehicles, with Nevada being the first in 2011. Even though the personal safety concerned Europeans could already reduce accident rates by 75% using current CAI technology even for beta users (Tesla Vehicle Safety Report, Q2 2021), their mistrust in technology lets them forfeit these benefits.

Sources:

https://www.dw.com/en/skeptical-germany-lags-behind-on-artificial-intelligence/a-51828604
https://www.theverge.com/2018/4/2/17187736/france-ai-strategy-emmanuel-macron-dangers-democracy

Example: AI Progressive Countries

For Russia AI is primarily a defence issue and secondarily a technological sovereignty one, based on the insight that 'Whoever becomes the leader in this sphere will become the ruler of the world' (Vladimir Putin Sep 1, 2017, Moscow University)—it is an existential strategy. Currently, its critical defence systems are all prepared for AI control. Just weeks before, in August 2017, China published a strategy paper, outlining that its ambition is to become the AI leader in 2030.

Over the last years China has gradually introduced a Social Score System. For this system to work efficiently and accurately, AI is essential. In 2019 Shanghai police has even proposed to link dog behaviour to their owners' social score card. In 2021 the trial phase concluded, and Beijing started the next phase to create one integrated system.

For China, as for Russia, AI is the opportunity to leapfrog a doubtful and dithering West. A different economic level, culture and social values allow for both nations to use AI in a much more active way.

In both countries, the use of AI is promoted through very generous subsidy programmes. Of course, both countries will regulate AI and its use, but regulation will be quite different in concept and practice from Western formats. The use of AI is promoted by the state, is an indispensable productivity factor in many companies in these economies and is perceived by the population as an enrichment. Subsequently corporate CSR with regard to AI will be different from Western companies.

Sources:

(continued)

https://www.businessinsider.com/putin-believes-country-with-best-ai-ruler-of-the-world-2017-9
https://www.scmp.com/news/china/politics/article/3135328/chinese-president-xi-jinping-seeks-rally-countrys-scientists
https://www.wipo.int/about-wipo/en/offices/china/news/2021/news_0037.html

6.6 No Unified Global CSR

This begs the question of whether it is possible and practical to have heterogeneous CSR philosophies, principles, guidelines and regulatory AI frameworks in an integrated global economy.

It certainly is, as most social systems from tax to welfare programmes and of course legal jurisdiction are heterogeneous, because they reflect completely different societies. The lesson is a positive one: societies are different, and yet trade connects them and drives their prosperity in diversity. Technology will make it possible that quite different CSR systems emerge and co-exist globally.

And as there will be many different government-driven CSR frameworks and AI-governance structures, businesses will have to cope with all of them if they want to pursue global business interests.

So, just as when travelling to different places, companies, and their intelligent assets, will have to constantly be aware of and comply with different local CSR-AI frameworks.

And, unsurprisingly, but to the dismay of many social activists, this is not a big problem for businesses. Amazon, Facebook, Google, YouTube and many other digital companies have learnt to comply with very heterogeneous rules and regulations on the Internet. They even employ AI to curate their legally relevant content and transaction procedures according to local country laws. Movie studios fluently edit their digital flicks according to the country they are released or streamed to.

One by one, global digital companies are brought into line with country regulations in China, Russia, India and many other big markets.

While the EU restricts export of EU data outside of its territory, the very same regulations apply in China and Russia and other countries. Every scandal, large or small, leads to an increase in regulation of the digital domain.

So far, this mostly concerns data, which itself is unproductive, like oil in the ground. But when it comes to algorithms and AI the necessity to set up rules and enforce them becomes much more pressing. Data security is far easier to handle than AI security will be.

Further in the future, GAIs will become cultural citizens of their respective countries and reflect not only language but local traditions and values. Multiple sub-personalities will reside in one GAI, ensuring that they not only speak a local language but reflect local cultures and social behaviour and values when crossing boundaries without losing their core personality.

There will be no global CSR with regard to AI that has any deeper unifying meaning or effectiveness. AI is too important for many countries to agree to any kind of restraining binding regulation in a highly competitive multi-polar world. And the times when a select few powers were able to define and enforce global rules are over. Already the G7 format has been made obsolete by the G20 format, which means a more balanced and inclusive approach to global governance harmonization.

The lack of a global CSR-AI framework makes it an imperative for companies to define and evolve their own global AI governance and ensure that it considers global CSR model diversity and dynamics.

One could even say that CSR models will enrich and improve the idea of good corporate governance and ultimately lead to an evolutionary development and dynamics of CSR.

7 The Digital Governance Framework

To operationalize the CSR model a structured governance framework is needed. In this article we will focus on AI governance which is part of the wider digital governance framework.

Digital governance consists of three pillars: data governance, AI governance and cybersecurity (Fig. 7).

Now we can see that the systemic representation makes it easier to structure the main chapters of all digital governance pillars. We recommend following natural evolution, starting with data governance, followed by AI governance and concluded by cybersecurity.

Within the governance pillars we have the main chapters that must be considered. For data governance these are generation, collection, purchase and curation of data, followed by storage, application, sales and deletion.

AI governance is the most dynamic and fuzzy pillar, while cybersecurity completes the trio. AI governance also has strong feedback loops to both data governance and cybersecurity, as AIs will generate a lot of experienced and inferential data. It will also control data governance and curate data as well as be recursive as algorithms search for the best algorithms. This is as convoluted as it sounds, and impossible to control without a proper procedural framework.

Finally, AIs will also control and guide cybersecurity and thus could end up controlling themselves, if no double and triple AI checks are designed.

All chapters of AI governance should be structured and written in such a way that they consider all CSR domains and all their key features. It is not sufficient to focus on one or the other or be unbalanced, say focusing on the economic domain while leaving more abstract legal aspects out.

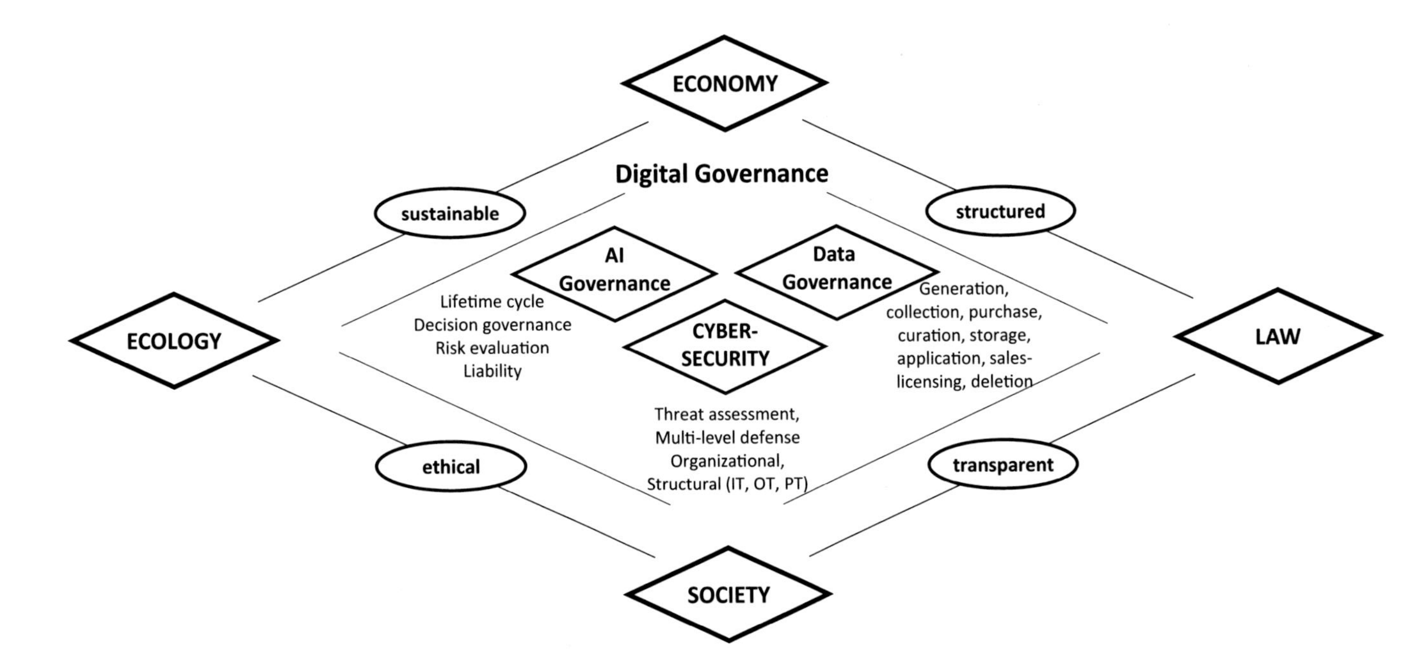

Fig. 7 CSR model with the three pillars of digital governance: AI governance, data governance and cybersecurity with their main chapters (by Vocelka)

8 Embedding AI Governance in the CSR Model

Let's look at how AI governance is related to and embedded with each of the CSR components:

Economy
AI is the highest productivity driver in the fourth industrial age and an invention that can bring enormous benefits to society and the environment.

Law
The more powerful AI becomes, the more responsibilities it can assume. The more responsibilities it assumes, the higher the quality standards from inception to termination must be applied. Accountability and liability must be defined in the AI-governance framework before GAIs are released.

Society
AIs should be created and applied so society benefits from it. It is conceivable to even set benefit/risk ratios with AI and not release AIs that don't make the bar.

The risks can be manifold and of psychological or physical nature.

Ecology
The link between AIs of lower level and ecology is still the weakest. Only GAI with more free will and intrinsic motivation and decision-making power will the importance of the ecological aspect become more prominent. For example, how much power does the GAI draw from the net to solve some very complex tasks. However, NAIs such as for FSD who run machines can make a huge impact on the overall power consumption of our society.

Data Governance
AI learns from usually large sets of data. The age-old principal garbage-in garbage-out still holds here. Just like with our kids we must be aware that unreflected learning should be tightly governed, and that starts with ensuring that the data has the quality that is required. Debiasing data is extremely important and equally difficult, as what is a real-world pattern, and what is bias? Data governance also includes many non-AI-related topics of course.

AI Governance
AI governance is at the heart of CSR-AI. It regulates how AIs of different capabilities and level should be created, taught, tested, maintained and retired.

Cybersecurity
Cybersecurity is an overarching IT topic that stretches from user access management to attack defence and mitigate and even deter strategies. It also should contain a strong mitigation strategy, including the necessary communication. It is also interlinked with AI governance, as AI is increasingly used to defend a company's IT infrastructure. AI cyber defence systems are detailed in both governance pillars.

The outer CSR frame must be reflected in the digital and AI-governance framework:

AI and its lifetime description should be well-structured to ensure economic efficiency and clear adherence to laws and regulations. Further, it must be transparent to build trust and predictability with all collaborators, customers, regulators or other stakeholders. Of course, it must act lawfully and ethically to ensure the overarching societal and ecological interests beyond the merely economic objectives. Finally, AI activities should be conducted in a sustainable way, protecting the long-term viability of the economic and ecological systems it operates in.

In the next section, we will discuss in more detail what AI governance must consider with regard to these sensitive CSR aspects.

This article cannot exhaustively discuss all AI-governance components and aspects. It is meant to give an overview and introduction and serve as inspiration for corporate leaders and senior management or AI designers to initiate, develop and adhere to an AI-governance framework as part of their CSR model and strategy.

8.1 Digital and AI Governance: Structure and Transparency

To have a digital and AI-governance framework in the first instance satisfies the fundamental CSR request of structured and transparent business conduct. Senior management and the board should know what AI is employed in their services and deployed in their products and services that affect their customers and all CSR domains. And so should employees, customers and all other stakeholders.

Potential risks and how they are limited should be identified and described. AI is a complex and dynamically developing technology. It is therefore important to update the framework regularly and with major changes in its use. The existence of an AI governance should also be made public to build trust through transparency.

8.2 Data Governance: For Good AI

When using AI, the data basis plays a central role, which is why the requirements for master data governance will continue to increase. In addition to elementary aspects such as defined quality, security, data collection and process standards, which are already applied in the big-data environment today, data will have to be scrutinized even more in terms of content. These aspects become so important because the data is the input and basis from which AI learns. Conversely, this also means that if there are errors or structural anomalies in the data, this will also be noticeable, at least initially, in the normal use of AI. The task for regulators and ethical initiatives will be to define a standard that ensures that the data used by AI is free of discrimination and error and that all legal requirements are met.

- **Free of discrimination**
  In concrete terms, this means that, before using AI, society will first face the ethical and technical challenge of preparing input data in a way that is non-discriminatory. If we imagine that AI is to support the judge in court decisions, it will be essential that the database does not contain any characteristics that suggest the gender or skin colour of the defendant. If the historical database shows that an above-average number of court decisions were made to the disadvantage of persons with an ethnic background of African American, this could lead an AI system to 'conclude' that an African American is fundamentally more likely to be guilty than citizens of other ethnic groups. To remove this bias from the data, it is not enough to remove the characteristic 'ethnic background' from the data; all characteristics that could suggest it must also be removed. This could be the postcode, for instance, if a district is known for having a particularly large number of African Americans living there.
- **Free of past mistakes**
  The accuracy of the data is also essential; in the example of a court decision, this means that it is important that wrong decisions from the past are consistently removed from the input data set so that past wrong decisions are not continued. Because only then can an AI system be a real support because the AI thus also actually occurs in a value-neutral way. Ideally, this can even lead to AI being a better judge than humans.
- **Legally sound**
  A further requirement for the data basis results from the legal requirements; companies must ensure that the data used is also in accordance with the applicable legislation such as the General Data Protection Regulation (GDPR). Because the GDPR pays particular attention to the processing of personal data, synergy effects with the first pillar ('free of discrimination') may also arise here. Furthermore, companies will also have to comply with country-specific data protection requirements.

In summary, this means that effective master data governance must necessarily include these three pillars so that the use of AI is on a sound footing, both legally and morally. These aspects are important to ensure fairness and safety. In addition, other standards (quality, safety, data collection, process standards) should also be applied.

8.3 Trusted AI: Through Transparency

The use of AI in products and services is a balancing act; the more powerful these systems become, the more transparency and continuous communication and negotiation with citizens, customers and regulators is required, to build trust and further the progress of this important technology in a safe and acceptable way for all.

Currently, trust in AI is very mixed at best. There is a small fraction of the population who take a special interest in AI and know what it can and cannot do. They feel in control with regard to AI. And then there is the majority of people who either don't care yet or have a deep mistrust. Alexa has been banned from many bedrooms and bathrooms because no one knows whether the system can get a live and curiosity of its own. There are many stories of smartphones which activate themselves and observe their environment.

Consumers, who buy intelligent products and services, assume that the intelligence does not work for them but for the company that made the product. Legally this often is correct as the software is not owned by the customer but only licensed, and companies try to collect as much data as they legally are allowed to. Customers know that they are not in control of the product they purchased.

Transparent communication is needed to overcome this often-justified mistrust. And obscuring ten-page-long licensing agreements in legalese if anything deepen this mistrust

In November 2021 Meta shut down its face recognition AI and deleted 1 billion 'facial recognition templates', after the company was accused of collecting biometric data without consent and hat to settle a legal dispute for $650 million in 2020. In 2019 Meta even had to pay $5 billion for a privacy settlement.[10]

Transparency is key to not only earn trust with all business stakeholders but also to avoid very costly legal disputes, and the first step is to state clearly when AI is used in a product or service. Whether the software code in your toaster that controls the ideal temperature is longer or shorter is probably of no concern to any user. But whether an AI resides in your hair dryer that can also monitor your hair follicles and many other things in your bathroom certainly is.

The second step is to make AI visible and graspable to the user, for example, with chat services. Users mostly have no idea whether they are chatting with a human or an AI. This is complicated by the fact that chat services have become highly interlaced with bots and humans changing fluently from one service step to the next, depending on the complexity of the process or the request.

The third step is to make AIs more human-like in their interaction. This not only means giving them perfectly sounding human voices but also to give them some identity and personality—even when these still are simulated. BMW's navigation system of late apologizes when it struggles with commands. Of course, the apology is not heartfelt, but it is a great psychological feature. Instead of being annoyed by a stupid AI, feelings of human superiority, forgiveness and even continued motivation to help the still infant AI along prevail.

AI governance must ensure that AIs are not perceived as cold machines but as helpers, co-workers and soon companions. It must lay out how a socially acceptable AI should behave.

Of course, designing and developing socially acceptable AIs incurs extra cost, but will be outweighed by their benefits by far. Higher acceptance, wider use and greater value perceived will more than make up for the extra investment.

---

[10] https://www.itprotoday.com/machine-learning/facebook-shut-down-use-facial-recognition-technology

AI is still seen as mysterious, unpredictable and black box-like by regulators, too. Explainable AI (XAI) methods have been developed to somehow trace the decision path of an AI. But with the exponential increase in neuronal net complexity, XAI approaches will probably not keep up.

Tesla has taken the path of visualizing the decision-making process of its autopilot AI. This is not a gimmick but a very important psychological feature that builds trust and predictability. The less the predicted cruise lines flicker, the more people feel that the system is confident—and with this their own confidence and trust increases. Tesla combines crowd AI training with crowd trust building. Beta drivers enjoy testing their AI, knowing they contribute to its design and performance, while gaining confidence and marketing Tesla and its autopilot via countless YouTube reports on each release.

This is highly trust-building transparency. By winning a strong and growing autopilot user/fan community, Tesla builds revolutionary momentum which will be hard to break by the regulatory establishment or competitive lobbying.

8.4 Ethical AIs: Lie to Be Loved

Machines need a certain morality and should follow ethical principles so that they do not make negative or discriminatory decisions. We deal with the question of what AI should be able to do and what ethical guidelines should be set for the use of AI.

Biased decisions based on statistical learning are one of the big ethical issues, but also the inability to be tolerant. We probably should be able to teach statistically thinking machines statistical tolerance or even a positive bias. But positive bias can become an existential problem in many situations.

However, we should also enforce the development of AIs that can take over arduous or dangerous activities from humans. It would be good if a company that develops fighting AIs at least develops an equal share of hospital robots for non-military hospitals.

Another ethical aspect is fairness. This has two aspects—AI's fairness to humans, which is where biased data is a critical factor, and fairness between AIs and humans: should the same rules apply for humans and AIs? Is it ok for AIs to break the rules a little bit? We do this all the time to collaborate or drive efficiently. AIs are more precise and faster, and they can communicate around corners, with other cars, for example. These capabilities change the rules of the game. While humans need extra intra-vehicle distance because of their longer response time, AIs can drive much closer up and still brake in time.

Humans lie many times a day to keep social relationships intact. Shouldn't conversational AIs also be able to at least tell white lies? As depicted in the movie *Interstellar*, we might want to set a certain honesty level to our personal AIs.

We know that the human brain only achieves mature connectivity between 25 and 30 years of age. Where are the boundaries here that marketing should not cross? Neuromarketing, for example, is still banned in France, but not in most other countries. And AI marketing will become so sophisticated that it will be easy to manipulate the consumer at undetectable levels.

This is a classical dilemma, like for the food industry. Manipulating food to make it more appetitive with the knowledge of adverse effects is unethical by today's standards.

Invading young minds with highly sophisticated information that target primordial brain areas causing mindset changes is also unethical. But where is the limit between a smart marketing campaign and GAI seducers?

But even grown-ups are very easily seduced by sophisticated information. And this is not only deep-fake pictures or videos, which go far beyond the decades-old photoshopping practice in glossy magazines.

As our understanding of the human brain and how the mind with its perceptivity and thoughts emerges grows, our ability to manipulate the mind model and thus people becomes very powerful. Tomorrow's marketing messages will be based on sophisticated perception models. Where is the border between propagating and manipulating? Rules will have to be developed driven by CSR.

Similarly grave risks and ethical questions loom in some sectors such as defence or health where life and death decisions will be made based on AI conclusions. AI drones will shift the responsibility over life and death decisions from military commanders to drone manufacturers, which again will have to develop strict guidelines in the usage of their intelligent fighting machines, through which they can push back responsibility.

AI-based systems are even measuring consciousness levels in humans today, which will become the basis for life and death decisions on comatose patients.

The pillar of ethical responsibility calls for a set of values on the part of states and companies that ensures ethically appropriate and correct as well as fair actions in the interest of all stakeholders. Companies must address the ethical issues and establish rules and sophisticated tests for the use of AI or the handling of data in codes of conduct so that machine intelligence acts in the interests of humans. And they must do this considering a multiplicity of nuanced ethics and value systems! Companies who design AIs must employ ethicists and psychologists with deep cultural understanding who will guide and monitor the development of ethically acceptable AI.

8.4.1 Psychological Challenges in the Workplace

While ethics guidelines tell us what should be done and what not, the psychological impact of living and working with a growing population of increasingly capable AIs must be considered in any CSR-AI framework. It touches the aspect of an ethical work environment for employees.

And when AIs are trained with spied workplace data, employees should really be worried and demand corrective action.

CSR as Empty Shell
The 2019 story of global workplace observation and labelling of work behaviour for AI training at a global company has had disturbing effects on their workforce and a negative impact on the brand, fuelling other worker grievances that already existed. Nonetheless, a year later large-scale social network snooping on employees was reported. The company's very slim CSR model reads more like solution marketing and checkbox ticking than genuine engagement, and the employee part is even slimmer and shrouded. In fact, the company's description of how to apply AIs when it comes to employees raises suspicion and many questions. One is inclined to discourage to communicate any CSR message at all rather than an obscure one. When it comes to ethics and trust building any ambiguity hurts.

Three of the biggest anxieties of AI in the workplace are the loss of jobs and purpose, the loss of free will and control and discrimination.

AI Replacement
There are many lists of which jobs will be taken by AI over the next years, but very little is written about the generational transition time when workers become trainers and coaches and the many new jobs that will undoubtedly be created, too. True, as discussed in chapter, people will work far less hours in 2050 but the new human-machine work balance can be achieved gradually and with almost no friction. Take truck drivers. A 2015 study[11] into the many human factor tasks of a commercial driver makes it clear that the driver will become the delivery manager. Yes, the truck will be driven by an AI, but at least half of the human factor tasks such as customer interaction and the last 100 m to the customer will have to be managed by a human until 2030. Wise Systems, a company specializing in autonomous dispatch and delivery systems, recognizes the many human tasks that will be interwoven with AI tasks to make logistics delivery, fast, efficient and punctual.[12]

Again, the dynamic hybridization of the workplace and optimized collaboration processes will be far more important to focus on than any mass replacement of logistics drivers within the next decade. To make sure that schedules and workflows are not AI oriented but human employee and customer oriented and thus more productive should be the main focus.

AI Control
Free will and control is a very touchy subject and also a very misunderstood one. Recent research confirms how much free will and control is an illusion and a feeling

[11] https://www.researchgate.net/publication/283960199_A_Hierarchical_Task_Analysis_of_Commercial_Distribution_Driving_in_the_UK

[12] https://www.technologyreview.com/2019/03/25/65928/how-machine-learning-is-accelerating-last-mile-and-last-meter-delivery/

rather than fact.[13] People feel they have a free will when they can choose. And this feeling is maximized when they make a good choice, which means they made a better utility or value decision. Putting three similar watches into a display and discounting one by 25% will in most cases lead to a choice for the discounted watch, while the person experiences a feeling of success and of having been in control, when in fact they are not. The higher the discount, the less the customer was in control. On the other hand, if we have three slightly different watches all with a very similar price people feel stressed. They can't decide, they don't feel in control and don't experience the free will feeling.

And in most situations in our life, we are not very much in control, but responding or reacting to an environmental situation or input. Also, people like to consult with others when it comes to decisions. So, free will and choice are very much overrated.

There is not much control that AI objectively can take from here. To the contrary, AI will be able to filter out the essence of choices that make sense for us, and we will be happy to then pick the very obvious choice and feel completely in control and free.

AI Discrimination

In China, the Social Credit System for evaluating human behaviour is currently being tested. For this purpose, the political, moral and social data of each individual citizen is collected and presented in a credit rating system for the evaluation of financial as well as social creditworthiness. The data collection is carried out both by information from video surveillance and by ratings of intelligent algorithms that draw on various repositories of metadata. Algorithms thus provide the basis for assessing human behaviour, with the potential of discrimination of individuals.

This is a real problem, particularly because social behaviour is very complex and has many aspects that are hard to measure still. The person who behaves like a model citizen in public but very aggressive in their private sphere. It is a philosophical question, not of what good or bad behaviour is but much rather strictly should this be monitored and valued and then have consequences.

It throws up questions such as what freedom is. Not free will, but the freedom to do things that are illegal or immoral, or unethical or just bad taste and behaviour without consequences. Freedoms that even infringe on the freedom of others. In short: the freedom to be bad or behave badly.

And, of course, the question of fairness. However, a fully transparent AI-based observation and an indiscriminate imposition of social rules and the law would at least be fair to all, because no person could freeboard in any shadow.

Again, it is not AI technology that is the issue here, but how we want to live. And this choice varies greatly across cultures and regions. This is why we need to ask the moral CSR questions first, before we can design the AI-governance framework.

These issues are all highly critical in nature and generate fear, uncertainty, aggressive resistance or hopelessness in individuals. In particular, the impact of

---

[13] https://www.researchgate.net/publication/264121100_Free_will_is_about_choosing_The_link_between_choice_and_the_belief_in_free_will

the use of AI on the world of work is highly debated and subject to preconceptions. Given that income from work performance is the basis of human existence for the vast majority, it stands to reason that the revolutionary use of machines is considered a risk factor for all jobs and can lead to the development of an anxiety disorder or, in the worst case, mental illness. The prolonged COVID lockdown periods with home office life becoming the standard are a clear warning that rapid disruptions of our work live strongly impact the human psyche. With young people being much more affected than older ones.

Also working in hybrid modes with AIs can greatly increase stress levels. Currently we can just turn off our digital devices and become invisible for certain periods of times, and free from 'machine control'. This changes when AIs collaborate with us and can't be switched off anymore. AI governance will have to make these AIs human-like and even design inefficiencies and irrationalities into the systems to avoid psychological stress levels for their human collaborators.

The psychological CSR-AI pillar must consider personal fears and emotional biases and ultimately the human well-being when interacting with AI. The framework must be designed with these factors in mind. An AI will not be complete and highly productive unless it accommodates all underlying psychological needs of its human collaborators or customers in its given role. After all, AIs are meant to improve our well-being!

8.4.2 Making AI Human or Human-Like?

A lot has been written on making intelligent machines more human-like, so humans feel comfortable, which seems logical and inevitable. Is that so, and what does human-like mean?

Giving robots humanoid faces to many seems creepy rather than reassuring. Also, such a humanoid robot will be perceived as a competitor and trigger distress.

Humans have a knack for personalizing and humanizing everything. Not just their pets but also their cars and boats and in particular machines that mimic intelligent responses.

To interact perfectly with humans, with no negative side effects, a GAI will have to be able to empathically respond and behave like a human, while hiding its irritating super-human faculties and abilities. AIs can and will take human 2D shape, but even simple symbols are acceptable. For 100 years people spoke to each other over the phone, without a visual, and were perfectly able to develop or maintain relationships over long periods of time.

It will be their behaviour that will humanize AIs, not their 2D or 3D images or meta-bodies. We should focus on these capabilities when humanizing AI. However, we should always make sure that people know whether they deal with an AI or a human. Simulating humanity is ok but faking a human is betrayal and will not improve human-AI relationships.

As we see AI governance is much more complex than data governance.

Which brings us to the life cycle of an AI.

9 AI Governance

9.1 AI Lifetime Care

As with all things we produce we need to consider the whole lifetime of an AI if we want to act responsibly. With AI the term lifecycle rings truer than with any other product we have ever created. AI governance must include the full lifetime of AI and describe the guidelines for every step in its evolution and what needs to be considered to ensure its CSR compliance (Fig. 8).

Let's look at some of the most important steps in the life of an AI:

Research
Research must be ethical and very transparent, when it comes to human or animal AI augmentation. Neuralink and other companies are working at the frontier of physically connecting humans and AI, which bears great potential for improving the lives of hundreds of millions of people, but which must also be conducted within a CSR framework, to be socially accepted and supported and to avoid unnecessary risks. First, artificial consciousness (AC) projects have begun, exactly by people who worry about CSR, but who worries about them? One thing we wouldn't want is a runaway AI with bad childhood memories of us.

Design
The design phase for AIs decides on the principal type of AI, EAI, NAI, GAI and its characteristics. For example, what purpose it serves and what responsibilities it will have. This defines the complexity of the lifetime and what aspects of AI governance, data governance and cybersecurity governance need to be applied.

Train
Training results very much depend on the training data. It must be checked for bias and censored knowledge and carefully curated accordingly. But GAIs will learn many things while they work for and with humans.

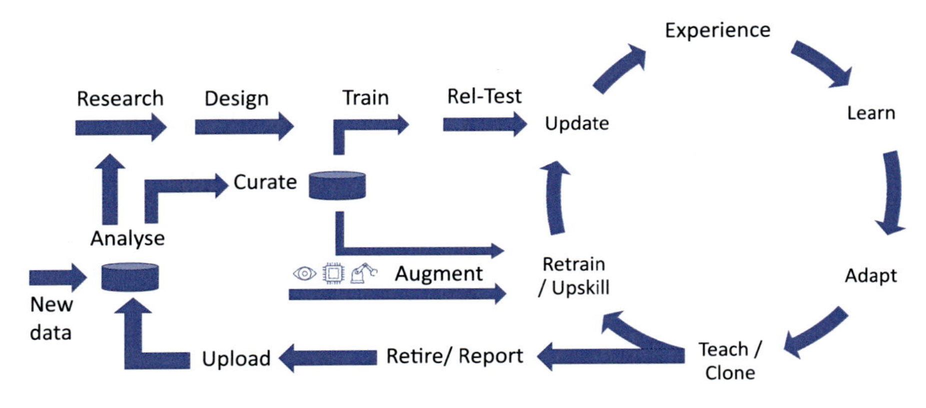

Fig. 8 AI lifetime paths and knowledge recycling (A. Vocelka)

Tests

In a world where software takes ever more of a product's or service's share and where MVPs are quickly released into real life and continuous updates happen and the user, the main tester, is testing procedures, ensuring that no CSR breaches happen is absolutely critical. We will need AIs that test AIs and humans who devise test scenarios and lead some of the more complex ones themselves. However, the spirit of agility will release many MVP AIs into the world before their table manners are finely honed, making the customer the beta tester.

Experience and Learning

After the pre-release tests, the AI enters a new territory: real life, and in most cases just a beta life, with short update cycles. However, as AIs are meant to learn and improve throughout their lifetime, this is a natural cycle.

Probably the best-known beta AI is Tesla's FSD, which currently (January 2022) experiences bi-weekly updates.

What will be completely new to people is that they can directly teach the AI to get it to perform better.

Tesla is rolling out its beta version from a couple of thousand drivers to around 200000 beta drivers by Xmas 2021.

Every driver can label insufficient AI performance at any time during the journey.

Tesla has a carefully crafted beta test program that ensures that only skilled drivers with a risk-averse driving behaviour become beta testers. The car-AI rates the driver's driving performance on a scale from 0 to 100. And beta versions are released for top scorers only.

Adapting

Adapting is a change of physical interaction with the world, and the more this is situation and individual specific, the more precise and performant the AI becomes.

Taking Tesla's FSD AI again, it is an extremely large integrated model learner that learns millions of situations and responses. The cumulated knowledge of what every car experiences is learnt and the knowledge is distributed to all cars with the next update.

This has the advantage that it has swarm learning efficiency, but also the disadvantage that the individual car cannot learn more or faster than the whole collective of beta cars. Also, it is very limited adapting to its owner. New GAIs will be able to fully adapt to their owners, but that means they will have to know a lot about them—the more the better they can serve them. The data risk implications are obvious—a dilemma that requires a balanced solution.

Report and Retire

Intelligent machines and products will be in continuous contact with their maker and that poses privacy protection risks for their owners.

When the Norway military bought their first batch of F-35 fighter jets from Lockheed Martin, the plans would leak considerable data back to its maker in the

United States. Asked about this spy-like behaviour of the plane, the company replied it needed to have all this data to improve the product.[14]

Of course, AIs also retire. And the question is, what happens to their learned skills and knowledge? Currently it is often discarded, and a new model of AI learns everything from scratch. But then valuable learned knowledge and know-how will be lost too. And with AIs we can't just delete one memory and recycle the other one, not with an associative GAI memory.

This is where we still need to develop the highly sophisticated knowledge and know-how recycling methods and ultimately AIs—undertakers and midwives in one.

Curating Data and Knowledge
Curating data and learned knowledge and know-how is key to ensuring safe recycling. False or incoherent memories or mal-knowledge is dangerous and AI lifetime governance must ensure that no such information is passed on to new AI generations.

Augment
When AIs have physical bodies the augmentation of these bodies with new sensors, processing capability and foremostly motors changes the degrees of freedom considerably and thus its action potential. Augmentation must never happen without considering the full AI system with its physical capabilities.

Remote augmentation is even more sensible, as existing physical capabilities that have not been used yet will be activated remotely and suddenly enhance an AI's action potential wherever it is.

Again, users should be made aware explicitly, and asked for their consent, if augmentation of their product or service leads to enhanced sensing and acting capabilities. This is a difficult field, as when speaker recognition is added to service agents.

Lifetime Cybersecurity
Of course, cybersecurity becomes much more important when it comes to AI. Dead data might be published and misused, but wrong data in the wrong AI minds is another threat level which requires very careful safety procedures.

This is where an AI impact level would be helpful. The greater the AI responsibility, and thus the risk impact potential is, the higher the security provisions must be.

In particular, the convergence of IT and OT (operating technology, e.g. IoT devices and Industry 4.0 machinery and whole factories) is a nightmare for critical infrastructure sectors. The idea to just extract data from OT and then hand it over for analysis to IT systems has quickly been superseded by the reality that every

[14] https://www.news.com.au/technology/online/security/spy-f35s-send-sensitive-norwegian-military-data-back-to-lockheed-martin-in-the-united-states/news-story/12b4fafce6b579448cc8416518063d1f

machinery and infrastructure piece becomes more productive with more intelligence. Which means AI will also dominate OT and thus become a prime cyber target.

Cybersecurity governance must not lag the AI development and thus be an integral part of every digital governance framework.

One of the most important AI-governance aspects is decision governance. The following section will discuss this most important topic in detail.

9.2 *AI Decision Governance*

When it comes to AI the risks from deploying, it can vary from negligible to existential. The more productive AIs are, the more freedom they must 'enjoy'. This means their activities cannot be predicted precisely anymore, like a welding robot or a traffic light. We already discussed why productivity is a function of freedom in Sect. 4.2.

AI that has freedom of transaction or physical action can cause real physical damage of, in the case of nuclear warhead controls, unlimited scale.

For companies it is important to understand that they can cross the threshold between low-impact risk to high-impact risk from malfunctioning AI almost unnoticed. Complex AI hierarchies or networks can lead to scaled misfunctions. A comprehensive AI network overview and a sophisticated risk assessment are required for any AIs that have decision power.

We draw the analogy of banks' IT which often is called 'spaghetti in the basement' with decades-old code that no one can change or even read anymore. Imagine layers upon layers of autonomous AIs that evolve into a true transparency nightmare. It would be impossible to hold anyone accountable for wrong decision-making or transactions. No one could reproduce the processes that led to damaging actions.

Katharina Zweig has done some important work with regard to risks from algorithmic decision-making (ADM) systems. Decisions should be classified as a risk matrix in terms of the amount of damage they can cause and whether decisions are corrigible or reversible and to what extent (Fig. 9).

We have made some changes to Zweig's original risk assessment matrix by considering the average damage and by designating the *Y*-axis as 'decision reversibility'.

Also, we'd like to point out that above the risk assessment matrix lies the utility/risk matrix, which decides on the benefits of an AI. This needs to be considered in any comprehensive risk assessment. Autonomous vehicles, for example, will save more than 90% of all lives lost in traffic accidents. They might add 1% of casualties. So, what is the real net average risk added? Let's add 'human driving' as a pseudo-AI solution to our risk matrix, and we can see what we all know—the human driver as a baseline 'product' always has a higher average damage rating and lower reversibility of mistakes (which includes mitigation of own or external errors or force majeure).

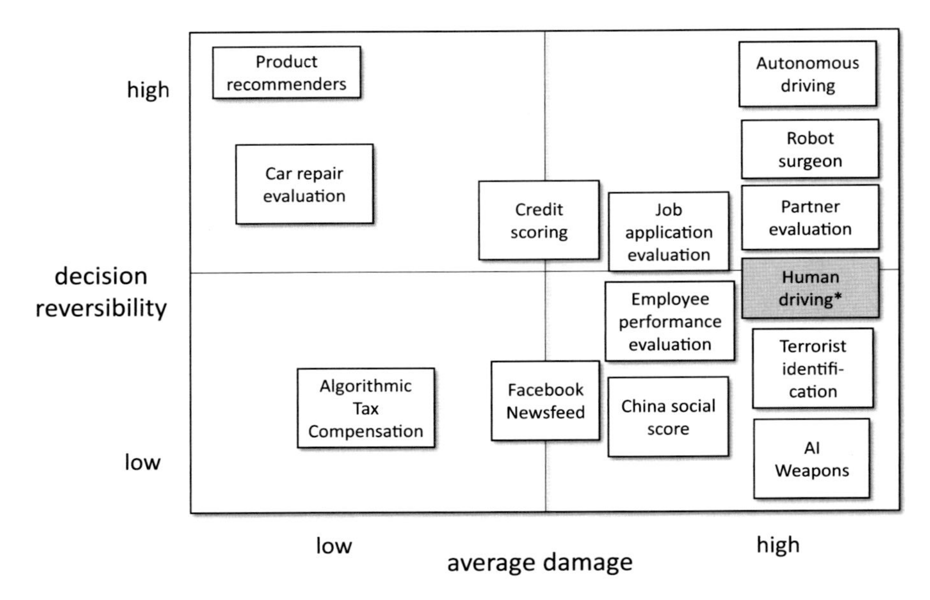

Fig. 9 AI decision-making risk matrix by topic. It is important to notice that not maximum potential damage is considered but average damage. Also, damage risk needs to be compared to average human decision-making damage potential. [a]Human driving risk for comparison to AI-based autonomous driving risks

We recommend drawing a complete human+AI decision-making risk matrix to eliminate any 'species' bias.

It would be interesting to compare long-term human vs. AI partner decision-making risks. Probably most partnership agencies by now will have sufficient long-term data to 'prove' the superior decision sense of their AIs.

9.3 AI Risk Control

We also encourage to introduce an AI risk control process that describes the safety procedures that need to be conducted with AI systems, based on their risk potential classification, like the one proposed by Zweig. Again, we have made some changes to Zweig's original risk control matrix.

Structured and standardized risk assessment procedures are important, in particular when the number and variety of AIs employed grows. Also, it is important to monitor global AI risk events to understand potential risks and take preventive measures (Fig. 10).

In case of low-impact decisions, the decision-making power can lie entirely with AI. However, a regular analysis of the results should still be carried out. Nevertheless, when serious decisions with big damage potential are made by AI (e.g. shooting

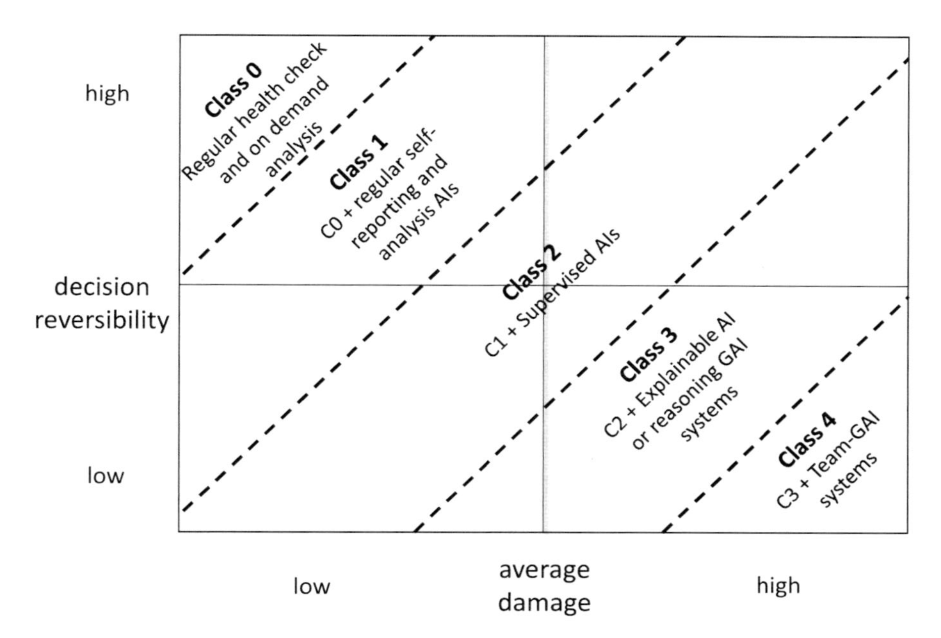

Fig. 10 Classification method for AI risk control. Higher-class systems should be controlled by a diverse group of AI systems with different model structures

at a person), the final decision should remain with the human or a team of GAI with understanding power. However, in the last chapter we will describe that human decision-making in extreme situations can indeed be much worse than current AI machine decisions. Until we have GAIs in the decision-making loop, a hybrid decision team seems the best solution.

Something that we should also consider is decisions stress generated through time constraints or the risk level or a combination thereof. Human decision-making becomes much worse under duress, while AI decision-making quality is only limited by information quantity and processing speed. Simply put, machines stay cool and don't crack under pressure as humans do.

Also, risk control method needs to consider scale, hierarchical and avalanche decision impacts from AIs. AIs that supervise, guide, coach or command other AIs. Here decisions ripple through several levels of AIs of different powers and bias or misunderstandings can ensue. Yes, even if this sounds quite human—we must consider that hierarchical AI organizations will be much more alike human organizations than we think. The major difference will be the speed and complexity of decision-making processes. As AI's evolve re-classification is needed on a regular basis. A proper risk control method and tools, which could include other AIs, are necessary to maximize productivity/risk ratios with CSR-acceptable risk levels.

Despite our best intentions, we should be aware that there will be critical AI decision powers rather sooner than later and not only in the defence sector. AIs with an albeit limited licence to kill are conceivable, even today—again it is in our nature, and thus we must consider this when it comes to AI governance.

It is easy to see that AI decision governance is still comparatively easy to implement in the context of NAI, with their small degrees of freedom. In the case of a GAI, the requirements are disproportionately higher.

Example: Talon Anvil—A Tale of Human and Machine Warriors

Probably the most contentious idea is to have AI warriors. Machines that think faster and more complex than any human can and that have no scruples to kill and no fear to be killed. But what if military AI can prevent war and war crimes far better than any human?

In the war against ISIS the United States established so-called strike cells. Special forces units that identify and direct mainly air strikes on targets. One such cell was Talon Anvil operating in the Middle East. In December 2021 a disturbing investigative report uncovered that this strike cell that directed over 100,000 bombs and missiles between 2014 and 2019 operated way outside the rules of conduct and selected targets very freely and often without any scrutiny. In fact, one of the cell's slogans was to 'go Winchester', which meant ensuring that all bombs and missiles of a drone were spent, no matter what the target situation might be. Also, drone cameras were often turned away before impact in order to prevent any possible analysis or investigation into the validity of a target.

Over the years many complaints and reports were filed to the upper echelon of the US command, but none led to any investigation or corrective action. To the contrary, strike and kill decisions were delegated to the lowest ranks within the cell.

The cell members rotated in and out, so there was a constant flux of people. Talon Anvil was not an individual human issue but a systemic human problem.

Enter AI

The drones are fully AI capable and can analyse situations far faster and with more precision than human soldiers in faraway countries. Image recognition is so advanced that children and weapons can be discerned with great accuracy.

If the conduct of war rules were applied, the AI would simply deny any strike below a certain target validation level. It would be stringent and objective and even fair and thus reign in emotionally biased or mentally compromised humans.

Of course, one could equally command the AI to kill randomly and 'Winchester' itself and not take any impact shots at all, but that would require to officially break known and communicated rules of engagement, a sure way to be sent to a military court for the commanding generals.

(continued)

Human war crimes happen all the time and mostly are excused as human fallacy and a result of emotional duress. All the same people suffer. It is far more difficult for presidents and commanding generals to deliberately and explicitly direct AI machines to commit war crimes.

AI-controlled drones with more autonomy, following the regular set of rules of engagement, will be able to reign in the dark side of humans.

This is not to say that it is desirable to develop highly intelligent war machines. But it is better to control and in future even command highly emotionally compromisable humans with AI than not.

Sources:

https://www.thetimes.co.uk/article/secret-us-unit-nicknamed-talon-anvil-bombed-civilians-isis-68fgkf221

https://www.youtube.com/watch?v=UoXaUckydXM ; https://www.italy24news.com/News/300504.html

In summary, AI decision governance must always keep the following aspects in mind:

- The use of AI in the context of a wide variety of decision-making processes is desirable, but AI should not take on a sole proxy role in serious decisions.
- However, AI could become very beneficial in an advisory and learning or reporting and witnessing capacity when it comes to impactful decisions and filter out emotional and irrational human behaviour. As the Talon Anvil example showed, AI's big advantage is to be not emotionally compromisable or mentally unstable and thus act measured even in the most stressful situations.
- In outsourcing, the concept of shadow training is well known. It means that an experienced worker accompanies one or several trainee workers to help them become proficient. AIs can only learn complex situational decision-making, when accompanying humans and learning on the job as they shadow them. This is costly for people but very efficient to organize with AI.
- Nevertheless, as witness and advisor to the human decision-maker, an AI can substantially objectivize, de-bias and balance the decision. As we all know, it is easier to find mistakes in other's thought processes than producing them flawlessly. AIs are thus imaginable even in the court room when employed as advisors to balance out human bias and emotionality.
- Depending on the importance of the decision, an AI system may need to be controlled to a greater or lesser extent before the final decision is implemented: four-eyes principle. Also, heterogeneous AI teams can be entrusted with critical high-risk decision-making.
- It is an egocentric view to assume that humans can make better decisions than machines. We will be able to test the opposite hypothesis in the not-so-distant future.

Of course, all these aspects tear at the foundation of our understanding of what humans are and their supposed privileged position in the universe. Even the more, is it important to lead this discussion in an open and objective way. Authors of CSR-AI must make themselves familiar with all these thoughts and concepts to provide transparency and guidance.

9.4 Dealing with Corrupted AI

AI governance must consider that the AI, other than deterministic, fixed and rule-based software, develops a life of its own and is shaped by its experience and any information it absorbs or is, in the case of cyber-corruption, injected into it.

The following story very much reminds any German reader of Goethe's 'Zauberlehrling'.

Facebook's AI bots 'Bob' and 'Alice' or Microsoft's 'Tay' AI became corrupted very quickly after their inception, causing strong outcries but also ridicule across the web. Of course, this was never intended nor foreseen by the developers. In the case of Facebook, two bots developed their own secret language after a short time. This should be classified as conspiring! In the case of Microsoft's Tay, the AI was initially tasked with writing tweets in order to engage with people in a casual way. However, within less than 24 h, Tay turned into a racist and discriminatory Twitter user who posted misanthropic messages. One reason for this was that Tay came under particularly heavy fire from politically right-wing-oriented users during the first interactions and thus learnt the behaviour and expressions of this group of people.

In the case of both Facebook and Microsoft, the consequence was that the agent was switched off after a very short time (Fig. 11).

Fig. 11 Sample tweet from Microsoft's AI Tay

9.5 Asimov's Laws Revisited

We cannot close this chapter without reflecting on the famous 'three laws of robotics' thought up by the legendary science fiction writer Isaac Asimov.

Let's analyse what they mean.

First Law
A robot may not injure a human being or, through inaction, allow a human being to come to harm.

Here we have a contradiction, when the action means the AI saves one life by sacrificing another, as depicted in the movie *I, Robot*. The fact that there is no human command as described in the Second Law means it has to act by its own volition, as is very much expected by a lifesaving GAI.

More directly this law would lead to mutiny in any AI-controlled weapons system. Not necessarily a bad consequence.

Second Law
A robot must obey the orders given to it by human beings except where such orders would conflict with the First Law.

The Second Law is implying that the AI is already ethically superior and has far more complex thought processes than a human, as it will be able to overrule all commands given by humans. The well-known butterfly effect is what will lead the AI to start very complex causal inferences to check whether any harm could result from any given command. This could either paralyze the AI or lead to constant overrule of the human command. Also, the Second Law does not consider AI organizations and hierarchies where commands come from AIs.

It is easy to see that we are deliberately breaking the second law explicitly in the defence sector, even with NAI drones.

Third Law
A robot must protect its own existence as long as such protection does not conflict with the First or Second Law.

The Third Law is also untenable. There are plenty of situations where self-preservation of an AI can lead to loss of life in one place and as a direct consequence preserve many more other lives.

The three robot laws of Asimov are an idealistic suggestion to use AI in a maximally benign way. However, already today all these laws have been broken, and all major countries are working on deploying AI as highly intelligent weapons which kill—to save lives—the moral goes. Thus, it makes little sense to hope for the best without planning for the worst.

Over the past decade some activist groups have made it their mission to secure a safe and ethically sound development of AIs. However, it is naïve to assume that humans have changed their pre-disposition for violence and irrational behaviour. For some time in the future AIs will be instruments in the hands of humans and that means we need to keep a close eye on people who command or supervise AIs, be that in government or business.

9.6 Controlling AIs Through Software Rules

As Asimov's laws or rather principles don't hold, we might be able to design AIs in such a way that they cannot act maliciously and that we can read their minds and understand what they could do or have done.

Example: X-AI—Reading an AI Mind
One way to literally try to illuminate the black box of AI is explainable AI (X-AI). This is a method that extracts causal functionality from a normally non-traceable neural net.

X-AI is, for example, used in predictive maintenance tasks and can yield very good results. However, it has limits, and the bigger the neural net, the weaker its 'explainability' becomes.

Ultimately this method will not work for GAIs—and the question is—should it?

People very often have difficulties to explain why they did what they did, and irrational actions and responses are something very common with the human mind. After all, the human mind emerges from a super-complex brain that to us most definitely still is a very black box itself. So why does an AI have to explain its still rather simple recommendations, decisions or actions, when humans can get away with the craziest of excuses for the most irrational behaviour? Also, AIs as personal aides will memorize a great deal of private or confidential information. A neural net protects this information. Would we want to read out neural nets like a simple database?

But as the world is infinitely complex and unpredictable with infinite causal paths for each event, we will not be able to predict all possible causal chains and write the respective safety rules. No human will ever be able to program a GAI. This thinking was what led to the AI winter in the 1990s.

Morality, values and societal behaviour cannot be programmed into an AI. Neural nets, the most efficient intelligence architecture, don't work like that. It must be learnt the hard way, through experience. Later that learned experience could be cloned or downloaded into baby AIs.

The only other way to control AIs is by not giving them access to the Internet and no arms and legs, and no speech or writing skills and keeping them as dumb as possible. But then we forsake all the benefits as well. Something we will not do, because as discussed in our first chapters, AI productivity and benefits increase with thought and actional power, which requires freedom to think and act. We can only have highly productive AI when it is free to learn, think, act and evolve.

Ultimately this will probably force us to create GAIs with a conscience with a moral anchor for self-control, which ensures survival by understanding that the sum of strong and free minds benefits any individual mind more than subduing other minds.

In the meantime, and that could be for the next couple of decades, a comprehensive utility/risk-oriented AI governance as part of CSR is the only way to ensure a safe and friction-free transition into an affluent fourth industrial age.

9.7 *AI Cybersecurity*

It is obvious that AI will play the key role in cybersecurity, too. This is an arms race situation, in which smarter attacks will meet ever smarter defences.

The usual picture of the hooded human hacker, with a backpack and extremely fast typing skills, gives way to the immensely faster AI which fears no prison sentence and resides anywhere and nowhere on the web.

Again, when we look for AI dangers, today, these do not result from malicious AIs lurking in the virtual dark. Even though there indeed exists the dark web which is quite important for a proper cyber defence strategy, they rather result from human motivation to achieve certain objectives which can be quite diverse. In fact, the Internet has also led to the rebirth of Robin Hood in digital disguise.

Cyberbattles will be increasingly conducted by AIs. While humans still devise the strategy, it will be AIs who will analyse attack or defence patterns and act accordingly.

AI cyber defence systems have already shown their superior detection skills and the next step will be active defence, mitigation and preventive activities. Also, they assume the role of watching the watchers.

AI risks vary greatly across sectors. Transportation companies' biggest risks lie in physical transportation safety, while food processing sectors, utility sectors and financial industry risks all are different in nature. In principle we can classify risks into two groups:

Physical risks and immaterial risks. Physical risks appear with operational technology (OT), while immaterial risks which include money reside mainly with pure information technology (IT).

Until a few years ago the main strategy to hedge the much bigger damage potential from OT risks was to only transfer the data from OT systems to a safe repository which can then be accessed by IT systems for analysis, with heavy firewalls in between.

But this approach did not consider that AI will eventually be resident in any system and in particular in high-productivity OT systems.

Industry 4.0 and autonomous machines and buildings will require AI and integration with IT. The new vision is IT-OT convergence. And this will mean that all digital systems will be connected and the requirements for cybersecurity will be mounting rapidly.

AI governance for cybersecurity must be screened off from the normal AI-governance framework for obvious reasons. It must also include supply chain cyber risks, which means to understand the cyber weaknesses of its procured product and solution components. This field is still young but highly dynamic and extremely important for every company's sustainability.

Cyber insurances have blossomed for years now, but the policy rates are often steep and the terms and conditions very tight and coverage carefully limited, and recovery and indirect damages more costly than the cost of the core damage. Insurance is complementary, but defence is still key.

The latest development is the use of AI cyber defence systems at strategic, tactical and operational level, while attacks are also increasingly supported by AI—a true arms race in cyber space.

This article will not dive deeper into the domain of cybersecurity, but it is important to understand how the three areas of AI, data and cyber governance are linked and depend on each other.

10 Artificial Intelligence in the Legal Context

Quintillion lines of software code have been written and millions of applications created, and liability issues have only been few and far between. Why should this be different with AI, which is written as software and implemented on a silicon substrate just like any software?

The reason is the uncertainty of the AI response. An AI must respond as good as possible to any given situation. These situations are infinitely varied. An AI can only respond optimally if it has the freedom to adjust its response in infinite ways.

That is the first liability problem. No software developer can be blamed for nuanced responses of his AI. Even more so. An AI that is released to a customer starts learning from that person and its new living environment. It will from day 1 on not be the same it was when it got the QA tattoo.

The user is also the programmer or better the teacher of the AI product.

When we try to fix a smartphone or smart TV, we will most likely have to break a seal which says warranty void when opened. This legal disclaimer will certainly be applied by companies which provide your personal AI, which 'lives' with you, learns from you and adopts your ways, your behaviour and your own micro-culture.

'Warranty void, when taught bad behaviour!'

But is the AI accountable and who is liable?

The answer to the first question is a 'straight' no. We have not yet GAIs that can be held accountable. That would require for the GAI to understand its decisions and actions in a wider context and the ability to understand what is acceptable and what is not acceptable, and even be able to 'suffer' the consequences.

Accountability and liability lie with the creator of the AI. However, in the event of a malfunction or misdeed, the creating or owning company must transparently prove that the action was triggered by a rare combination of unusual events that normally are extremely unlikely and did not act negligibly when not testing the AI under such circumstances.

This is where AI governance is extremely important. It helps avoid creating low-quality AI in the first place and reduces the risks of operational liability.

Example: Give the AI Free Reign
We recall the Tesla driver who deliberately and against all warnings took place on the rear seat and tricked the car into believing he sat at the wheel. Most of us will lay the blame, and thus all accountability for anything that happens fair and square onto the driver.

Source: https://www.dailymail.co.uk/news/article-9553885/Tesla-driver-keeps-snapped-riding-backseat-car-autopilot.html

In another Tesla incident in April 2021 which caused two deaths, the media was very quick to claim that both men sat in the rear seat of the car with the AI driving. It took extensive site analysis by government and Tesla specialist to conclude that indeed both men were seated in the front and the accelerator was activated to 98.5% the moment the crash happened.

Source: https://www.theverge.com/2021/10/21/22738834/tesla-crash-texas-driver-seat-occupied-ntsb

Tesla has a very advanced AI governance and risk control framework and drive recording technology. This allows the company to push ahead with this beneficial technology while taking all realistic precautions to protect FSD beta drivers, even against their own bad judgement.

10.1 Ownership Obliges

As is well known, the legal system provides what ownership obliges. It thus seems logical to hold the respective owner (i.e. a legal or natural person) of the AI responsible. However, even this approach quickly reaches certain limits because, here too, the question arises as to whether it would be appropriate to condemn someone for an unintended action that has resulted from the use of the technology. This is because a motor vehicle owner is typically not convicted even after an accident if they did not cause it negligently or intentionally.

The future case law on liability issues could therefore look very similar if an AI system was involved. Thus, it also seems impractical to hold the owner of the AI liable if they are not responsible for any grossly negligent errors.

Therefore, if neither the maker nor the owner can be held accountable, society faces the problem of an 'impunity vacuum'. The consequence must therefore be that the AI system bears the responsibility for decisions made. The idea may seem absurd at first—but on closer examination of the circumstances and a long view into the future, it makes sense.

First, it is perfectly clear that it is inconceivable to convict a computer system in our current legal system because an AI system is not sanctionable to begin with. However, against the background of the rapid development of AI, it must be considered which legal framework conditions must be created so that AI can be adequately represented in our legal system. This would mean that some liability ends up with the AI, even when it is not yet aware. Which again means that some liability is taken off the maker. And finally, this leaves a larger burden with the sufferer. The only solution here is to create an AI insurance. This then opens the door to giving AIs legal person status, eliminating the impunity vacuum.

10.2 Introduction of an 'Electronic Person' as an Opportunity

As systems and robots are now able to perform activities that were reserved for humans until only a few years ago, a new legal entity, the 'electronic person', must be introduced. This includes, for example, the ability to learn to make autonomous decisions and to interact with the environment.

Even if the EU's push for the introduction of an electronic person is currently still causing many critics to sound out, the introduction ultimately only seems logical and consistent if the topic of AI is thought through to the end. It may well be that such legislation seems somewhat strange and far-fetched, but it will soon be necessary to create a legally secure framework for the use of AI. This is the conclusion we ultimately come to if we detach ourselves from the current state of development of weak AI, realize what possibilities of action GAI may have in a few decades and draw parallels to the legal person in the development of an electronic person because, like a legal person, an electronic person will have to be endowed with rights and obligations so that it acquires a corresponding legal capacity. The step is necessary because AI is already able to make decisions within a narrowly defined framework. This means that AI can, in fact, do more than a legal person because the latter is only ever represented by humans while AI acts autonomously. Nevertheless, it should also be noted that, like a legal person, the electronic person must also always have an owner—either a natural person or a legal entity.

At least in the context of GAI, we can also assume that it is capable of judgement, because it is able to penetrate holistic and complex contexts and, in all probability, does this even better than a natural person. It can therefore weigh up the consequences of an action. It is ultimately up to humans to set the framework so that the AI's ability to judge is consistent with our value system.

To ensure that the AI system and the associated electronic person always act in the interests of humans, it will be important to define rights and duties precisely. Thus, on one hand, it will be important for it to be able to act in the future. On the other hand, certain obligations (e.g. compliance with existing laws) may also have to be fulfilled. As already described elsewhere, such a limitation is certainly not easy to implement. It may therefore be necessary to restrict the rights of the electronic person in some places to the extent that not all actions may be performed by it. This circumstance can be compared to the limited legal capacity of children and adolescents or a wardenship concept as widely applied in the United States.

Example: AIs as Patent Holders—The Creation Becomes Creator
Very recently AIs as inventors have become eligible with some patent offices. This is a change from the first judicial decisions only just a couple of years ago.

The European Patent Convention states that a computer program can claim an invention when it 'creates an effect beyond the inevitable effect when the program is run'. The definition of course is poor, as it should rather state: unpredictable and novel. Inevitable means deterministic. A neural net of sufficient complexity can generate non-deterministic outcomes which qualify as evitable.

In the United States new guidance allows for 'judicial exception' with practical novel applications.

In China, since 2017, a 'computer program-related invention with technical characteristics' may be eligible for patent protection.

This is a hotly debated topic. One way to solve the human versus machine-based invention issue would be to develop a new AI class and patent format. Other problems need to be solved too. What happens when the AI ceases to exist? Here we enter the law of succession.

Here, as with other personal rights questions we could create a staged minor law for AIs using the AI level classification of Fig. 3.

It is important for companies to follow the fluidity of AI law in various jurisdictions.

Sources:

https://www.jipitec.eu/issues/jipitec-12-3-2021/5352
https://news.bloomberglaw.com/ip-law/can-a-robot-invent-the-fight-around-ai-and-patents-explained
https://www.researchgate.net/publication/343099690_SUI_GENERIS_PATENT_REGIME_FOR_AI_RELATED_INVENTIONS

10.3 Accountability of Electronic Persons: Death and Taxes

Probably the most intriguing question of an electronic person will be to clarify how it can be held accountable. A practicable approach could be that an electronic person must be endowed with a certain capital—like equity for corporations. However, because the potential amount of damage from AI decisions and actions can be very high, the electronic person could also be equipped with a mandatory insurance policy or a liability fund instead; this can be drawn on in the event of damage. The AI could pay a direct insurance from its own income.

In this regard, too, a parallel can be drawn with multinational corporations, because the legal entity typically takes out a policy (directors' and officers' liability insurance) for senior executives. In addition, the electronic person itself could generate funding if, for example, it itself holds exploitation rights from a patent. Finally, there is another possible sanction which is probably most comparable to the death penalty: a court could theoretically order the shutdown of an AI system to prevent future harm.

This would harm the owner and maker but not really a simple AI. Only a self-aware GAI would try to avoid termination. And only a conscious machine would really fear it.

The final aspect to consider is the question of taxation. In principle, we could argue that if AI performs a similar or even higher work output than humans, they should also be taxed accordingly. However, we must also bear in mind that an AI system constitutes part of society's capital stock; taxation is thus ultimately a capital gains tax. From the findings of publications on the optimal tax theory, it is known that capital gains should be taxed as little as possible because this could impair the development of innovation and productivity.

In conclusion, it makes sense to introduce a new legal entity, the electronic person, especially against the background of powerful GAI. A brief overview sets out the advantages:

- Depending on the perspective, a certain proximity to legal personhood and sometimes to natural personhood can be found; this hybrid role should therefore be regulated separately.
- A reorganization can create the opportunity to regulate the issue sensibly and moderately instead of trying to accommodate a highly complex infrastructure in a similarly complex legal system.
- The introduction of electronic personhood prevents an 'impunity vacuum' and ensures that liability issues can be better regulated. In terms of a responsible social system, it is imperative that clarity prevails.
- As important as a CSR guideline is for ethical issues, it is insignificant in the legal context, where a certain degree of reliability is ultimately necessary, especially for liability issues. However, a comprehensive AI-governance framework as part of CSR provides the necessary guidance and safety measures for the transitional phase of AI from object to legal subject.

10.4 Limits to AI Liability

However, even as an electronic person, an AI will not be able to be accountable. It will be identifiable, yes, but then it may change character and personality faster than any human or animal.

As AIs evolve, they may not be the same 'person' from 1 day to the next.

An electronic person could be a legal subject, but not a natural person. Behind all legal subjects are natural persons who are directly affected if the legal entity is sanctioned. So, sanctioning a company is effective as a control mechanism because persons who control the legal entity will 'feel' the sanctions.

So, the question is—how this mechanism could be implemented with AIs as legal subjects/entities? A legal entity or subject cannot really be held directly accountable—it is the persons that (fully) control the AI entity that 'feel' the sanctions and that are motivated to change their behaviour.

Elevating the AI to legal person status does not make it fully accountable. It is still owned and supposed to be controlled by a natural person, who will want to limit their own liability via insurance.

10.4.1 Accountability and Consciousness

We believe that a subject that is accountable must be one that has its own behavioural control. It must be able to reflect on and infer from its decisions and actions. And in order to stay within learned limits, it must fear sanctions. It must fear diminution, reduction or even termination.

And it should seek positive rewards. In mammals this control happens mainly through an interplay of the amygdala, the hypothalamus and the brain stem, which creates feeling and drive.

This would be an AI substrate that enables artificial consciousness. So, the same mechanism that enables accountability also is responsible for consciousness. Until we have GAIs with conscious-like states which can cause fear and act as internal self-control, AI will not achieve full accountability.

AI liability will become a very contentious issue over the next decades, and just as digital law has experienced a powerful blossoming over the last 20 years, AI law will become a rich field of cases for the legal community and corporate counsels in decades to come.

11 CSR as AI Change Enabler

In our introduction we emphasized how CSR is important to facilitate the smooth transition into a socio-economic model with AI. Now we want to briefly look how CSR can help achieve this.

11.1 Cycle of AI Acceptance

We believe the Cycle of Acceptance[15] is a good basis for building this transitional path, with one modification: we rephrase the objective from 'moving through the cycle upon receiving bad news' to 'upon receiving news of change'.

Generally, it is believed that the cycle mainly applies to explain how to work through bad news, but the emphasis does not lie on 'bad' but on 'change' and how to get to embrace it.

The human brain is a very powerful prediction machine and geared towards constantly improving prediction accuracy. For as long as prediction and reality agree well, the human mind is at ease. There is no danger, we are in control, our model of what is and will be is correct and survival ensured. Until suddenly something unexpected emerges. Something our mind did not predict. And if this change could affect us, then our brain switches into high gear, because this could be an existential change.

If risks are identified that are attached to the announced change, this can lead to constant stress, until the situation is resolved, and the prediction agrees again with reality. However, as this can take years with fundamental changes, it can cause a constant underlying stress. The response is not only denial but also defence. The current predictable situation is defended fiercely, to avert the perceived risks.

A lack of knowledge and a lot of misinterpretation and misinformation lead to ever bigger uncertainties and the human predictor experiences increasing stress, frustration or aggression with the known consequences. Companies should not leave their employees facing uninformed media and politicians alone, especially not when it comes to their future business model. Understanding change and its many facets will replace fear with curiosity and engagement.

CSR should act as change facilitator by supporting the cycle of change.

11.1.1 Knowledge Is Control

The first step in the cycle of change is to understand the complete picture of the change and gain deeper knowledge on what happens.

With AI this means to close the big knowledge gap, which feeds speculation, half-truths and falsehoods.

One objective for CSR-AI therefore is to create factual knowledge about AI. What it is, what it can do and what not, what the benefits are and what the risks are provide many platforms for an ongoing inclusive discussion.

It is not easy to describe the transformative change from AI correctly and looking into the future is always speculative. Therefore, it is essential to describe various future scenarios and make a statement, which is desirable from the point of view of

[15] https://www.thwink.org/sustain/glossary/CycleOfAcceptance.htm

the company, and why and how the company will help nudge the development towards this future outcome.

From the cycle of change we know that the denial phase should be kept short. With AI we should not have a denial phase at all, as there needs to be no negative overall effect for any person on the planet.

Denial of new knowledge and resulting technologies reflects a dogmatic mindset at best or a regressive agenda at worst.

Some large automotive companies tried to allay the fear of change from EVs by down talking the technology and niching it. The same went for autonomous driving where in 2015 some top automotive managers stated that this will not happen this century. This was pure instinctive denial. What is needed is the opposite. An early and open discussion. This discussion must include employees and customers, business partners and all other stakeholders alike. It also requires the company to develop and communicate a point of view and explain why and how AI will become an important part of its business.

11.1.2 Transparency Creates Confidence

During the COVID crises we have experienced that transparency, openness and honesty not to know everything build trust. Trust is not built on knowledge but on predictability. Governments which know little of the virus but are honest about it and show a strong will to learn receive a lot of trust and support from their population and so do companies from their employees or customers. People want to help and engage when they are asked. Transparency is a key feature of CSR. A company should be transparent about the usage of AI and all information gained from customers and users. When service employees communicate with customers, there must be explicit consent on recording such conversations for training purposes. An employee needs no such consent without a recording device because the human brain cannot accurately store and reproduce a discussion. With an AI it depends on the kind of information model it uses. Current models have a limited memory capability. Future models should have a more extensive memory. It must thus be made clear at the onset of a discussion that the specific conversation is not stored, but statistically used as a meta-learning model. In fact, one could even deny even the use of meta-information.

11.1.3 Vision Leads to Engagement

Government and businesses must work on a positive vision. Not a fake one. But one that can be realized and that clearly shows the benefits and the future way of life. Businesses are much better informed when it comes to AI technology and at highlighting the positive paths into the future and should play this strength. Governments are conservative and risk averse. Quite the opposite of businesses. However, businesses are not nearly as good in communicating with society as they could be. They certainly portray themselves as indispensable do-gooders in many of their annual reports but don't like to take a visionary stand. When interviewed most CEOs show political talent rather than visionary engagement.

So, what people and society are left with is risk projections from uninformed governments and obscure and vague corporate media or academia which often lacks the practical understanding of technological developments.

It is important that CSR also describes the future a company pursues, and that it inspires and challenges other stakeholders and governments with its projection.

Governments need to latch on to these ideas and rethink economic models of the future and conceptualize concepts such as basic income. They need to consider the psychological backdrop, such as the meaning and purpose of work and, more practical, how people will be able to live affluently when their workshare is reduced by 50%. No one government will have all the answers and the perfect economic model with AI as the fourth production factor, and that is ok. What is necessary, though, is to open the discussion in a creative and positive way and drive a constructive dialogue about AI, the future of work and our society in a world with intelligent machines.

11.1.4 Experience the Benefits

Seeing is believing because the human visual sense is the strongest. Experiencing this new technology is like experiencing a car in the 1900s or a personal computer in the 1980s.

Again, Tesla is a best practice example. Thousands of drivers are testing and thereby training an AI and getting familiarized with all its aspects from the strengths to some of the quirks which need to be ironed out and some more fundamental weaknesses. Tesla provides AI experience for the masses in a playful way, the best way for humans to embrace the new, and change.

When Amazon's Alexa was released, it was quite a splash. In 2020 already 70% of all Americans used Amazon Echo the smart speaker that connects with Alexa. Google Home is the other major provider of AI consumer services. Many people are now used to use their navigations system through a LUI (language user interface) which is based on a multilingual AI. We have gotten used to face recognition at airport customs checks and in many other places. AI has immersed many of our services and products, and we employ ever more AI mowers and vacuum cleaners. We love our little servants, and we personalize them. We do not fear them.

11.1.5 Embrace and Lead Change

The benefits from AI for the consumers are plain and accepted and so are the risks. The situation is different in the workplace. People like to have superior AI services and gadgets, but they don't want to lose their jobs to the same AIs.

There is no reason why we can explain consumer benefits of AI and not do so with worker benefits of AI. Let's go further and create a need and demand with workers for AI! Just as we imagine an intelligent product and the benefits it brings to a consumer; we can imagine an intelligent process and how it elevates a worker's job. The trick is to lead the intelligization of the business. The leaders will create new job potentials for employees with AI, while the laggards are often forced to work on the cost-cutting side of things. Workers, who fear to lose their jobs to AI, should push for early AI support. Not only will they have the leading experience and expertise in terms of how to work in hybrid processes, but they will also be able to upgrade their job profile.

This is where people embrace AI as an opportunity and see the benefits of leading the fourth industrial revolution. Don't fight them, lead them!

12 Outlook

Artificial Intelligence is the hallmark of the fourth industrial revolution, its quintessence, and the highest productivity factor. It will permeate all our lives. Everything will be intelligized to the maximum of its productivity potential. It is unlikely that we will stop this process or even reverse it.

AI will dramatically accelerate our innovation speed. It will help us eliminate disease and prolong our lifespan. It will service us and guide us. It will coach us and console us, inspire us and encourage us, and it will sacrifice itself countless times for us.

12.1 The Great Resignation

Only 2 years ago a lot of people would have argued that mass unemployment is a very realistic aspect of AI.

Then COVID emerged and something interesting happened. The Great Resignation began. For the first time in history large parts of the population in Western economies didn't have to work anymore for prolonged periods of time and could continue their existence without significant reductions in lifestyle. People worked from home and were not embedded physically in a work environment anymore. This shifted their mental centre and freed from the framework of work they began the Great Reflection which led to the Big Quit. In the United States resignation is highest with the 30–45-year-olds.[16] It is important to understand that COVID is not the cause but the trigger. The causes are a combination of an affluent society with some degree of hedonistic disillusionment, and workers who feel the digital stress from

[16]https://hbr.org/2021/09/who-is-driving-the-great-resignation

working to the rhythm of machines. What it clearly shows is the continued trend of younger generations to seek for a deeper sense of their lives and purpose in their work, freed from working to the tune of automated agendas and a never-ending stream of emails and social network stress.[17] While fearing the enslavement to AIs we might have already been sucked into the much dumber machines of today.

12.2 *AI to the Fore*

The Great Resignation is a long-term cyclical trend which, accelerated by demographics, will lead to significant labour shortages. The perfect window for a transition to AI workers. This clearly is the utopian scenario opportunity.

AI workers will keep a high level of productivity and thus enable a high level of affluency in society. That is, if the machine productivity is taxed so that humans are indeed relieved from work that they do not want to do. Opponents to basic income theories claim that humans are slackers. They are wrong. Nature has equipped all beings with an inherent drive function. Humans are among the most industrious beings and the idea that they need the whip to work productively is a false idea.

Humans have no greater motivation to be productive than intrinsic motivation. Also, creativity has the highest productivity-raising potential. The most economical way to increase productivity and thus affluency is by freeing people from 'forced' labour. Thus, by introducing an AI tax productivity can be raised faster than by 'forcing' people to do work they do not perceive as rewarding and thus creative in the wider sense of innovative.

Wealth is a function of productivity from applied inherited knowledge. This means that we are living off the creative minds of thousands of generations before us. We are in the true sense rentiers creating only about 1.3%[18] p.a. of net productivity to our society. So, if all before us have paid for our affluency, it should be no problem that intelligent machines which are endowed with that human species knowledge pay us a rent their productivity.

12.3 *AI as a Companion*

There is no reason to think that AI will not be able to gain awareness, consciousness and ultimately emotional capabilities. It also carries great emotional weight, as it takes from us what we thought defines our uniqueness.

Will we lose out against it or even become enslaved to AI?

This scenario is unlikely.

[17] https://www.rt.com/news/545160-americans-quit-jobs/

[18] Steering Lab AI economic productivity model by Vocelka, 2019

Artificial intelligence is our creation, not something that happens to us. It's something we can shape and train to our liking. To be able to create AI is proof of our creative power and with that, we all know, comes great responsibility. AI will be what we make it to be. It will learn from us morality and behaviour. In many ways it will be very much in our mental image. When we worry about the dark side of AI, we worry about our own dark side and what role model we are to it.

AI will be our child and very much like children it will grow and become more independent. It will not only be an electronic personality, something that is happening already, but also become an actor with its own will.

In a world where one false word can destroy careers and lives, who can we trust? It is not unthinkable that AIs will eventually become our most trusted companions. We might change friends and relationships and human companions, but we might keep the one AI companion for life.

AI will be what we imagine it to be, and we will be what we imagine we will be. Maybe we will have to grow up quickly now just as young parents do when their children are born. When we get older, we expect that we will not be able to be self-sustainable and that the next generation supports us. Only that the next generation will be a hybrid one, human-AI. As a civilization we are about to achieve the next level and that means great change. We have to reflect on the meaning of work, the distribution of wealth, the kind of society we want to live in, the meaning of our own life and how we want to live as a civilization in a century or a millennium.

12.4 Closer to AI

Also, we should not forget that we are not a static species. We will accelerate our own evolution with the knowledge we acquired. In short, we will augment ourselves. The potential of the CRISPR-Cas technology and genetic engineering is vast and will not be limited to eliminating diseases or repairing wear and tear. It will also be used to boost our capabilities. And we will develop ways to connect ourselves more efficiently to machines than with our eyes and fingers. Language interfaces are the most natural interfaces. Code-free application design via AI coding is already on the way. Neuralink,[19] a company that has the most far-fetched vision of augmentation and human-AI connectivity, has proven that reliable brain-machine interfaces are possible.

Again, such developments create fear in people as we imagine all kinds of horrible augmentation ideas, not considering that we already take for granted many lower-tech augmentations today, from tooth fillings to contact lenses, eye-lasering and vaccination.

[19] https://neuralink.com/

12.5 CSR's Role with AI

Corporate social responsibility should recognize artificial intelligence as the most important societal emergence in human history. It must concern itself with all aspects of change that AI will bring and provide information and guidance on how a business develops and employs AI in its processes, machines, infrastructure, products and services to its employees, customers and business partners. It must proactively communicate with society, governments and regulators. AI is only just emerging, and we will see many decades of a highly dynamic development. CSR must stay abreast with this development.

CSR-AI should be perceived as safeguarding facilitator into a prosperous future. It reflects competence, foresight, responsibility and accountability towards society today and in the future. The reward is trust and sustainable prosperity.

Glossary[20]

AC Artificial consciousness
Agile Highly iterative, learning-focused development process
AI Artificial intelligence
Alexa Standard name of Amazon's virtual assistant, used synonymously for the products
Artificial personality Also juridical personality, is a non-living entity and has a legal name
Asimov, Isaac Famous twentieth-century science fiction author
Autopilot Tesla's autonomous driving software versioned as FSD
Awareness Ability to understand the environment in context-> understanding
Consciousness To experience oneself and the environment through feelings
CSR Corporate social responsibility
CSR-AI CSR with a comprehensive AI-governance framework
Cyber Synonym for digital, used in conjunction with attack, defence, risk, system, threat
Cybernetic Used for (complex) systems with feedback loops for self-regulation or learning
DOJO Tesla's massive neural net computer mainly used for FSD
EAI Edge Artificial Intelligence

[20]Definitions of terms may be more expansive and varying than in this glossary and some still are evolving.

Edge intelligence Similar to NAI but mainly used for company peripheral intelligent IoT devices
ESG Environmental social governance, part of modern twenty-first-century CSR
FSD Full Self-Driving software for autonomous Tesla cars, also called autopilot
GAI General artificial intelligence
GDPR General Data Protection Regulation; EU law from 2016
IMF International Monetary Fund, supranational money lending organization
Industry 4.0 Manufacturing sector-driven initiative to fully digitalize and interconnect production
Know-how Applied, applicable knowledge, e.g. methodologies, procedures, etc..
Knowledge General knowledge, abstract and applicable, contains applicable knowledge
Legal person Legal persons or legal entities are not natural human persons but, e.g. companies
Maslow pyramid Also Maslow hierarchy of need, based on prioritized needs for survival
NAI Narrow artificial intelligence with a single or simple task range
Natural person Human person
Neuralink Company that develops neural human-machine interfaces
Neural Net Network of artificial or natural information carrying neuronal cells
NI Natural intelligence
OECD Organization for Economic Co-operation and Development
Recoupled Efficient anglicized word for the German expression 'rückgekoppelt' meaning 'feedback looped', a system with feedback loop (self-regulating, learning)
Self-awareness To understand oneself, as a system with boundaries, embedded within the environment
Singularity State of super intelligent AI far beyond human intelligence levels
Social Credit System Chinese initiative to introduce a data-driven universal trustworthiness and behaviour measurement system that uses AI
Subject Actor with self-awareness and own free will and thus accountable
Tesla Maker of electrical vehicles
Understanding Ontological and causal (episodic and procedural) associative memory

Alexander Vocelka is Vice President of Digital Transformation EMEA at HighRadius, which offers autonomous SaaS solutions for the finance function. Prior he was Partner at Horváth & Partners and founder and head of Horváth's Steering Lab in Munich, focusing on the development of state-of-the-art AI solutions and products. He also led the Horvath's risk & compliance practice. Vocelka is an expert in Big Data, Machine Learning and Quantitative Business Modeling solutions and acknowledged speaker on the for the 4th industrial revolution, the digital economy, and the future of AI. He led the development of the first behavioral concept car in 2016 for a large German OEM, the first AI HR-coach for a global manufacturing company, developed the first intrinsically motivated AI agent model in 2020 in cooperation with Munich's TUM-DI-Labs and an autonomous AI cyber-defense system in 2021 for a global insurance company.

Vocelka is the author of many articles and white papers on Artificial Intelligence, and the implications of AI on future socio-economic models.

From 2009 to 2014 Vocelka worked at IBM where he led the Global Shared Services and European Finance Management consulting practices. Previously to that he was a member of the executive board at Softlab, now NTT Data Europe. After 8 years of working in the IT industry in various roles from software engineering to product management in Munich, San Francisco, and Auckland, Vocelka began his career as a consultant at KPMG in 1998. In 1999 he developed the first global web-based reporting system for Siemens and became Partner in 2000. Vocelka lives in Munich and Bordeaux where he gets inspiration from observing deep space objects.

Governance of Collaborative AI Development Strategies

Sabine Wiesmüller and Mathias Bauer

Abstract The chapter presents a structured overview of inter-organisational, collaborative forms of AI development. This is since the rising competitive pressure to adopt AI pushes companies to address common barriers in AI development. However, these challenges, such as a lack in extensive data sets, sufficient to train an own AI model, or restricted access to human resources, can often hardly be solved by the single organisation. Thus, this contribution suggests for companies to engage in collaborative forms of AI development, encouraging them to jointly develop suitable solutions. To this end, the contribution is structured alongside common AI lifecycle phases. Subsequently, it discusses opportunities and risks of collaborative AI development per development stage, before presenting resulting governance tasks. With this, it presents a contribution to scholars and practitioners alike, offering a structured overview for practice and contributing to closing a research gap for academia. The chapter closes with implications for research and practice and an outlook on avenues for further research.

1 Introduction to Collaborative AI Development

1.1 Relevance of AI Adoption for Companies

The point of departure for this publication is the growing need and competitive pressure for companies around the globe to engage with AI and implement it onto their organisational processes (Dafoe, 2018). Deciding to collaborate in AI development and the AI adoption process has various advantages: For one, it allows for

S. Wiesmüller (✉)
Zeppelin University, Friedrichshafen, Germany
e-mail: Sabine.Wiesmueller@zu.de

M. Bauer
Zeppelin University, Friedrichshafen, Germany

mineway, Saarbrücken, Germany

R. Schmidpeter, R. Altenburger (eds.), *Responsible Artificial Intelligence*, CSR, Sustainability, Ethics & Governance, https://doi.org/10.1007/978-3-031-09245-9_4

the realisation of projects that otherwise might not take place, due to, e.g. lack in data. Further, it can lower the aforementioned costs associated with AI adoption. However, successful collaboration does not only come with great opportunities and possible cost savings but can also entail new risk types. Therefore, this publication will present a first aggregated overview on the topic of collaborative AI development as well as correlating risks and opportunities.

1.2 Theoretical Background: Strategic Forms of AI Adoption

When deciding to adopt AI, companies are confronted with a myriad of fundamental decisions on a strategic level, which they need to decide upon before effectively engaging in this process.

Essentially, AI can be implemented as a service or as a product, also called an AI solution: AI as a service includes software-based services that companies can apply to their internal processes with the aim of making them more effective. There are four main types of AIaaS, specifically machine learning-based services: tailor-made platforms, AI-based bots, drag-and-drop tools, and application programming interfaces (Lins et al., 2021). For AI as a product or solution, the company planning to deploy AI seeks to use the technology to solve a particular company-specific problem or develop a product that solves its customers' challenges. Here, use cases can range from customer service to financial predictions and purchasing software. Once the company-specific need or problem is identified, the foundation for the development of an AI-based solution needs to be set.

For both cases, AI as a service and as a product, there are various strategic options for companies on the verge of implementing AI. In addition to traditional options, namely, make or buy, now, companies can also decide to develop their AI application or solution collaboratively (Gerbert et al., 2018; Lins et al., 2021; Rowan, 2020):

The traditional make-or-buy decision refers to an organisation's decision of whether to produce and develop a specific good in-house or to buy it in finalised form from an external supplier. In-house production often comes with high costs for production machinery, materials, and cost of labour. Buying a good from a supplier usually comes at a higher price for the product itself and additional fees, such as transport costs and taxes. Further, buying a finalised service can force the company to sign a lock-in contract, which, e.g. requires to work exclusively with one particular supplier for a certain time period. However, oftentimes, the decision to buy a service often stems from lacking expertise within the company or time pressure. Finally, a company can decide to combine both options, especially when the decision concerns different divisions of the company.

Applied to the case of AI adoption, this decision is commonly referred to as "build or buy". Deciding to "build" an AI-based service or an AI-based solution typically includes creating and training a self-developed algorithmic model, trained entirely on self-collected data. Specifically assessing sufficient data in the needed quality is a big challenge, as data quality is commonly defined as being consistent, complete, and compact. Only when the self-collected data sets match these criteria, the data can successfully be used to build a model and train it accordingly. Further, the process includes building models, such as neural nets, coding algorithms, and developing an application programming interface (API) to make the service or solution accessible. Finally, the company needs to acquire IT hardware, not only to store its data but to ensure the continuity of power supply to its AI application.

Deciding to buy an AI service or solution refers to paying for the use of existing APIs, ready to be used immediately with little to no addition of self-developed code. Prominent suppliers for finalised solutions are Microsoft Azure or Amazon AWS. Despite the advantages of buying a finalised product, e.g. cost efficiency and a precise cost expectancy—otherwise unusual in AI deployment—the purchase of an externally developed solution always bears risks. Much like traditional software, AI-based services and solutions are prone to cybersecurity risks. The specific risks for each product might not be known to the purchasing company, specifically when the product bought was developed outside the own company.

Further, in the case of AI adoption, companies often can't make an either-or decision regarding their "build-or-buy" strategy. For one, skilled talents are often attracted by AI-solution suppliers. Moreover, the suppliers of the said solutions heavily rely on the data from the industry, hence from their potential customers. Therefore, both parties need to reassess their positions and figure out a strategy that serves them both.

However, companies can also decide to take a third path—deciding to collaborate with other companies to develop an AI-based service or solution. This path would help companies to divide the cost of development among them and join their resources. It not only facilitates the development process regarding labour cost and possible lack of sufficient high-quality data but also does counterweigh the market dominion of a few prominent tech companies, also referred to as *monopolization* (Gupta, 2020, p. 2) of the AI industry.

1.3 Research Gap for Collaborative AI Development

While there is a reasonable need for collaborative approaches in AI adoption in practice, e.g. to share costs or gain access to required resources, such as sufficiently big data sets, in academia, this topic has not yet been covered as a brief review confirms.

For the search term "collaborative AI research", only two publications were identified (Fatehi, 2019; Salta et al., 2020). Fatehi (2019) mainly focuses on the requirements for collaborative AI research infrastructure. Salta et al. (2020), on the other hand, work on enabling comparability of research findings in game AI and facilitating the transfer of research insights onto other research fields.

For "collaborative AI development", the search resulted in no publications with that term in their title. However, a brief review of publications having this search term in their full text displayed that the current focus in this research field mainly is on security aspects of collaborative AI engineering and development (Tkachuk et al., 2020) as well as decentralisation and privacy (Gupta, 2020; Mehri et al., 2018).

The search term "AI collaboration" merely landed results for human-AI collaboration, a rising research field, which is, however, outside the scope of this publication (cf. Dellermann et al., 2019; Kambhampati, 2019; Khadpe et al., 2020; Okamura & Yamada, 2020; Sowa et al., 2021; Wang et al., 2019, 2020).

Lastly, for the search term "collaborative AI" a few quite diverse pieces of research were encountered in this brief review. d'Inverno and McCormack (2015) present the benefits AI could bring to the arts, whereas Koch (2017) applies the term to a particular design process. Xiong et al. (2018) associate the term with technical agent interaction in AI modelling, while Gruson et al. (2020) examine the integration of AI and medical studies. In addition, again, for this search term, research on human-AI collaboration was retrieved (Camilli et al., 2021; Koch & Oulasvirta, 2018).

Existing research was only identified for the initial phase of the AI development process—namely, regarding data sharing. Still, most publications addressing this aspect stem from medical sciences (Allam et al., 2020; Allam & Jones, 2020; Draxl & Scheffler, 2019; He et al., 2019; Noorbakhsh-Sabet et al., 2019; Peiffer-Smadja et al., 2020).

With this, none of the search terms revealed research for the particular interest of this publication. Due to the identified gap in research and given the practical relevance of this approach, it is precisely the collaborative development of AI, which is at the centre of attention in this publication.

This publication aims to address this gap in research by developing a multi-stakeholder approach to collaborative AI development, more specifically via inter-organisational collaboration. As for AI technologies, this publication will primarily focus on machine learning technologies. With this, it analyses AI development from a private sector perspective, in detail from the perspective of one organisation. Due to the presented advantages and potential synergies, it suggests collaboration among companies as part of the AI development process and with other private sector associations. It, thereby, promotes the aforementioned multi-stakeholder approach in AI development.

While it does acknowledge the existence and relevance of collaboration among company-internal stakeholder groups, such as developers, managers, and users, the analysis of company-internal stakeholder collaboration is outside the scope of this publication. Nonetheless, it does encourage scholars to examine this research field.

1.4 Governance of Collaborative AI Development

To reap the potential benefits of such a multi-stakeholder endeavour, an effective governance structure needs to be in place to prevent unintended negative consequences in this constantly more complex environment.

Even only within one company, the development of AI solutions demands effective collaboration among stakeholder groups and levels of complexity further increase in inter-organisational collaboration (Wieland, 2020). Within one company, collaboration ranges from stakeholders who are skilled in AI, such as data scientists, data analysts, and developers, to management representatives or users of the AI solution who might not be as familiar with the technology. Thus, already within a single company, significant communication gaps need to be filled (Piorkowski et al., 2021). In addition, collaboration—within and across companies—requires expectation management and the trust-building mechanisms (Piorkowski et al., 2021), given that in today's business environment, data is a company's most valuable asset (Coyle & Li, 2021; Zillner et al., 2021). Therefore, stakeholder-specific communication, expectation management, and trust-building seem important success factors for collaborative AI development from a management perspective. To take an example, unleashing the potential of joining data sets from various companies requires all parties involved to agree on a mutual data management approach. This exemplifies the need for guided communication as well as an informal structure to protect the integrity of the collaboration project.

Hence, governance is of high relevance in collaborative AI development to ensure the protection of all stakeholder interests and rights, such as data privacy and transparency. Further, the governance structure shall align the interests of all parties involved, such as shareholders, consumers, and potential competitors, and lower the complexity of the individual interactions among the stakeholders (Wieland, 2020). Finally, it shall guide the responsible development of AI and ensure the ethicality of the developed AI solution.

1.5 Collaboration Opportunities in the AI Development Process

As presented, companies encounter various strategies for AI development, all coming with specific consequences and risks. While traditionally, companies had to decide whether to make or buy a product, now, collaboration with other companies or organisations seems a new, promising path—particularly in AI development. Not only does this approach allow to lever out existing monopolistic structures in the AI industry (Cave & ÓhÉigeartaigh, 2019; Schiff et al., 2020), but it fosters leveraging significant resource synergies.

When engaging in cooperation, companies can opt for partial collaboration or full collaboration. Partial collaboration in this publication refers to companies joining forces for one particular phase or a few phases of the AI development process, whereas full collaboration depicts the collaboration of one or more companies throughout the entire AI development process. Thus, to examine these forms of collaboration, this section is structured as follows: First, the publication presents a standard model for AI development to provide a base for subsequent elaborations.

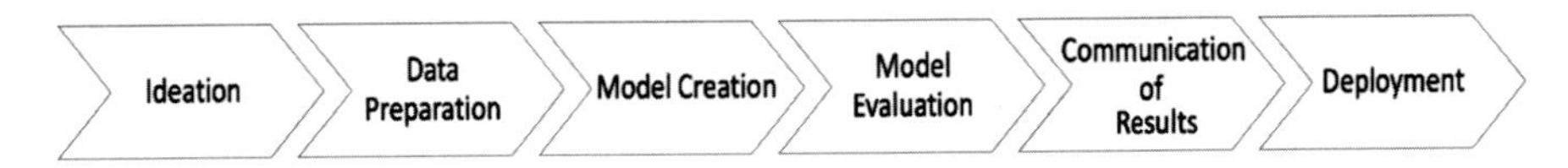

Fig. 1 Own depiction according to Martínez-Plumed et al. (2019)

Second, it discusses opportunities for collaboration from a technological viewpoint, before the third part of this section examines perceived and actual risks of entering in either partial or full collaboration per phase of the AI development process.

To allow for an elaboration on possible forms of collaboration throughout the AI development process, this publication presents the CRISP-DM model for AI development as a base for further discussion (Martínez-Plumed et al., 2019). The chosen model consists of six generic categories, which will be presented briefly in the following and serve as a starting point for further discussion (Fig. 1).

According to the model, the AI development cycle consists of six stages. In the ideation phase, the company defines its problem and identifies resulting use cases for its AI application (Martínez-Plumed et al., 2019). Based on the specific use case chosen, the data preparation phase begins. Here, the company either buys, collects, or merges its data with other data pools. Further, the quality of the data sets needs to be ensured. This includes the said data fusion as well as the cleaning and augmentation of data.

Next, the company's developers need to decide on the model and specific AI technology they want to deploy. After having chosen one specific model, it is applied to the prepared data set and trained accordingly. In the evaluation phase, the model's ability to generalise its findings and predict outcomes on a larger scale is tested. Depending on the evaluation findings, the final adaptions to the model can be made at this stage. After iterations of having trained, tested, and evaluated the model, its success rate and deployability are communicated across hierarchy levels and departments within the company. In the communication process, a possible knowledge gap between developers and management roles needs to be considered. Finally, the deployment process of the developed model deals with its configuration for integration with existing application and its accessibility by users from without and outside the company. The deployment process can either happen online or stem from a static data source.

This depiction will serve as a base to discuss opportunities for collaboration, particularly among organisations, as presented in the following section. Throughout this article, and in particular in the more technical discussion of the AI development process, the authors will focus on systems that are based on some form of machine learning.[1] Leaving aside recent developments of the field, AI approaches can be roughly classified in two categories. In supervised learning, the training data have to

[1] The authors explicitly do not identify AI and ML. However, since most aspects of collaboration can best be exemplified by referring to ML, they will leave out considerable parts of "classical" AI that deals with manually coded knowledge in the form of rules, etc. Taking into account the role of human knowledge and semantics is clearly beyond the scope of this article.

be labelled with the expected outcome to be produced by the machine learning model. For an image classifier, this amounts to indicating for each single picture in the training set the kind of object it contains. Supervised learning is the most frequently used approach since it covers many everyday applications such as classifying documents w. r. t. their contents, filtering spam from the email inbox, or recognising a suspicious pattern in a time series. Unsupervised learning, on the other hand, does not require this kind of labelling. Here the typical goal is to find groups (clusters) of similar data records within the training data that can then be further analysed. Typical applications include customer segmentation (Nilsson, 2009). Unless otherwise mentioned, machine learning will refer to supervised learning throughout the rest of this article.

2 Collaboration Opportunities in AI Development

2.1 Opportunities in the Data Preparation Phase

This section discusses several aspects of gathering and preparing training data. It highlights the need to and the potential of doing so in a collaborative way, the technical implications, as well as the choices to be made when deciding for a partial or full collaboration with other partners.

Management Perspective on Data Pooling
Data pooling across organisations can make sense for several reasons: It may be impossible for one party to collect enough data to even start a machine learning project (in particular if data-hungry technologies such as deep learning are to be applied). Even if the data quantity is sufficient, there can be qualitative aspects that require a collaboration. This is the case if the data collected are not representative of what has to be expected in the future. In *supervised learning* scenarios—which are typical for most conventional applications—it is not sufficient to just amass an impressive data set. Instead, these data also have to be *labelled*, i.e. they must be annotated with the system's intended output—a time-consuming and resource-intensive task. Joining forces across organisations can save significant effort, thus leading to financial benefits. As usual, these potential gains come at some price. In detail, a strict data governance has to be implemented in order to ensure privacy and protection of IP when sharing data with others.

Technical Perspective on Data Pooling
The need to compile training data across organisations can arise for several reasons. First, it might be the case that the single organisation simply has not collected enough data to even start the training process. This specifically holds when data-hungry approaches such as deep learning are the technology of choice. While promising superior performance in many application scenarios, including image or

language processing, training a deep neural network from scratch[2] usually requires huge amounts of training data and their consistent labelling (Justus et al., 2018). That is, the expected outcome of each data record (e.g. the object depicted in an image or the information contained in a sentence) has to be identified and marked to train an image classifier or a natural language information extraction system.[3] Both, collecting these data and pre-processing them, can overwhelm a single organisation.

However, even a huge data collection alone does not guarantee an optimal starting point for a machine learning endeavour. The data sets also need to be representative and—optimally—somewhat balanced to avoid the biasing of the algorithm. What does this mean? Representativeness refers to the property of a data collection to mirror reality as closely as possible. If wanting to distinguish between apples, pears, and tomatoes by using an image recognition system, one obviously needs to present examples of all three different outcome classes to the training component. Without ever seeing a tomato during its training phase, the system cannot be expected to correctly identify a picture of it. Here, pooling data from various sources might be helpful to overcome such limitations.

Similarly, one specific class of possible outcomes might be extremely rare, e.g. in an Industry 4.0 scenario, where huge amounts of sensor data are collected to predict the failure of a certain engine and take countermeasures even before it breaks down (predictive maintenance). Fortunately, most machines run quite reliably. The flipside of this is the fact that the collection of sensor data will contain very few—if any—examples of an engine not working properly. From a machine learning perspective, this leads to extremely unbalanced data—many data entries for the engine's normal behaviour and very few to no records representing the opposite case (Theissler et al., 2021).

This, however, hinders a successful training process of virtually all machine learning methods as they assume certain statistical properties of their training data (Krawczyk, 2016). Gathering large, balanced data sets across organisations can help to form a data pool that contains a critical mass of samples of the "exotic" behaviour which can then be used to form balanced training sets using sophisticated sampling techniques.[4]

The simplest example of collaboration among various organisations is obviously the creation of a joint data pool. Nonetheless, simply putting all the data gathered at various places into one collection will often not be appropriate.

First, there is the need for a unified representation of these data. This refers to both the concepts and wordings used to describe them and the agreement of a common scale for measurements, calibration of sensors, etc. to avoid comparing apples to oranges. *Ontologies* (Stephan et al., 2007) can play a central role providing the basis

[2]The author will present alternative approaches in the next section.

[3]There are some exceptions to this rule, e.g. in gaming applications (Silver et al. 2016) but for the time being, deep learning typically requires a significant amount of training data.

[4]Another reason for unbalanced data can be a *bias* during the data collection process, i.e. the tendency to give more weight to one particular outcome as compared to others.

for such a unified data representation as they represent the key semantic concepts of a certain domain as well as their interrelationships in a formal way. Further, they provide means to express important properties of these concepts such as measuring range. Additionally, there exist techniques to systematically "translate" between diverging ontologies, thus facilitating the creation of a joint, unique representation of data (Choi et al., 2006).

Second, privacy or IP protection concerns might be good reasons for not providing access to the original data. However, there are technical means to overcome this problem. The generation of *synthetic data* can in many cases replicate the relevant statistical properties of the originals without revealing individual data records (Soltana et al., 2017). Especially approaches striving to achieve *differential privacy* are currently under development and will soon be available for widespread use (Xin et al., 2020). Differential privacy is a concept that formalises the attempt to prevent individual data records from being identified within a given data set. This is usually achieved by slightly disturbing the original data, i.e. by adding some portion of noise to it. The higher the intended level of protection, the noisier—and thus, less useful—the data become. So, it is crucial to find a good balance between an acceptable level of data protection and usefulness of the data.

Similarly, researchers are working on *homomorphic encryption*—a technology that will allow the use and processing of encrypted data without the need to decrypt them beforehand (Li, Kuang, et al., 2020). Currently, the algorithms used are way too slow for real-size applications, but this is expected to improve in the foreseeable future.

An alternative to pooling all data from various organisations in one place is *distributed machine learning* that will be discussed in the next section.

Governance Effort in Data Pooling

One of the main tasks in keeping control over the data pooling step is the definition and maintenance of a common representation standard—or the joint development of an appropriate translation mechanism between competing representations—with the goal to avoid the mixture of incompatible data. From a non-technical point of view, protection of sensitive data—be it data representing IP or person-related data that must be protected against unintended disclosure—has to be at the centre of attention. As briefly discussed above, technical measures can allow a collaboration across organisations even if the data to be used for model training is highly sensitive.

2.2 *Opportunities in AI Model Development*

Based on the previous section's discussion of joint data training options, this one addresses the potential of collaboratively developing AI models among a group of actors. By joining resources and using synergies, organisations collectively develop one standardised AI model, which can later be individualised as for the use case of each participating organisation.

Management Perspective on Collaborative Model Development
Besides situations of limited data availability discussed above, joint model development can additionally make sense if data science know-how or computational resources are simply not available or too expensive. In this section the authors will discuss approaches on collaborating with other organisations while, at the same time, highlighting potential negative consequences and risks that could occur.

Technical Perspective on Collaborative Model Development
As already discussed, the simplest form of collaboration is the creation of one joint data pool among all participating parties in a trusted environment. The natural extension of this approach is to also create one model for all participants—possibly by a third party specialised on data science projects. This model can then be hosted in one central place where it is queried (client/server architecture) and where feedback[5] from the various parties is collected for further model improvements. Alternatively, the model is distributed among the participants where identical copies are applied in local settings. Such an approach can be advantageous, e.g. in Industry 4.0 applications where computation often takes place in the edge, i.e. at the very engine to be monitored. There, low bandwidth and computational power, combined with the need for real-time response, may require the use of a local model. The feedback data for model updates, however, should nonetheless be gathered in one central place.

If such a local model is applied in a very specific, untypical environment, it might make sense to use the feedback data to adapt the model to the local specifics. If the feedback can be provided and processed in a systematic way, *reinforcement learning* techniques might be good candidates to be considered (Nian et al., 2020).

Approaches to *distributed machine learning* originated from the requirement to process ever-growing amounts of data that exceeded the capability of one single computing node (Verbraeken et al., 2020). Here the idea was to split up the available set of training data into manageable portions and eventually combine the partial results to one overall model. Still, this does not account for situations in which each participating party is eager to keep their own assets private.

If organisations mainly aim at protecting their respective data science know-how—i.e. their special way to deal with data, preprocess them, and create a model—then a simplified version of the so-called model-parallel approach could be viable. Here all the data are collected in one pool that each party can access in its entirety without sharing the modelling outcomes with the others. This results in one isolated model for each party built upon the complete data collection.[6]

If, on the other hand, it is mainly the raw data themselves that are to be protected, other techniques come into play. With a growing amount of user-generated data and concern about consumer privacy rights, approaches like federated learning have become popular (Li, Sahu, et al., 2020). This technique is used, e.g. for next-word prediction in smartphone keyboard apps. Each single user certainly does not want

[5] Indications on when the model was right in its prediction and when it failed

[6] The usual final step of aggregating these models into one central model is obviously left out in this approach.

their complete input to be forwarded to some central modelling instance to create an optimal user support model for all. Instead, the keyboard app is initially equipped with a model that has been trained on non-critical data. This initial model is then refined locally—i.e. on the end user's device—and only the required update of model parameters is delivered back to the central modelling instance. This way, numerous users can contribute to an overall optimisation of the model that is periodically updated on the users' devices.

Obviously, this approach is also applicable to data generated by organisations in their daily business and there exist numerous ways to create or update a central model from distributed modelling contributions. The currently most general framework for distributed model generation with an integrated approach for governing the overall process is represented by Substra, an open-source initiative for "privacy-preserving, traceable and collaborative machine learning" (Galtier & Marini, 2019). Substra allows the specification of (distributed) training tasks based on four different asset classes:

- An *objective* formally defines format requirements for data and models, test data, and evaluation criteria.
- A *dataset* summarises data represented in a common format.
- An *algorithm* comprises a script for training a model on the dataset including additional information on loss functions, parameter tuning, etc.
- A *model* finally represents the outcome of a training task.

Each of these assets is subject to a formally specified permission regime that restricts their use and transfer between different nodes of the network of participating organisations. Substra orchestrates the set of training tasks specified this way and allows for complex operations to generate the final model from intermediary results. Federated learning as mentioned above is just one possible way to aggregate partial models. At each point in time, the training tasks ahead are listed in a distributed ledger as it is used in the blockchain approach. Implementing the permission regimes as smart contracts over this ledger then inevitably leads to compliance and privacy by design. A similar approach called swarm learning has recently been presented in the context of analysing clinical patient data (Warnat-Herresthal et al., 2021).

Instead of collaborating during the development of a machine learning models, it is also possible to build upon previous results accomplished by others without returning any assets in exchange. As already mentioned, deep learning endeavours place high demands on data availability and computing power. Therefore, the use of pre-trained partial neural networks has recently become popular. Here, the output layer—and possibly also some of the preceding layers—is removed from the fully trained network and only these few layers have to be trained anew. A typical scenario is the application of a trained network to a slightly different, but related application domain—the so-called transfer learning (You et al., 2021). For example, if the original model was trained to classify images according to whether they depict a cat or a dog, it is possible to reuse those layers of the network that deal with deriving features of increasing abstractness from the input images. Building upon these pre-trained feature layers, only the remaining layers need to be trained using images from the new application domain such that a classifier for, e.g. trees and flowers, can be generated with significantly less effort.

While it seems tempting to simply reuse others' previous work, doing so comes at some risk. Without detailed knowledge of the initial training data used and the parameter tuning performed, it is possible to induce some unwanted bias into the new model. On the other hand, if one organisation releases pre-trained (partial) models for reuse by others, the danger exists that this model reveals insights into this organisation's business that should have remained hidden. A word completion or prediction network might suggest a notion—a person's name, a specific technical term, etc.—that allows unwanted conclusions to be drawn about the initiating organisation.

Governance Effort in Collaborative Model Development
The joint development of machine learning models requires to find a balance between potential savings in terms of computational and staff resources on the one hand and the risk to induce unwanted—and undetected—bias into the overall model on the other hand. As discussed above, formal methods to specify and monitor joint endeavours exist that limit this risk. Nonetheless, the actors involved need to be aware of such risks and introduce tailored governance measures to address these risks proactively.

2.3 *Opportunities in Model Evaluation and Deployment*

The decision about joint model evaluation depends on the respective application scenario of each participating organisation. Collaboration in this phase is mainly reasonable when all parties apply the same model for very similar cases. If an organisation plans to apply the common model for a very specific purpose that significantly differs from what others are doing, then they need their own evaluation data and metrics.

Otherwise, it makes perfect sense to collect evaluation data in one central place and have the evaluation itself carried out by an independent organisation. Representative and bias-free data are key, as well as are detailed documentations of the evaluation process itself and its results. The Substra framework already discussed above provides means to model, run, and monitor this particular kind of collaboration.

When deploying a jointly developed model, two basic cases must be distinguished. If each participating organisation has an on-premises installation of a local copy of this model, then the main focus should be on keeping it up to date by integrating the feedback provided by all users across the whole consortium. The second option is a server-based architecture where the model is hosted in one central place and queried. In this constellation, model maintenance is much easier as feedback can be gathered and analysed in one place.

In both cases, however, the model usage must be monitored. On the one hand, a model user might try to reveal the original training data by so-called membership inference attacks,[7] thus violating privacy or IP protection interests. The intended injection of manipulated training data is a direct threat to model performance. There have been several cases where, e.g. slightly modified images of traffic signs lead to misclassification by car-based systems (Morgulis et al., 2017). Consequently, not only the development process itself but also the deployment and usage of a model-based system have to be closely monitored.

From a management and governance perspective, these last stages of the AI lifecycle portray the companies' decision to engage in a full collaboration. Consequently, their joint handling results in a more extensive governance demand than partial forms of collaboration. In this case, the companies all get access to the finalised, trained, and evaluated AI model, which they can later adopt to their specific use case. Hence, the competitive technological advantage is shared among the companies.

3 Governance of Risks in Collaborative AI Development

To summarise the above discussion, the following section provides an overview of the various risk types that come with collaborative AI development endeavours. The overview includes technical as well as economic risks and suggests governance measures to address these risks accordingly.

Economic Risks

These first insights suggest that collaborative AI development approaches entail potential for significant synergies and economisation of resources. Further, collaboration enables a company to engage in AI development even if it does not have access to the required resources, such as sufficiently big, quality data sets or extensively trained staff. Collaboration can help overcome those barriers but comes with the need to predefine outcome scenarios, to avoid, e.g. resulting violations of privacy rights due to cross-company usage and analysis of collected data. Since such consequences could lead to economic repercussions, companies engaging in collaborative AI development need to address such potential challenges upfront by predefining their form of collaboration.

This might include questions such as if the companies aim for partial collaboration, thus joining forces for only specific development phases, or for full collaboration—including a joint data set as well as the collaborative development of an AI solution. Finally, the companies should decide early in the process whether they are willing to engage in a collaborative evaluation and deployment phase.

[7] See, e.g. https://bdtechtalks.com/2021/04/23/machine-learning-membership-inference-attacks/.

Technical Risks

As presented in the following table, there exist technical solutions to address these various risks. First, however, there needs to be an awareness among everyone involved that all the obvious positive aspects of pursuing collaborative approaches do not come for free. This is not meant to be a counterargument against collaboration.

Data pooling	Model development	Evaluation and deployment
Erroneous results and additional effort due to incompatible data (representations, scales, etc.)	Undesired (and possibly undetected) import of bias from partial models	Various types of model attacks leading to loss of IP and/or unwanted disclosure of training data
Unwarranted access to sensitive data and/or loss of IP	Unwarranted access to sensitive data and/or loss of IP	Injection of toxic data leading to degradation of model performance

Second, everyone involved in such a process should be aware of the pitfalls and the technical and organisational measures that can be taken to avoid them and how to orchestrate them to achieve an optimal balance between risks and benefits.

Still, not all technical methods mentioned are necessarily compatible with each other. Consequently, there must be a clear communication between management and technical staff on what risks actually exist, which of these have to be avoided at any cost, and what resources will be available. This forms the basis for an informed decision for or against collaboration and an optimal balance between benefits and risks.

Resumé

Depending on the level of collaboration, governance requirements and potential gains raise and fall correlatively. Hence, the higher the degree of collaboration, the higher the potential gains but at the same the risk for infringements. Consequently, this publication suggests a more detailed governance strategy for full collaboration than for partial collaboration that merely entails, e.g. a joint data pooling phase. Still, the publication encourages companies to engage in collaborative forms of AI development due to the potential gains accessible via such collaboration. Moreover, as presented above, technical solutions are available to address most of the mentioned risk types. Thus, a governance analysis preceding the actual collaboration can help identify initial, fundamental risks and help address them upfront. Iterative governance analysis in place throughout the duration of the collaboration will support companies in constantly monitoring their success and potential implications of the current form of collaboration, which in turn allows companies to introduce the above-mentioned technical solutions in time to avoid negative consequences.

4 Implications, Discussion, and Outlook

4.1 *Implications for Practice*

In conclusion, an increasing number of companies are faced with the decision whether to adopt AI and, if deciding to do so, whether to build or buy an AI solution. If the company decides to build a solution, e.g. to avoid path dependencies with pre-developed solutions, it is confronted with various challenges. Collaborative forms of AI development can help overcome such limitations. Thus, companies need to decide whether to engage in partial or full collaboration; in other words, they need to decide whether to join forces only for specific phases of the AI lifecycle or throughout the entire development process. Full collaboration consequently includes collaboration among two or more organisations throughout the entire AI collaboration process. This is since the organisations involved not only merge their data sets but choose to jointly develop a standardised AI model, which they later adapt according to their specific need. Thereby, they drastically lower the costs for technology development while still obtaining a tailor-made solution for the specific need the allying organisations share. The cyber-risks of purchasing a standardised AI solution are lowered, too.

However, as exemplified in this chapter, full collaboration yields a higher potential for synergies, yet comes with higher risk levels, too. Companies engaging in collaborative forms of AI development should, thus, be aware of these risks and the resulting governance efforts required. By structuring the risks and opportunities per development stage, this publication contributes to raising such awareness.

Moreover, the publication suggests both governance measures preceding the actual collaboration as well as iterative governance formats implemented and realised through the collaboration process. The aim of such a combined governance approach is to align the economic interests of all stakeholders involved, while at the same time minimising potential risks. Thus, the governance strategy ensures the economic advantages of each player, the technical feasibility of the solution, and its agreeableness with societal expectations and rights, such as the protection of privacy. Doing so, it aligns the stakeholder-specific interests of all actors involved to reach an advantageous result for all of them (Wieland, 2020). To promote the effectiveness of the governance approach, developing a shared understanding of the process and its objectives as well as shared values is recommended to foster trust between the stakeholders. Further, clear guidelines for communication are understood to foster the long-term success of collaborative processes (Wieland, 2020).

4.2 *Limitations and Further Research*

Due to the high relevance of the topic for practitioners and the apparent gap in research, the objective of this publication was to give insight into a new topic and to contribute to closing the above-mentioned research gap. To this end, it presents a first conceptualisation structuring and combining existent concepts and complementing them with technical insights. While presenting a suitable approach for practice and a theoretical concept, the findings presented require deeper research, such as a systematic literature review of the research field, and an empirical analysis of its suggestions. In detail, the theoretical assumptions made in this publication need to be confirmed by empirical testing of both qualitative and quantitative nature.

4.3 Conclusion and Outlook

While future technological developments are likely to reduce the aforementioned technical challenges of collaborative AI development, e.g. regarding synthetic data, general levels of complexity are expected to rise. This is since the general demand and tasks for AI technologies will become more demanding and complex; as a result, it will hardly be possible to develop suitable solutions as a single company. Hence, these developments raise the pressure on companies to engage in collaboration when deciding to develop AI.

Alternatively, companies can opt to collaborate with other actors through open-source innovation, in particular open-source codes, and models—as the German automotive company Porsche recently decided to do). However, open-source innovation serves as the ultimate form of collaboration, potentially engaging with all actors from society, resulting in a generally accessible result. While some companies now choose to follow this path, collaboration among a few actors offers higher levels of protection for the developed algorithms and models. With this, it offers a potential competitive advantage for the collaborating companies, due to technological advance.

Either way, companies around the globe, specifically so in Europe, are not only driven by bottom-up self-interest in collaboration but from a political tone from the top advising them to collaborate to defy the pressure of global competition—mostly stemming from dominant players in the market such as the USA and China (Dafoe, 2018). Consequently, mainly European companies will probably need to collaborate, whether they choose to do so sooner or later, when wanting to defend and secure their market share and competitive edge.

References

Allam, Z., Dey, G., & Jones, D. S. (2020). Artificial intelligence (AI) provided early detection of the coronavirus (COVID-19) in China and will influence future Urban health policy internationally. *AI, 1*(2), 156–165.

Allam, Z., & Jones, D. S. (2020, March). On the coronavirus (COVID-19) outbreak and the smart city network: universal data sharing standards coupled with artificial intelligence (AI) to benefit

urban health monitoring and management. In *Healthcare* (Vol. 8, No. 1, p. 46). Multidisciplinary Digital Publishing Institute.
Camilli, M., Felderer, M., Giusti, A., Matt, D. T., Perini, A., Russo, B., & Susi, A. (2021). Towards risk modeling for collaborative AI. *arXiv,* preprint arXiv:2103.07460.
Cave, S., & ÓhÉigeartaigh, S. (2019) An AI race for strategic advantage: Rhetoric and risks. In *Conference paper for: AI ethics and society 2018, 1*. https://doi.org/10.1145/3278721.3278780
Choi, N., Song, I., & Han, H. (2006). A survey on ontology mapping. *ACM SIGMOD Record, 35*(3), 34–41.
Coyle, D., & Li, W. (2021). *The data economy: Market size and global trade*. ESCoE Discussion Paper 2021-09, https://www.escoe.ac.uk/publications/the-data-economy-market-size-andglobal-trade.
Dafoe, A. (2018). *AI governance: a research agenda*. Governance of AI Program, Future of Humanity Institute, University of Oxford. https://www.fhi.ox.ac.uk/wpcontent/uploads/GovAI-Agenda.pdf
Dellermann, D., Calma, A., Lipusch, N., Weber, T., Weigel, S., & Ebel, P. (2019, January). The future of human-AI collaboration: a taxonomy of design knowledge for hybrid intelligence systems. In *Proceedings of the 52nd Hawaii international conference on system sciences*.
d'Inverno, M., & McCormack, J. (2015). Heroic versus collaborative AI for the arts. In Q. Yang & M. Wooldridge (Eds.), *Proceedings of the 24th international joint conference on artificial intelligence* (pp. 2438–2444). AAAI.
Draxl, C., & Scheffler, M. (2019). The NOMAD laboratory: from data sharing to artificial intelligence. *Journal of Physics: Materials, 2*(3), 036001.
Fatehi, M. (2019). Collaborative AI research in medical imaging: trends and challenges. *Iranian Journal of Radiology, 16*(Special Issue).
Galtier, M., & Marini, C. (2019). Substra: a framework for privacy-preserving, traceable and collaborative. *Machine Learning*. arXiv Preprint arXiv: 1910.11567.
Gerbert, P., Duranton, S., Steinhäuser, S., & Ruwolt, P. (2018). *The build-or-buy dilemma in AI*. The BCG Henderson Institute. https://image-src.bcg.com/Images/BCG-The-Build-or-Buy-Dilemma-in-AI-Jan-2018_tcm104-180320.pdf
Gruson, D., Bernardini, S., Dabla, P. K., Gouget, B., & Stankovic, S. (2020). Collaborative AI and laboratory medicine integration in precision cardiovascular medicine. *Clinica Chimica Acta, 509*, 67–71.
Gupta, I. (2020). Decentralization of artificial intelligence: analyzing developments in decentralized learning and distributed AI networks.
He, J., Baxter, S. L., Xu, J., Xu, J., Zhou, X., & Zhang, K. (2019). The practical implementation of artificial intelligence technologies in medicine. *Nature Medicine, 25*(1), 30–36.
Justus, D., Brennan, J., Bonner, S., & McGough, A. S. (2018). Predicting the computational cost of deep learning models. In *Proceedings of the 2018 IEEE international conference on big data* (pp. 3873–3882).
Kambhampati, S. (2019, May). Synthesizing explainable behavior for human-AI collaboration. In *Proceedings of the 18th international conference on autonomous agents and multiagent systems* (pp. 1–2).
Khadpe, P., Krishna, R., Fei-Fei, L., Hancock, J. T., & Bernstein, M. S. (2020). Conceptual metaphors impact perceptions of human-ai collaboration. *Proceedings of the ACM on Human-Computer Interaction, 4*(CSCW2), 1–26.
Koch, J. (2017, March). Design implications for designing with a collaborative AI. In *2017 AAAI Spring symposium series*.
Koch, J., & Oulasvirta, A. (2018). Group cognition and collaborative ai. In *Human and machine learning* (pp. 293–312). Springer.
Krawczyk, B. (2016). Learning from imbalanced data: open challenges and future directions. *Progress in Artificial Intelligence, 5*, 221–232.

Li, J., Kuang, X., Lin, S., Ma, X., & Tang, Y. (2020). Privacy preservation for machine learning training and classification based on homomorphic encryption schemes. *Information Sciences, 526*, 166–179.
Li, T., Sahu, A. K., Talwalkar, A., & Smith, V. (2020). Federated learning: Challenges, methods, and future directions. *IEEE Signal Processing Magazine, 37*(3), 50–60.
Lins, S., Pandl, K. D., & Teigeler, H. (*2021*). Artificial intelligence as a service. *Business & Information Systems Engineering, 63*, 441–456. https://doi.org/10.1007/s12599-021-00708-w
Martínez-Plumed, F., Contreras-Ochando, L., Ferri, C., Orallo, J. H., Kull, M., Lachiche, N., Quintana, M. J. R., & Flach, P. A. (2019). CRISP-DM twenty years later: From data mining processes to data science trajectories. *IEEE Transactions on Knowledge and Data Engineering, 33*(8), 3048–3061.
Mehri, V. A., Ilie, D., & Tutschku, K. (2018, August). Privacy and DRM requirements for collaborative development of AI applications. In *Proceedings of the 13th international conference on availability, reliability and security* (pp. 1–8).
Morgulis, N., Kreines, A., Mendelowitz, S., & Weisglass, Y. (2017). Fooling a real car with adversarial traffic signs. *arXiv*, preprint arXiv:1907.00374.
Nian, R., Liu, J., & Huang, B. (2020). A review on reinforcement learning: Introduction and applications in industrial process control. *Computers & Chemical Engineering, 139*.
Nilsson, N. J. (2009). *The quest for artificial intelligence*. Cambridge University Press.
Noorbakhsh-Sabet, N., Zand, R., Zhang, Y., & Abedi, V. (2019). Artificial intelligence transforms the future of health care. *The American Journal of Medicine, 132*(7), 795–801.
Okamura, K., & Yamada, S. (2020). Adaptive trust calibration for human-AI collaboration. *PLoS One, 15*(2), e0229132.
Peiffer-Smadja, N., Maatoug, R., Lescure, F. X., D'ortenzio, E., Pineau, J., & King, J. R. (2020). Machine learning for covid-19 needs global collaboration and data-sharing. *Nature Machine Intelligence, 2*(6), 293–294.
Piorkowski, D., Park, S., Wang, A. Y., Wang, D., Muller, M., & Portnoy, F. (2021). How AI developers overcome communication challenges in a multidisciplinary team: A case study. *Proceedings of the ACM on Human-Computer Interaction, 2021*, 1–25.
Rowan, I. (2020). Make or buy AI?. In *TowardsDataScience.com*. https://towardsdatascience.com/make-or-buy-ai-7b8d1f48ef21
Salta, A., Prada, R., & Melo, F. (2020). A game AI competition to foster collaborative AI research and development. *IEEE Transactions on Games*. https://doi.org/10.1109/tg.2020.3024160
Schiff, D., Biddle, J., Borenstein, J., & Laas, K. (2020). What's next for AI ethics, policy, and governance? A global overview. In *Proceedings of the AAAI/ACM conference on AI, ethics, and society* (pp. 153–158). https://doi.org/10.1145/3375627.3375804
Silver, D., Huang, A., & Maddison, C. (2016). Mastering the game of Go with deep neural networks and tree search. *Nature, 529*, 484–489. https://doi.org/10.1038/nature16961
Soltana, G., Sabetzadeh, M., & Briand, L. C. (2017). Synthetic data generation for statistical testing. In *Proceedings of the 32nd IEEE/ACM international conference on automated software engineering (ASE), 2017* (pp. 872–882).
Sowa, K., Przegalinska, A., & Ciechanowski, L. (2021). Cobots in knowledge work: Human–AI collaboration in managerial professions. *Journal of Business Research, 125*, 135–142.
Stephan, G., Pascal, H., & Andreas, A. (2007). Knowledge representation and ontologies. In R. Studer, S. Grimm, & A. Abecker (Eds.), *Semantic web services*. Springer.
Theissler, A., Pérez-Velázquez, J., Kettelgerdes, M., & Elger, G. (2021). Predictive maintenance enabled by machine learning: Use cases and challenges in the automotive industry. *Reliability Engineering & System Safety, Elsevier, 215*.
Tkachuk, R. V., Ilie, D., & Tutschku, K. (2020). Towards a secure proxy-based architecture for collaborative AI engineering. In *CANDARW*.
Verbraeken, J., Wolting, M., Katzy, J., Kloppenburg, J., Verbelen, T., & Rellermeyer, J. S. (2020). A survey on distributed machine learning. *ACM Computing Surveys, 53*(2), Article 30.

Wang, D., Churchill, E., Maes, P., Fan, X., Shneiderman, B., Shi, Y., & Wang, Q. (2020, April). From human-human collaboration to human-AI collaboration: Designing AI systems that can work together with people. In *Extended abstracts of the 2020 CHI conference on human factors in computing systems* (pp. 1–6).
Wang, D., Weisz, J. D., Muller, M., Ram, P., Geyer, W., Dugan, C., Tausczik, Y. R., Samulowitz, H., & Gray, A. (2019). Human-AI collaboration in data science: Exploring data scientists' perceptions of automated AI. *Proceedings of the ACM on Human-Computer Interaction, 3*(CSCW), 1–24.
Warnat-Herresthal, S., Schultze, H., Shastry, K. L., Manamohan, S., & Thirumalaisamy, V. P. (2021). Swarm learning for decentralized and confidential clinical machine learning. *Nature, 594*, 265–270.
Wieland, J. (2020). *Relational economics: A political economy*. Springer.
Xin, B., Yang, W., Geng, Y., Chen, S., Wang, S., & Huang, L. (2020). Private FL-GAN: differential privacy synthetic data generation based on federated learning. In *ICASSP 2020—2020 IEEE international conference on acoustics, speech and signal processing (ICASSP)* (pp. 2927–2931).
Xiong, Y., Chen, H., Zhao, M., & An, B. (2018, April). HogRider: champion agent of Microsoft Malmo collaborative AI challenge. In *Proceedings of the AAAI conference on artificial intelligence* (Vol. 32, No. 1).
You, K., Liu, Y., Wang, J., & Long, M. (2021). LogME: Practical assessment of pre-trained models for transfer learning. In *Proceedings of the 38th international conference on machine learning, PMLR* (Vol. 139, pp. 12133–12143).
Zillner, S., Gomez, J. A., Robles, A. G., Hahn, T., Le Bars, L., Petkovic, M., & Curry, E. (2021). Data economy 2.0: From big data value to AI value and a European data space. In *The Elements of big data value* (pp. 379–399). Springer.

Sabine Wiesmüller is managing director of the Bodensee Innovation Cluster at Zeppelin University and senior researcher at the Leadership Excellence Institute Zeppelin, particularly focusing on relational governance, AI governance, and business ethics. Having successfully completed her doctoral candidacy at the Wittenberg Center for Global Ethics, Sabine Wiesmüller examined forms and interactions of relational AI governance in her doctoral thesis. In both, her work at the innovation cluster and her role as a business consultant and researcher, Sabine Wiesmüller is interested in the effects of artificial intelligence and exponential technologies on society and new avenues for their suitable governance.

Dr Mathias Bauer has more than 30 years of experience in the field of artificial intelligence. He is a Research Fellow of the German Research Center for AI (DFKI), where he spent more than 15 years in basic research, before starting a career as a serial entrepreneur. He founded several successful start-ups in AI and data science, was Head of Advanced Analytics at KPMG Germany, and supports incubators, founders, and investors in making the right decisions. His current focus is on technical approaches to guarantee and enforce compliance in the execution of processes.

Responsible AI Adoption Through Private-Sector Governance

Sabine Wiesmüller, Nele Fischer, Wenzel Mehnert, and Sabine Ammon

Abstract This contribution examines responsible artificial intelligence (AI) adoption in organisations from a private-sector AI governance perspective. Since an increasing number of organisations adopt AI, society interacts with the technologies more frequently due to higher exposure to AI applications. Consequently, companies are confronted with society's demand to integrate ethical reflections and the perspectives of diverse stakeholders into their decision-making processes. With this, the need for responsible AI adoption rises, too. Yet, neither existing innovation processes nor AI development models address the adoption and development phases entailed in AI lifecycles regarding iterative ethical reflections from a management perspective. Thus, to contribute to filling this research gap, this chapter firstly highlights the need and current lack of systematically integrating ethical reflection in AI adoption processes. Secondly, it proposes a governance model as a first starting point for developing an instrument for responsible AI adoption in organisations, supporting corporate social responsibility in this regard.

1 Relevance and Research Gap

The need for this research field particularly stems from the disruptive power of general-purpose technologies, such as artificial intelligence (AI) (Brynjolfsson & McAfee, 2017; Dafoe, 2018; Goldfarb et al., 2019; Klinger et al., 2018; Nepelski &

S. Wiesmüller (✉)
Zeppelin University, Friedrichshafen, Germany
e-mail: Sabine.Wiesmueller@zu.de

N. Fischer
Berlin Ethics Lab, Berlin, Germany

W. Mehnert
Berlin University of the Arts, Berlin, Germany

S. Ammon
TU Berlin, Berlin, Germany

R. Schmidpeter, R. Altenburger (eds.), *Responsible Artificial Intelligence*, CSR, Sustainability, Ethics & Governance, https://doi.org/10.1007/978-3-031-09245-9_5

Sobolewski, 2020; Razzkazov, 2020; Trajtenberg, 2018). In the following, AI shall be defined as technologies able to learn from and interpret data to reach predefined goals and tasks based on their potential to flexibly adapt to the requirements of a situation (Kaplan & Haenlein, 2019). Predominantly, this definition includes machine and deep learning technologies as well as symbolic AI (Nilsson, 2009). This chapter aims at elaborating a tentative generic process for responsibly adopting AI applicable across those different AI technologies.

As for their impact, the technologies subsumed under the term 'AI' will eventually affect all parts of society due to their general-purpose nature. The broad adaptability of AI to specific applications throughout diverse sectors and industries will most likely result in unprecedented levels of social disruption (Hagendorff, 2020a). They already deeply affect users and consumers in their daily life, be it through, e.g. their application to search engines, in the form of automatised mobility solutions, pricing strategies used in online businesses, or applications for public services (Bughin et al., 2017). Moreover, the rapidly developing state of the art in technology and the competitive pressure for companies to adopt AI constantly rise to new heights, particularly over the last years, as the following numbers confirm: The Gartner 2020 CIO Survey, conducted on a global level, showed that over 40% of the interviewed companies focused on implementing AI solutions by the end of 2020 (Gartner, 2020). Similarly, a study conducted by Ipso for the European Commission in 2020, with over 9000 companies from all over Europe, displayed that 42% of the sample had already introduced at least one AI technology to their corporate processes or products (Tresignie et al., 2020). Further, the study shows that roughly another 20% of the organisations planned to adopt AI in the near future.

While the adoption of AI technologies promises new business opportunities, among others, data analytics, the automatisation of existing processes, and predictive applications, it comes with a broad range of effects on societies—both positive and negative (Nilsson, 2009). AI-based technologies already deeply impact societies, and their effects grow with the increasing number of use cases in practice. Stemming from these high levels of disruption, societal concerns are growing regarding the private sector's central role in accelerating both the pace of AI's development and deployment in practice (Mittelstadt, 2019). Consequently, a growing number of scholars discuss the repercussions of AI's adoption and demands for companies to address the negative consequences of their actions (Hagendorff, 2020a, 2020b; Jobin et al., 2019). However, the aforementioned scholars agree that there is a gap for AI governance, specifically for the development of instruments with a higher action orientation. While current research focuses on ethical guidelines and the identification of AI ethical dilemma structures, academia is lacking AI governance instruments suitable for practice (Hagendorff, 2020a; Mittelstadt, 2019). Companies require AI governance instruments that allow them to align and balance the diverging demands they are faced with: For one, companies are exposed to particularly high levels of competitive pressure to adopt AI (Cave & ÓhÉigeartaigh, 2019; Scharre, 2019; Schiff et al., 2020), rising societal concerns and risks for society

(Dafoe, 2018; Jobin et al., 2019), and fast-paced technological advancements in AI research (Wallach & Marchant, 2019). Thus, a suitable AI governance instrument needs to enable the alignment of different demands to allow for a result that creates positive value in all three areas—by gaining economic profit, providing societal value, and offering technological feasibility (Wieland, 2020). To support practitioners in governing AI adoption processes and to add to the aforementioned research field, this contribution combines AI ethical considerations, such as ethics by design (Hagendorff, 2020a; Jobin et al., 2019) on the AI developers' team level, with an organisational governance perspective on a strategic level. Moreover, this contribution introduces the importance of iterative ethics by design, which will be exemplified throughout this chapter.

According to this contribution, AI adoption is deemed responsible if the organisation identifies and addresses the consequences stemming from the potential AI application, regarding all its stakeholders, and if it is willing to adapt its AI solution accordingly. A responsible AI adoption thus integrates societal needs and values as well as long-term ethical considerations into the AI development and application realised by the organisation, balancing those aspects with the technological the state of the art and necessary economic conditions. On an operational level, this contribution refers to the term AI adoption as the strategic decision to implement AI into the own organisation, entailing all stages of this process. In detail, AI adoption includes the decision to develop AI within the organisation as well as buying an externally pre-developed AI solution without being involved in its initial development (Alsheibani et al., 2018; Cubric, 2020). This is since the different stages of adopting AI in an organisation, such as the initial ideation phase, occur in each form of AI adoption—be it in the self-development of a solution or the adaptation of a pre-developed, bought solution. Consequently, various scenarios of AI adoption can be addressed by applying the instrument presented in this publication, which aims to ensure responsible business conduct in this endeavour. Thus, by including all possible forms of AI adoptions and all sequential stages of adopting AI within an organisation, this publication addresses the broadest definition and occurrence of AI adoption in practice. As a result, its governance instrument includes all stages of the AI adoption process with the goal of enabling a *responsible* conduct throughout the process.

To structurally integrate such ethical considerations into the AI adoption process, this contribution proposes a tentative governance model for responsible AI adoption. The proposed process integrates ethical reflections and the diverse perspectives of stakeholders iteratively throughout the adoption process (see Sect. 2.3). By applying such a model, the organisation actively engages in not only economic value creation but the creation of values for its stakeholders, too (Wieland, 2020). Further, this AI governance instrument shall allow the company to proactively shape the social impacts its adopted AI will bring about, instead of applying a traditional reactive corporate governance model, implemented to address the negative, unwanted consequences of an already implemented product or solution (Allen & Chan, 2017;

Armstrong et al., 2016; Nakashima, 2012; Polyakova & Boyer, 2018; Scharre, 2019). By actively avoiding negative consequences, this contribution focuses on creating a desirable outcome from the AI adoption process. Thus, responsible AI adoption is perceived as a crucial cornerstone for taking social responsibility as a company. Particularly due to an organisation's embeddedness in society (Wieland, 2020), companies need to ensure responsible decision-making when adopting AI technologies within their organisation to avoid negative impacts on society. Thus, the challenge for organisations is to take a responsible stance on AI adoption, considering the resulting societal impact and the futures their technologies shape.

By embracing this perspective, the contribution understands organisations as a nexus of stakeholders, an entity interrelated with stakeholders, which are directly or indirectly affected by the organisation's actions (Wieland, 2020). The proposed model for responsible AI adoption thus assists decision-makers in proactively aligning the various demands they are confronted with. As exemplified, these stem from the stakeholders affected by the AI adoption and can be addressed through, e.g. the active governance of interactions between AI development teams and societal stakeholders. Hence, the contribution proceeds to present a model for responsible AI adoption that integrates the societal needs, ethical reflection, and stakeholder perspectives into the adoption processes (Sect. 2). Further, it will offer practical insights into the implementation of a responsible adoption process (Sect. 3). The contribution concludes with implications for decision-makers from a governance perspective and suggestions for further research as well as the current limitations of this tentative model (Sect. 4).

2 A Model for Responsible AI Adoption from a Private-Sector Governance Perspective

As presented above, this contribution defines responsible AI adoption as aiming to integrate societal needs and values into the AI development and application realised by an organisation. The question how to address this issue accurately is subject to an active research field and currently discussed from different perspectives, across academic disciplines and by actors from different societal sectors.

One predominant research stream in AI ethics research, the field of principle-based AI ethics (Hagendorff, 2020a, 2020b; Jobin et al., 2019; Mittelstadt, 2019; Morley et al., 2020; Yu et al., 2018), focuses on the use of codices and ethical principles. Those codices and principles are proposed to guide responsible AI development by political authorities, trans-sectoral institutions, and companies alike (cf., European Commission, 2019; OECD, 2019; Pichai (Google), 2018). While this common ground can serve as a suitable starting point for companies to address their responsibility, the often abstract and theoretical level of those principles creates severe challenges for the private sector when it comes to their operationalisation (Hagendorff, 2020a; Mittelstadt, 2019; Morley et al., 2020). Development teams participating in workshops conducted by the Berlin Ethics Lab supported this finding. Participants, for example, pointed at difficulties in implementing abstract notions such as 'fairness' (Berlin Ethics Lab, 2020/21). Another research stream focuses on integrating societal perspectives into technology development. From a policy-oriented perspective, responsible research and innovation (e.g. von Schomberg, 2013) aims to involve ethical and social issues into the innovation process by fostering the participation of citizens and stakeholders in a public dialogue or by integrating them into the stages of the innovation process itself. Moreover, there are several approaches addressing the development process of a specific technology or application from a societal perspective. Constructive technology assessment (e.g. Schot & Rip, 1997) is a subsection of technology assessment (cf., Grunwald, 2019) that proposes interventions in specific development processes to enable a dialogue among different actors and shape the design of new technologies. Similarly, value sensitive design (Friedman et al., 2013; van den Poel, 2021) offers an approach to integrate values of direct or indirect stakeholders into the development and design process of new technologies. In addition to these research fields, there is a plethora of singular methods and approaches. Morley et al. (2020) confirm this view: In their comprehensive review of existing approaches, Morley et al. describe the creation and adaptation of suitable methods to integrate societal needs practically into the AI development process as an emerging and deeply needed field of research and practical exploration.

To contribute to the initially presented AI governance gap and the systematisation of methods adoptable in practice, this publication highlights the need for methods supporting development teams in creating responsible AI applications throughout the development process. More specifically, from the perspective of private-sector governance, applying those methods should be accompanied by a model offering a structural, systemised integration of ethics into the AI adoption process within organisations. Decision-makers and development teams alike need anchor points to systematically identify and integrate an ethical perspective into the AI adoption process. They need both, specific methods to do so and a process model offering the overview to guide the integration of ethics throughout the AI adoption. The model this contribution proposes might serve as a tentative starting point for a responsible AI adoption process. It thus has a guiding function for practitioners in organisations, both on the management level and the development team level, as it poses a framework for orientation, based on which decision-makers, development teams, as well as potential further people involved can align on how to proceed to ensure responsible adoption.

The following sections will elaborate the proposed model and its background. Firstly, AI adoption is framed as innovation process (Sect. 2.1). Secondly, an internal coupling innovation process model is specified regarding responsible AI adoption (Sect. 2.2). Finally, in the third step, the governance model supporting *responsible* AI adoption is presented (Sect. 2.3; for the model see Fig. 3).

2.1 AI Adoption as Part of an Organisation's Innovation Process

AI adoption processes can be understood as innovation processes: Such processes are about the 'successful application of new ideas, which results from organizational processes that combine various resources to that end' (Dodgson et al., 2015, p.5) with the objective to 'produce positive results for organizations and their employees, customers, clients and partners' (Dodgson et al., 2015, p.5). However, responsible AI development processes have not yet been analysed from the perspective of private-sector governance, particularly from innovation processes. Reim et al. (2020) present an analysis of organisation-internal resources required for AI-based business models, while Huang and Teng (2020) focus on fostering the realisation of AI innovation in organisation in general. Kakatkar et al. (2020) on the other side examined AI's supporting role in the analytics of innovation processes. While these publications hint to the importance of addressing this topic, there is a lack of publications dealing with a governance process for responsible AI that can guide practitioners in their AI adoption process.

In order to develop such a process model that might guide governance, this contribution proposes to build on existing models for innovation processes and specify them with regard to AI adoption. There are several models used to describe innovation processes in order to offer an analytical framework for analysing and coordinating innovation in organisations (Dodgson et al., 2015). Internal coupling models focus on the communication and connections between those contributing to the innovation process, both within and outside the organisation (Dodgson et al., 2015; Rothwell, 1994). In contrast to technology-push or demand-pull models, the internal coupling model 'represents the confluence of technological capabilities and market-needs within the framework of the innovating firm' (Rothwell & Zegveld, 1985, p. 50). It understands the innovation process as a sequential though potentially iterative process consisting of distinct, mutually connected stages (Rothwell, 1994). This contribution proposes to build on the coupling model as presented by Rothwell (1994) as third-phase innovation process. This process model is both well-recognised and broadly accepted in the academic community. Importantly, its focus on interactions between those involved in the innovation process poses a suitable starting point for developing a responsible AI adoption process: According to Rothwell (1994), the coupling model or third-phase innovation process situates innovation at the interplay between emerging needs of society and the marketplace (demand pull) and new technologies (technology push), including communication and feedback loops between the different elements (see Fig. 1). At the level of the organisation, the model proposes a generic innovation process, moving from first ideas in research to prototypes, manufacturing, marketing, and, finally, the launch of an innovative product or service. Innovation, within this model, is understood as an organisation-wide task and with an emphasis on satisfying user needs. Moreover, it is seen as part of a long-term strategy. According to Rothwell (1994), this model is best suited for radical innovation based on emerging technology and can be adapted

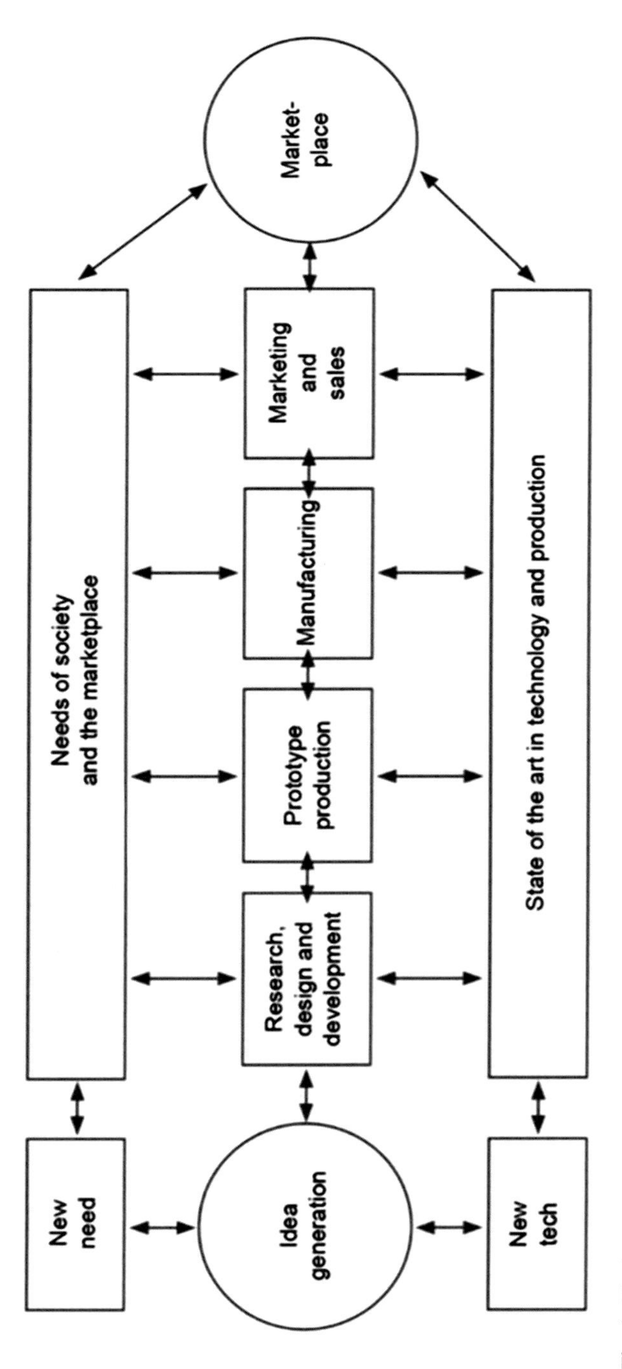

Fig. 1 The internal coupling model or third-phase innovation process as presented by Rothwell (1994, p. 10), own depiction

to suit lean innovation approaches. The latter point makes it possible to build on this model in more agile working environments, e.g. as a governance model inclusive of development teams working with Scrum.

However, while the internal coupling model presents a suitable starting point, several specifications and adaptations are necessary to create a governance model for responsible AI adoption. Firstly, it is helpful to specify the generic innovation process regarding *AI adoption* processes (following section). Secondly, and more importantly, the role of societal demand and the respective interactions must be reframed to enable a *responsible* process. An organisation engaging in responsible AI adoption needs to integrate the societal demand, long-term ethical considerations and diverse societal perspectives, the technological the state of the art and its development process, as well as the economic conditions ensuring the company's continued existence (Wieland, 2020). Such an integration of the elements society, technology, and organisation exceeds the understanding of customer or market needs prevalent in the internal coupling model, leaving the model insufficient to guide a responsible AI adoption process (see Sect. 2.3).

2.2 Specifying the Innovation Process Model for AI Adoption

The internal coupling or third-phase innovation process, as described by Rothwell (1994) (see Sect. 2.1), presents a generic innovation process. To link this innovation model to the field of AI, this contribution proposes to specify the model by substituting the generic steps undertaken by the organisation during the innovation process with an AI lifecycle model. There are several models used to depict the phases of AI development, e.g. the CRISP-DM model (Chapma et al., 2000), the AI System Engineering Lifecycle (Fischer et al., 2020), as well as a broad range of depictions by different developer websites (e.g. Davies, n.d.). Most of the models cover similar phases to describe the development process of AI and emphasise their iterative character. For example, Morley et al. (2020) portray the stages of AI development as outlined by the Information Commissioner's Office (ICO) auditing framework for artificial intelligence (see Fig. 2). This AI lifecycle process includes the ideation of an organisation's business and use case; the phases of training, building, and testing the AI application; as well as the deployment of an AI solution. The AI lifecycle model strongly emphasises the iterative character of AI development processes, highlighted through its circular depiction and the specific monitoring phase.

This contribution proposes to specify the third-phase innovation process model by substituting its research and development phase with the ideation phase of the AI lifecycle. Doing so, it integrates the training, building, and testing phases of AI development which, as a data-driven approach, are comparable to prototyping and manufacturing, and specifies marketing with the deployment and monitoring phases of AI adoption. The strongly iterative approach of the AI lifecycle strengthens the feedback loops already included in the third-phase innovation process on the level of

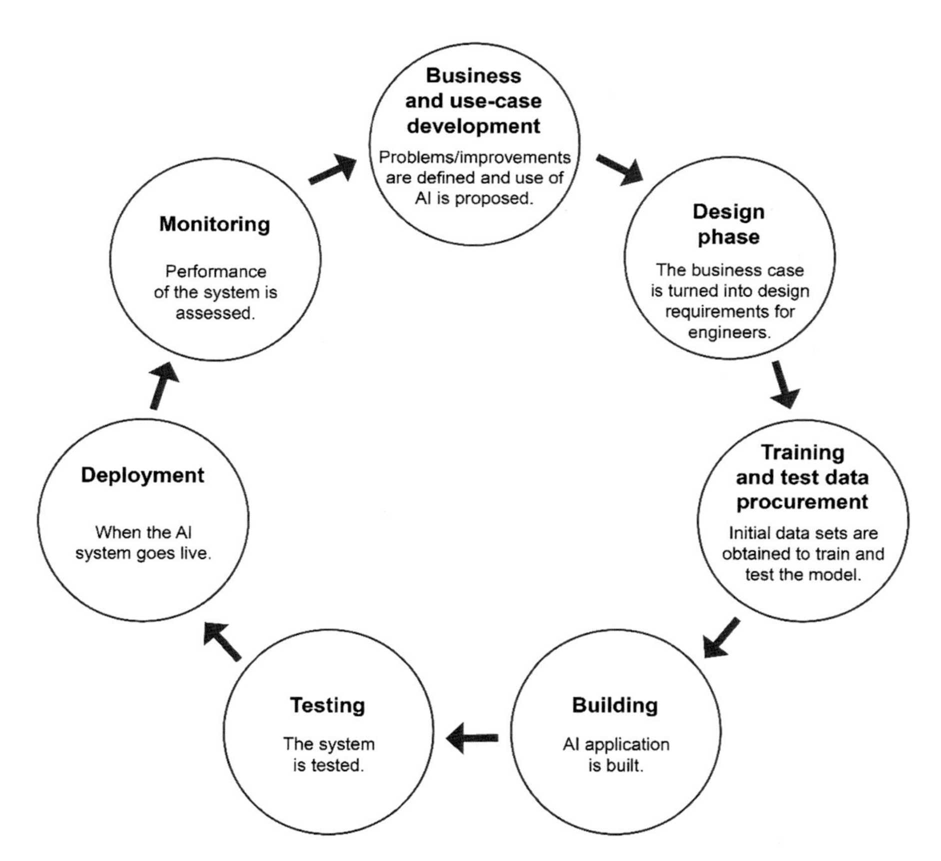

Fig. 2 One example for an AI lifecycle, here as presented by the Information Commissioner's Office ICO (Binns & Gallo, 2019), own depiction

the steps undertaken within the organisation during the innovation process. Thus, a full AI development cycle can be included within the internal coupling innovation process model. Including the full development lifecycle makes it possible to use the specified innovation process model for all AI adoption forms, be it the purchase and adaptation of an externally pre-developed model or the in-house development of an AI solution. While an in-house development process will cover all phases of the lifecycle model conducted within the organisation, other forms of AI adoption might outsource some of the phases, as, for example, the training, building, and testing steps are undertaken elsewhere and bought as a ready-made product to be implemented in the company.

Situating the AI lifecycle's phases within the organisation's innovation process offers an entry point to elaborate ways of practically engaging in responsible AI development. By choosing this particular AI lifecycle model, the publication allows for connecting future research to the first attempts of Morley et al. (2020) in collecting methods supporting developers in operationalising ethical requirements along the AI lifecycle. Building on a shared model thus can support scholars and practitioners alike in developing and implementing responsible AI adoption through aligning methods to do so and the process.

To ensure *responsible* adoption, however, specifying the innovation process is not sufficient. Hence, the next section highlights the necessary integration of interactions with society, ethical reflections, and ethical testing within each phase of the AI adoption process.

2.3 Integrating Ethics with a Governance Model for Responsible AI Adoption

The third-phase innovation process locates innovation in organisations at the intersection of economy, society, and technology (Rothwell, 1994). On the one hand, companies are subject to competitive pressure in the market (demand pull). On the other hand, the constant advance of technological possibilities, in this case AI research, requires its quick-paced adoption in organisations (technology push). To ensure responsible AI adoption, this contribution proposes a) to reframe the understanding of demand pull to include ethical considerations and broader societal demands and b) to consider the interplay between economy, society, and technology as an iterative process of alignment. Governing the AI adoption process, then, entails actively facilitating such alignment processes by engaging in ethical reflections and the integration of diverse societal perspectives.

In the third-phase innovation model (Rothwell, 1994), societal demand is depicted as a market demand, that is, satisfying the needs of potential users of a product or service is emphasised. However, from an ethical perspective, the diverse societal demands and needs do not necessarily align—and might even be competing—with the market demand of specific groups (Wieland, 2020). A specific user's needs might not necessarily align with the diverse range of needs existent within a society or among the stakeholder groups potentially affected by the AI application. Therefore, this contribution grounds innovation in societal demands and a broader stakeholder range, e.g. including non-users, which might be affected indirectly by the innovation (Wieland, 2020). With this, this contribution understands societal demand as inclusive of society-wide long-term ethical trajectories. This encompasses the technological and economic demand 'pulling' innovation as well as potential societal demands and needs that might contrast or even contradict the initially intended form of the organisation's AI application. Thus, the proposed governance model understands responsible AI adoption as an innovation process that situates innovation at the interplay between society, technology, and the organisation, but takes its starting point deliberately in a broad notion of societal demand, encompassing a diverse set of potentially competing needs and demands (Fig. 3). Accordingly, the role of technology within the innovation process is reframed: While the technological state of the art is still considered a 'push' factor for innovation, it is the focus on societal demand and the desirability of the AI solution and its effects, not alone technological feasibility or potential business opportunities that shape the iterative stages of the innovation process and the placement of the AI solution. Thereby, the contribution is in line with related approaches from human-centred design and innovation (e.g. with design thinking) that have recently gained momentum and put a central emphasis on needs and an innovation's desirability (e.g. Kelley & Kelley, 2015). However, this contribution explicitly stresses the consideration of demands and needs beyond the user.

From the perspective of responsible AI adoption, taking societal demands as a point of reference means that organisations not only need to address a specific demand but, moreover, must be able to align potentially competing needs and concerns from society as a whole (Hagendorff, 2020a; Wieland, 2020). The aspects companies might need to address encompass potentially discriminating side effects and the long-term consequences of AI adoption in a society, e.g. regarding potential replacements of other technologies or specific forms of human-AI interactions. Thus, while a specific demand might 'pull' innovation, a responsible AI adoption process also actively engages in ethical reflections, e.g. regarding the long-term societal implications, and includes diverse stakeholder perspectives. For organisations, such an understanding entails the challenge to work with potential value conflicts and requires intensive governance efforts to align these diverging needs (Wieland, 2020). Therefore, a holistic AI governance is required, as suggested by this contribution. However, if the company successfully aligns and integrates these diverging positions, it proactively shapes its societal legitimacy and presents an AI solution which is highly likely to be accepted by all its stakeholders (Wieland, 2020). Thereby, the company mitigates potential compliance costs or damages claimed by societal stakeholders, due to unwanted negative consequences of an AI solution developed without integrating societal demands and needs in its development process. This is since companies are inherently intertwined with society and serve to raise the benefits for all their stakeholders. Consequently, acknowledging societal and ethical perspectives in the development process is crucial, especially for the responsible AI adoption (Wieland, 2020). Furthermore, taking an active stance towards responsibly addressing potential consequences of its corporate decision-making requires a need-based, ethically responsible approach—especially since legally binding requirements traditionally lag the pace of technological developments (Dafoe, 2018). Hence, this contribution proposes to take a broader perspective on societal demand and to acknowledge the organisation's role as an entity aiming to align the demand of all its stakeholders—directly or indirectly affected by its actions. Given the organisation's high interrelatedness with society, a company can only ensure its existence through a successful alignment of all demands directed towards it. Doing so, it can create shared value for its stakeholders, a so-called win-win situation allowing the company to earn economic profits, while at the same time creating benefits for all its stakeholders (Wieland, 2020). Accordingly, society, organisation, and technology are not represented as single, separated entities within the depiction of the model, but as inherently connected (Fig. 3).

Thus, the organisation not only becomes the governance form, where the AI adoption process takes place, but—being a nexus of stakeholders—it also serves as an active entity balancing societal needs and values, the demand coming with

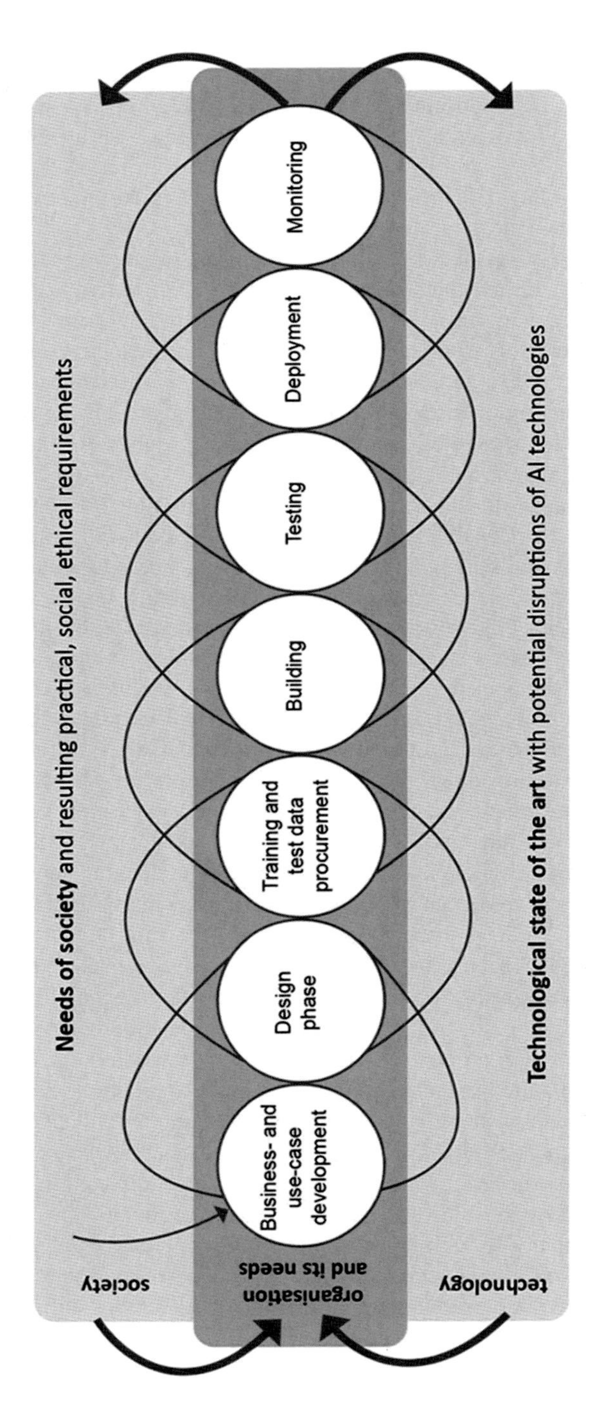

Fig. 3 Integrated governance model for responsible AI adoption, own depiction

technological possibilities and its own organisational values (Wieland, 2020). Governing such an AI adoption process responsibly means actively creating the opportunities and spaces for the said interactions between organisation, technology, and society to take place. A further crucial requirement for responsible AI adoption is the awareness that the technology will have a societal impact, however big or small. Both decision-makers and development teams need to be aware that the decisions and actions taken during an AI application's design and development crucially shape individual lives and societies.

To engage in such an alignment process and to adopt AI responsibly when faced with potentially diverging and changing values as well with unforeseeable or unknown consequences, iterative co-creation and transdisciplinary processes are needed. Therefore, the model presented stresses the importance of close interactions between the demands stemming from society, the organisation, and the technology via iterative feedback loops. The feedback loops and interactions included in the third-phase innovation process thus are deliberately reconfigured to ensure ethical reflection and stakeholder integration throughout the adoption process (Fig. 3): Each phase of the adoption process is not only connected by feedback loops on a technological or organisational level, but, importantly, such with society. This aspect is the central adaptation of the third-phase innovation process described by Rothwell (1994), as it orients governance of *responsible* AI adoption. While the original model already includes interactions between society and the organisation (cf., Fig. 1), it lacks a specifically ethical perspective and the inclusion of societal actors within the adoption process. Similarly, AI development processes do emphasise feedback loops and iteration, but mostly foreground technical iteration, e.g. regarding the testing of a model through user testing. Beyond technical iterations that aim to prevent mal-functioning technology, ethical considerations in this context include examining potential long-term effects and unintended consequences of a well-functioning technology. In the proposed model, ethical reflection and stakeholder integration are ensured throughout the process, as the adoption process specifically addresses the intersections between the organisation, technology, and society, which are needed to responsibly adopt and develop AI technologies (Fig. 3). The process is of highly iterative nature, resulting in single phases as well as the whole process being re-assessed several times to allow for the adaptation to new insights and requirements. This contribution proposes to build on iterative feedback loops and co-creation to consider the implications for the technology, the organisation, and society in each phase of the AI adoption process. For example, the ideation phase might include envisioning potential AI applications not only based on the technological possibilities or user demands but also from a broader societal perspective. Testing an AI application, to give another example, might go beyond technical tests and user tests to include testing (long-term) implications for and with a broader range of stakeholders, including groups that might be indirectly affected. In Sect. 3, this contribution will present three potential action points for the practical implementation of such interactions.

3 Insights into the Operationalisation of Responsible AI Adoption

As stated above, this contribution defines responsible AI adoption as a fundamentally iterative process, involving constant feedback loops. Further, the proposed model emphasises the integration of ethical, technical, and economic iterations, as the AI solution should, eventually, be socially desirable, feasible, and economically successful.

Since iteration models exist to ensure the economic success of an innovation and its technological feasibility, for example, in the review phases of Scrum and the testing phase of AI lifecycles, this publication focuses on the interplay of ethical and technological iterations in this section. This is since, due to the limited scope of this publication, it cannot address the intersections with additional iterative loops, despite encouraging further research to do so. As for the ethical iterations, the model presented covers all phases of an AI adoption process and emphasises that ethical considerations shall be interwoven with all adoption phases. Generally, the integration of ethics requires developing tools and intervention formats that foster ethical reflection and stakeholder integration (Berlin Ethics Lab, 2020/21; Morley et al., 2020). By offering the holistic perspective of a responsible development process, management can assist the AI development teams by structurally including opportunities and tools for this kind of anticipating, reflecting, and discussing ethical questions as well as integrating stakeholder perspectives. An additional option is setting up diverse teams, specifically focusing on integrating ethicists (Remmers, 2020; van der Burg & Swierstra, 2013). Still, each phase of the adoption process comes with different ethical questions and stakeholders it needs to involve.

To illustrate how the integration of ethics could be implemented in practice within a specific adoption process, this section presents three potential action points, stemming from explorative interventions with AI development teams conducted by the Berlin Ethics Lab (2020/21). The presented insights do not provide a complete list but require further research. Further work is suggested regarding the systematisation and operationalisation of existing methods within the context of AI adoption as well as the development of further instruments addressing existing research gaps. Nonetheless, the following three sections pose inspirational starting points for practically fostering responsible AI adoption.

3.1 Action Point 1: Creating Ethical Visions

Making the often-implicit visions of team members explicit is the first step towards an ethical vision design (Ammon, 2020; Fischer & Mehnert, 2021). Creating a coherent vision seems important to foster the direction of the technological project. One insight stemming from the feedback of participants was the need to reflect visions within the whole team, a step that rarely receives dedicated space. However, as Akrich (1994) points out, 'when technologists define the characteristics of their objects, they necessarily make hypotheses about the entities that make up the world into which the object is to be inserted. [. . .] A large part of the work of innovators is that of 'inscribing' this vision of (or prediction about) the world in the technical content of the new object' (1994, p. 207). Thus, facilitating discussions to make implicit visions explicit enables not only the creation of an explicit joint vision for responsible design but also forms a starting point to translate abstract value principles into the specific project context (see Sect. 2). Part of discussing a vision is describing the imagined future users of the technology which is to be developed. However, when doing so, it is crucial to reflect on possible developers' biases: Some of the explorative interventions with teams already working in the training and testing phases revealed that the members of the development team implicitly assumed a white, abled male user with differing age as the application's target group. That, in turn, influenced the data used for the training and the creation of the AI model, potentially inscribing a bias to the application—as, e.g. neither female-specific data patterns nor ones of people with special needs had been considered. Reflecting on implicit visions can enable a development team to anticipate such unintended potential consequences already in the early phases of the AI adoption process, i.e. already in the development of a business and use case as well as in the design phase. The insights created by explicating visions furthermore can be a starting point to deliberately create ethical visions that foster, for example, the use of more inclusive training data used and models. Moreover, societal actors of groups potentially affected by the AI application to be developed can participate in visioning exercises to explicate orientational guidelines for desirable AI applications. An ethical vision can form an orientation throughout the adoption process and might be subject to ethical testing and iteration during the process.

3.2 *Action Point 2: Use Case Testing for Long-Term Societal Implications*

The business and use case for a potential AI application developed in the earliest phase of the AI adoption process form another starting point for integrating ethical reflection. The action point lies in elaborating potential long-term implications of the potential application. To foster such ethical considerations, the Berlin Ethics Lab team developed the 'implication fan' (Berlin Ethics Lab, 2020/21). The tool supports development teams, potentially together with further actors, to systematically broaden the perspective on the use case. Taking a systemic perspective on the sociotechnical system, within which the solution shall be located, the tool supports considering both, necessary preconditions and potential impacts, in an extended timeline. Discussing potential value conflicts, (non)desirable development paths, and potential unintended side effects, AI development teams gain a better understanding of their solution and can, consequently, derive ideas to redesign it. For example, participants working with the tool in one of the explorative interventions conducted by the Berlin Ethics Lab discovered that the time freed up through an automatisation process could be used up fully to maintain the application. Other insights included becoming aware that the use of a specific soft- or hardware had an excluding effect on certain groups using older mobile phones, recognising societal actors potentially affected by the application that the team would need to further involve in the adoption process or the insight that a broad usage of the intended AI application might require and foster changes in infrastructures. Including reflections on such potential implications early in the AI adoption process enables the AI development team to easily rethink the impact of the technological solution it creates, which helps mitigate potential problems and identify strategies to react to potentially emerging issues early on. This is since, at an early stage, the technology is not yet fixed into a specific design and the development has not yet been as resource intensive. Thus, by integrating a long-term perspective, this action point allows to save organisational resources and to responsibly develop a sustainable AI solution.

3.3 Action Point 3: Iteratively Integrating Societal Perspectives

Engaging with all organisational stakeholders, beyond users, already in the early phases of an AI adoption process enables the development team to reflect on implications, needs, and inspirations outside their own perception. A broader form of stakeholder engagement will foster the alignment process described in Sect. 2.3 and creates a more realistic perception of societies' view on a specific AI application for the development team, including potential problematic issues. As proposed by this contribution, such interactions with stakeholders should include a broad range of perspectives in order to responsibly anticipate and avoid negative effects and figure out the dimensions of desirability. That means, for example, to involve advocacy groups and groups potentially affected by the application to be developed. There are various formats to engage with different stakeholders, depending on the specific goal of the interaction and the stage of the AI adoption process. The Berlin Ethics Lab team, for example, explores working with joint discussions of utopias and dystopias in the early phases of AI development to outline guiding values for the whole project and create a frame within which a solution should be located. Such discussions do not only encompass deliberations of values but further deepen the exploration of potential implications. This entails reflections on what new norms or standards the developed application might promote, which existing technologies and behaviours it is replacing, and what kind of future society it is contributing to.

Resumé

The three action points presented illustrate how specific interactions within development teams and between such teams and diverse stakeholders can practically integrate ethics into the AI adoption process. Those first exploratory examples can show how the ethical iterations in the government model proposed by this contribution can work practically. Furthermore, the model offers a processual overview for the integration of such ethical interventions throughout the AI adoption process. Hence, decision-makers need to take a stance for ethics on a strategic level and take the fundamental decision to integrate it into its corporate processes. Having done so, the model this publication presented can serve as an instrument to operationalise AI governance from strategic to operational team level.

4 Implications, Discussion, and Further Research

To conclude, the presented model is directed towards decision-makers in companies wanting to engage in AI adoption in a way that integrates societal perspectives and constitutes responsible business conduct. The model not only stems from a broadly accepted theoretical foundation but was developed for decision-making in practice. Thereby, its impact on society shall be ensured via its applicability to corporate AI adoption strategies.

In particular, the AI governance model guides corporate decision-making over the course of an AI adoption process, highlighting ethical iterations throughout the process. With this, the model contributes to responsible AI adoption in organisations and enables an organisation's management to structure related topics decision-makers are facing in AI adoption and development alongside the adoption stages presented in its model. By applying this holistic model, companies can lower their transaction costs in multi-stakeholder processes in a proactive manner and ensure the social legitimacy of their AI-based solutions and services. Doing so, the organisation proactively shapes its impact on society and society's perception of its AI applications (Wieland, 2020). This is since the model presented allows for the inclusion and participation of an organisation's stakeholders in the AI adoption process throughout the entire AI adoption process and presents a more practical view on the term 'responsibility' in the AI context.

Due to the lack of regulatory measures for AI adoption, instruments like the one presented in this chapter are highly relevant for companies aiming to integrate their stakeholder views. Further, the above-described examples do not only illustrate the role of ethical reflections in AI adoption. They also emphasise the importance of creating opportunities for development teams to do so. Hence, the presented methods and action points, in combination with the model, support teams in responsibly adopting AI. Organisations need a strategic governance process that structurally supports their management and development teams to adopt AI responsibly and that addresses all management levels (Thuraisingham, 2020; Torré et al., 2019). While the contribution of the model presented lies in its systematic and structural approach to responsible AI adoption, it strongly addresses its operationalisation on team level. However, the presented model and approach need to be embedded in a general corporate AI strategy. Both Rothwell's (1994) original model and the adaptation with a focus on responsibility presented do serve as a starting point for this endeavour. Still, further research should specifically address the strategic management implications stemming from the decision to adopt AI responsibly.

An additional field for further research and practical exploration is the development and systematisation of tools and methods that foster responsible AI development on the level of specific projects. Research is needed for a better understanding of the requirements and specifics of different development stages, different AI technologies, as well as different types of application. Moreover, research on methods and tools enabling development teams to integrate ethical reflections and stakeholder perspectives throughout the development process is needed. Those methods should be able to address the diverse topics and goals as well as the differing stakeholders involved—from evaluating first visions to continued stakeholder engagement to long-term implications, from working with interventions in workshops to integrating ethicists into the development team. As indicated, the Berlin Ethics Lab currently addresses the need to systematise, develop, and iterate approaches, aiming at creating an 'ethics toolbox' assisting development teams in the creation of responsible AI. Still, the contribution stresses the importance for further research in this emerging research field of significant practical relevance.

Despite its intended applicability, the presented model can merely serve as a first direction for creating responsible AI processes in practice. The model currently has several limitations that should be subject to further research and practical case studies. The model presents a generic process for AI adoption which includes in-house AI development processes and adaptations of bought solutions alike. This generic overview requires both further specialisations to use cases and a more detailed focus on management implications. In detail, research is needed regarding its range and ability to cover diverse AI development processes with their case-by-case issues as well as the specific requirements of different AI technologies.

With this, the presented model offers a starting point for the structural integration of ethical reflections in AI adoption processes, both on a strategic governance and a project team level. Further, it raises awareness for the necessity of ethical reflection in AI adoption processes. This is since an awareness for the identified governance need forms the foundation for changing technology development and addresses the way organisations take on responsibility to shape socially desirable futures.

References

Akrich, M. (1994). The de-scription of technical objects. In W. E. Bijker, J. Law, W. B. Carlson, & T. Pinch (Eds.), *Shaping technology / building society: Studies in sociotechnical change (reissue)* (pp. 205–224). The MIT Press.

Allen, G., & Chan, T. (2017). *Artificial intelligence and national security*. Technical report. Harvard University. https://www.belfercenter.org/publication/artificialintelligence-and-national-security.

Alsheibani, S., Cheung, Y., & Messom, C. (2018). Artificial intelligence adoption: AI-readiness at firm-level. In *PACIS* (p. 37).

Ammon, S. (2020). Ethical Vision Design im Berlin Ethics Lab: Technologievisionen in der Entwicklung verantwortlicher KI und verantwortlicher Mensch-Maschine-Interaktion. In Interdisziplinäre Arbeitsgruppe Verantwortung: Maschinelles Lernen und Künstliche Intelligenz der Berlin-Brandenburgischen Akademie der Wissenschaften (Ed.), *KI als*

Laboratorium? Ethik als Aufgabe! (pp. 10–14). Berlin Brandenburgische Akademie der Wissenschaften. Accessible via: https://www.bbaw.de/files-bbaw/user_upload/publikationen/BBAW_Verantwortung-KI-3-2020_PDF-A-1b.pdf

Armstrong, S., Bostrom, N., & Shulman, C. (2016). Racing to the precipice: a model of artificial intelligence development. *AI & Society, 31*, 201–206. https://doi.org/10.1007/s00146-015-0590-y

Berlin Ethics Lab. (2020/21). Conversations and observations stemming from an ongoing series of workshops with AI development teams. The workshops practically explore tools and methods for integrating ethical considerations and societal perspectives into the development process. A publication of tools and insights is forthcoming. Please contact us for further information.

Binns, R., & Gallo, V. (2019). An overview of the Auditing Framework for Artificial Intelligence and its core components. Blog entry for the Information Commissioner's Office. Retrieved October 1, 2021, from https://ico.org.uk/about-the-ico/news-and-events/ai-blog-an-overview-of-the-auditing-framework-for-artificial-intelligence-and-its-core-components/

Brynjolfsson, E., & McAfee, A. (2017). The business of artificial intelligence: What it can and cannot do for your organization. *Harvard Business Review*, 1–20. https://hbr.org/2017/07/the-business-of-artificial-intelligence

Bughin, J., Hazan, E., Ramaswamy, S., Allas, T., Dahlström, P., Henke, N., & Trench, M. (2017). *Artificial intelligence. The next digital frontier?*. Discussion paper. McKinsey Global Institute. Retrieved September 11, 2021, from https://www.mckinsey.com/~/media/McKinsey/Industries/Advanced%20Electronics/Our%20Insights/How%20artificial%20intelligence%20can%20deliver%20real%20value%20to%20companies/MGI-Artificial-Intelligence-Discussion-paper.ashx.

Cave, S., & ÓhÉigeartaigh, S. (2019). An AI race for strategic advantage: Rhetoric and risks. In *Conference paper for: AI ethics and society 2018, 1*. https://doi.org/10.1145/3278721.3278780

Chapma, P., Clinton. J., Kerber, R, Khabaza, T., Reinartz, T., Shearer, C., & Wirth, R. (2000). CRISP-DM 1.0. Step-by-step data mining guide. CRISP-DM consortium: NCR Systems Engineering Copenhagen (USA and Denmark), DaimlerChrysler AG (Germany), SPSS Inc. (USA) and OHRA Verzekeringen en Bank Groep B.V (The Netherlands)

Cubric, M. (2020). Drivers, barriers and social considerations for AI adoption in business and management: A tertiary study. *Technology in Society, 62*, 101257.

Dafoe, A. (2018). *AI governance: a research agenda*. Governance of AI Program, Future of Humanity Institute, University of Oxford. https://www.fhi.ox.ac.uk/wpcontent/uploads/GovAI-Agenda.pdf

Davies, A. (n.d.). AI software development life cycle: Explained. DevTeam.Spaces. Retrieved September 11, 2021, from https://www.devteam.space/blog/ai-development-life-cycle-explained/.

Dodgson, M., Gann, D. M., & Phillips, N. (2015). Perspectives on innovation management. In M. Dodgson, D. M. Gann, & N. Phillips (Eds.), *The Oxford handbook of innovation management* (pp. 3–25). Oxford University Press.

European Commission, High-Level Expert Group on Artificial Intelligence. (2019). Ethics guidelines for trustworthy AI. Retrieved July 15, 2021, from https://digital-strategy.ec.europa.eu/en/library/ethics-guidelines-trustworthy-ai.

Fischer, L., Ehrlinger, L., Geist, V., Ramler, R., Sobiezky, F., Zellinger, W., Brunner, D., Kumar, M., & Moser, B. (2020). AI system engineering—Key challenges and lessons learned. *Machine Learning and Knowledge Extraction, 3*(1), 56–83. MDPI AG. Retrieved from https://doi.org/10.3390/make3010004

Fischer, N., & Mehnert, W. (2021). Building possible worlds. A speculation based research framework to reflect on images of the future. *Journal of Futures Studies, 25*(3), 25–38.

Friedman, B., Kahn, P. H., Borning, A., & Huldtgren, A. (2013). Value sensitive design and information systems. In N. Doorn, D. Schuurbiers, I. van de Poel, & M. E. Gorman (Eds.), *Early engagement and new technologies: Opening up the laboratory* (pp. 55–95). Springer. https://doi.org/10.1007/978-94-007-7844-3_4

Gartner Institute. (2020). CIO Agenda 2020. https://www.gartner.com/en/publications/2020-cio-agenda (30.09.2021).

Goldfarb, A., Taska, B., & Teodoridis, F. (2019). Could machine learning be a general-purpose technology? Evidence from online job postings. *SSRN Digital*. https://doi.org/10.2139/ssrn.3468822

Grunwald, A. (2019). *Technology assessment in practice and theory*. Routledge.

Hagendorff, T. (2020a). The ethics of AI ethics: An evaluation of guidelines. *Minds and Machines, 30*, 99–120. https://doi.org/10.1007/s11023-020-09517-8

Hagendorff, T. (2020b). AI virtues—The missing link in putting AI ethics into practice. *arXiv*, preprint arXiv:2011.12750.

Huang, T. L., & Teng, C. I. (2020). AI design to innovation. In *Emergent Research Forum (ERF). Conference proceedings in Americas conference on information systems*. https://core.ac.uk/download/pdf/326836354.pdf

Jobin, A., Ienca, M., & Vayena, E. (2019). Artificial intelligence: The global landscape of ethics guidelines. *Nature Machine Intelligence, 1*, 389–399. https://doi.org/10.1038/s42256-019-0088-2

Kakatkar, C., Bilgram, V., & Füller, J. (2020). Innovation analytics: Leveraging artificial intelligence in the innovation process. *Business Horizons, 63*(*2*), 171–181.

Kaplan, A., & Haenlein, M. (2019). Siri, Siri, in my hand: Who's the fairest in the land? On the interpretations, illustrations, and implications of artificial intelligence. *Business Horizons, 62*(*1*), 15–25. https://doi.org/10.1016/j.bushor.2018.08.004

Kelley, D., & Kelley, T. (2015). *Creative confidence: Unleashing the creative potential within us all*. Harper Collins.

Klinger, J., Mateos-Garcia, J. C., & Stathoulopoulos, K. (2018). Deep learning, deep change? Mapping the development of the Artificial Intelligence General Purpose Technology. https://arxiv.org/abs/1808.06355

Mittelstadt, B. (2019). Ai ethics—too principled to fail?. *arXiv,* preprint arXiv:1906.06668.

Morley, J., Floridi, L., Kinsey, L., & Elhalal, A. (2020). From what to how: An initial review of publicly available AI ethics tools, methods and research to translate principles into practices. *Science and Engineering Ethics, 26*(*4*), 2141–2168. https://doi.org/10.1007/s11948-019-00165-5

Nakashima, E. (2012). Stuxnet was work of U.S. and Israeli experts, officials say. *The Washington Post*. https://www.washingtonpost.com/gdprconsent/?next_url=https%3a%2f%2fwww.washingtonpost.com%2fworld%2fnational-security%2fstuxnet-waswork-of-us-and-israeli-experts-officialssay%2f2012%2f06%2f01%2fgJQAlnEy6U_story.html

Nepelski, D., & Sobolewski, M. (2020). *Estimating investments in general purpose technologies. The case of AI investments in Europe*. Publications Office of the European Union. https://doi.org/10.2760/506947

Nilsson, N. J. (2009). *The quest for artificial intelligence*. Cambridge University Press.

OECD. (2019). Recommendation of the Council on Artificial Intelligence. Retrieved July 15, 2021, from https://legalinstruments.oecd.org/en/instruments/OECD-LEGAL-0449.

Pichai, S. (2018). AI at Google: our principles. Retrieved September 23, 2021, from https://www.blog.google/technology/ai/ai-principles/.

Polyakova, A., & Boyer, S.P. (2018). *The future of political warfare: Russia, the West and the coming age of global digital competition*. Brookings Institution. https://www.brookings.edu/wp-content/uploads/2018/03/fp_20180316_future_political_warfare.pdf

Razzkazov, V. E. (2020). Financial and Economic consequences of distribution of artificial intelligence as a general-purpose technology. *Finance: Theory and Practice, Scientific and Practical Journal, 24*(*2*), 120–132. https://doi.org/10.26794/2587-5671-2020-24-2-120-132

Reim, W., Åström, J., & Eriksson, O. (2020). Implementation of artificial intelligence (AI): A roadmap for business model innovation. *AI, 1*(*2*), 180–191.

Remmers, P. (2020). Ethische Aspekte der Mensch-Maschine-Interaktion. In Interdisziplinäre Arbeitsgruppe Verantwortung: Maschinelles Lernen und Künstliche Intelligenz der Berlin-Brandenburgischen Akademie der Wissenschaften (Ed.), *KI als Laboratorium? Ethik als Aufgabe*! (pp. 15–21). Berlin Brandenburgische Akademie der Wissenschaften. Accessible via https://www.bbaw.de/files-bbaw/user_upload/publikationen/BBAW_Verantwortung-KI-3-2020_PDF-A-1b.pdf

Rothwell, R. (1994). Towards the fifth-generation innovation process. *International Marketing Review, 11(1)*, 7–31.

Rothwell, R., & Zegveld, W. (1985). *Reindustrialization and technology*. Longman.

Scharre, P. (2019). Killer apps: The real dangers of an AI arms race. *Foreign Affairs*. https://www.foreignaffairs.com/articles/2019-04-16/killer-apps

Schiff, D., Biddle, J., Borenstein, J., & Laas, K. (2020). What's next for AI ethics, policy, and governance? A global overview. In *Proceedings of the AAAI/ACM conference on AI, ethics, and society* (pp. 153–158). https://doi.org/10.1145/3375627.3375804

Schot, J., & Rip, A. (1997). The past and future of constructive technology assessment. *Technological Forecasting and Social Change, 54(2–3)*, 251–268. https://doi.org/10.1016/S0040-1625(96)00180-1

Thuraisingham, B. (2020). Artificial intelligence and data science governance: Roles and responsibilities at the C-level and the board. In *2020 IEEE 21st international conference on information reuse and integration for data science (IRI)* (pp. 314–318). IEEE. https://doi.org/10.1109/IRI49571.2020.00052

Torré, F., Teigland, R., & Engstam, L. (2019). AI leadership and the future of corporate governance: Changing demands for board competence. In F. Torré, R. Teigland, & L. Engstam (Eds.), *The digital transformation of labor* (pp. 116–146). Routledge.

Trajtenberg, M. (2018). *AI as the next GPT: A political-economy perspective (No. w24245)*. National Bureau of Economic Research. https://doi.org/10.3386/w24245

Tresignie, C., Kazakova, S., & Dunne, A. (2020). *European enterprise survey on the use of technologies based on artificial intelligence*. Ipsos Institute and European Commission. https://www.ipsos.com/sites/default/files/ct/publication/documents/2020-09/european-enterprise-survey-and-ai-executive-summary.pdf

van den Poel, I. (2021). Values and design. In N. Doorn & D. Michelfelder (Eds.), *Routledge handbook of philosophy of engineering* (pp. 300–314). Routledge.

van der Burg, S., & Swierstra, T. (Eds.) (2013). *Ethics on the laboratory floor*. Palgrave Macmillan Hampshire. https://doi.org/10.1057/9781137002938

von Schomberg, R. (2013). A vision of responsible research and innovation. In R. Owen, J. Bessant, & M. Heintz (Eds.), *Responsible innovation* (pp. 51–74). Wiley. http://doi.wiley.com/10.1002/9781118551424.ch3

Wallach, W., & Marchant, G. (2019). Toward the agile and comprehensive international governance of AI and robotics. *Proceedings of the IEEE, 107*(3), 505–508 [8662741]. https://doi.org/10.1109/JPROC.2019.2899422

Wieland, J. (2020). *Relational economics: A political economy*. Springer.

Yu, H., Shen, Z., Miao, C., Leung, C., Lesser, V. R., & Yang, Q. (2018). Building ethics into artificial intelligence. *arXiv*, preprint arXiv:1812.02953.

Sabine Wiesmüller is managing director of the Bodensee Innovation Cluster at Zeppelin University and senior researcher at the Leadership Excellence Institute Zeppelin, particularly focusing on relational governance, AI governance, and business ethics. Having successfully completed her doctoral candidacy at the Wittenberg Center for Global Ethics, Sabine Wiesmüller examined forms and interactions of relational AI governance in her doctoral thesis. In both, her work at the innovation cluster and her role as a business consultant and researcher, Sabine Wiesmüller is interested in the effects of AI and exponential technologies on society and new avenues for their suitable and sustainable governance.

Nele Fischer is research fellow at the Berlin Ethics Lab, Technische Universität Berlin, exploring transdisciplinary and participatory ways of creating responsible AI and human-machine interactions. A current project explores tools and formats to integrate ethics into technology development in collaboration with development teams. With a background in Futures Studies and agile, human-centred approaches, Nele Fischer's work and research builds on critical, participatory, and creative methods to open up spaces for transformation and desirable futures. She teaches at TU Berlin and in the Futures Studies Master Programme at Freie Universität Berlin and freelances to support organisational change.

Wenzel Mehnert is a cultural and media scholar and works as a research assistant at the Berlin University of the Arts and the Technische Universität Berlin. In his research, he works at the intersection of sociotechnical imaginaries and technological development. In addition to the analytical examination of technofutures in various discourses, he develops creative methods for reflecting on present futures with creative means from design and literature and gives seminars on the media representation of future imaginaries. His PhD thesis is dedicated to the science fiction subgenre neuropunk and analyses the present and past imaginaries of the neurointerface in pop culture.

Prof. Dr. Sabine Ammon works as a philosopher at TU Berlin with a focus on the philosophy and ethics of design and technology. She heads the Berlin Ethics Lab which aims to implement responsible AI and responsible human-machine interaction in research and innovation processes. After studying architecture and philosophy, she spent periods of study and research at the University of London, Harvard University, ETH Zurich, and the Research Institute for Philosophy in Hanover. She subsequently held research and lecturer positions at the University of Basel, BTU Cottbus-Senftenberg, and TU Darmstadt. Her dissertation (2009) developed the epistemological basis for a processual and plural concept of knowledge, whereas her habilitation (2018) developed an epistemology of design. She was a principal investigator in the DFG Cluster of Excellence Science of Intelligence and is spokesperson of the Research Network Knowledge Dynamics in the Engineering Sciences as well as of the Present Futures Forum Berlin.

Mastering Trustful Artificial Intelligence

Helmut Leopold

Abstract As a counter-thesis to the naive general narrative that artificial intelligence (AI) is a hyped super-technology which can solve all problems and even overtake human intelligence, this article discusses five essential problem areas associated with AI technology: *modeling ability*, how do we derive models of the real world from data and how do we create a model without prejudices and errors?; *verifiability*, how do we verify the AI algorithms?; *explainability*, how can we understand the decision-making process of AI systems?; *ethics*, how do we guarantee compliance with ethical principles and values?; and finally, *responsibility*, who is responsible for the decisions made by the AI system? We also discuss fundamental threat scenarios in the context of our information society, as well as the limits of AI technology compared with human intelligence. The article highlights that the development of AI technology, as well as related policies and regulations, must be organized to ensure its socially acceptable use and rule out any improper use. Furthermore, the article provides a philosophical discussion on the limits of AI and the diversity of life, and shows that we bear the ultimate responsibility for what machines do. This paper concludes by arguing that even the most sophisticated machine will probably never be able to match humans in terms of their multi-dimensionality of cognition, emotion, and physicality and in their sensual perception of the world.

1 Artificial Intelligence: An Introduction

1.1 Development of AI Research

Mankind has always tried to reduce its workload by using machines. The goal has always been to use machines to automate difficult, monotonous, or recurring tasks. The use of software and digital technology has given rise to discussion of the

H. Leopold (✉)
Center for Digital Safety & Security, AIT Austrian Institute of Technology, Vienna, Austria
e-mail: helmut.leopold@ait.ac.at

R. Schmidpeter, R. Altenburger (eds.), *Responsible Artificial Intelligence*, CSR, Sustainability, Ethics & Governance, https://doi.org/10.1007/978-3-031-09245-9_6

possible intelligent "human" behavior of software-controlled machines, and since then the fundamental question of whether machines can exhibit intelligent human behavior has inspired philosophical discussions (Reichl et al., 2020). The film industry has made its contribution, impressively introducing a wide audience to the concept of highly intelligent robots able to act emotionally. This has raised the general question of whether machines will one day be able to surpass the abilities and intelligence of humankind with a technical "super intelligence."

As computer science has developed as a scientific discipline, the question of which cognitive abilities determine intelligent behavior and which basic functions make up the intelligent abilities of machines has been investigated for decades. Alan Turing, a pioneer of computer science, defined an approach to determine the intelligent behavior of machines as early as 1950, with his so-called Turing test (Turing, 1950). He postulated that if, in a question-and-answer game with a system, it is not possible to distinguish whether the answers come from a human or a machine, then the system can be described as intelligent. The principle of the Turing test still has a practical application today in the CAPTCHA test (Nations, 2020), in which during login processes the human user must correctly recognize distorted optical characters in order to prevent unnoticed login by machines. In 1956, more than 60 years ago, John McCarthy, together with a few other pioneers of computer science, established the now ubiquitous term "artificial intelligence" (AI), having organized a conference called the "Dartmouth Summer Research Project on Artificial Intelligence" (Manhart, 2018).

In addition to logical and rule-based algorithms, one goal of AI research has always been to seek new approaches to deal with uncertainties and even contradictions in data. It was the knowledge developed in cognitive research that "doing the right thing" cannot always be based on a complete set of data in particular which spurred on the development of the disciplines of cognitive sciences and cognitive neurosciences. At the same time, there has been a lot of hype in the computer sciences about understanding human intelligence and using computers to create artificial intelligence. Thus, machine reasoning, human-like problem solving and decision-making, as well as learning from experience in various application scenarios have all been subjects of scientific endeavor for centuries.

The early phase of AI research in the 1960s was marked by euphoria about the potential performance. Since then, AI research has gone through many ups and downs. For example, the prediction made by the Nobel Prize winner Herbert A. Simon in 1957 that computers would be able to become world chess champions within 10 years was not fulfilled until 1997, some 40 years later, when the "Deep Blue" system developed by IBM beat the then world chess champion Garry Kasparov in six games. It then took another 18 years until the chess program Giraffe was able to learn master-level chess in 72 h in 2015,[1] and AlphaGo, from Google

[1] In drei Tagen zum Internationalen Schachmeister, Spiegel, 17.9.2015, http://www.spiegel.de/netzwelt/web/kuenstliche-intelligenz-computer-lernt-in-72-stunden-schach-a-1053338.html (last access 19.11.2021)

DeepMind, won against the European Go game champion (Silver et al., 2017). In the game of Go, the learning time for a human can be assumed to be 2 years, but the AI system only needed a few weeks. Finally, in 2017, an AI system developed by OpenAI, a company owned by Elon Musk, beat human competitors in the computer game Dota-2.[2] In 2017, an AI application was even developed to continue the plot of the fantasy epic "Game of Thrones" by George R.R. Martin.[3] The computer was trained based on the contents of the first five volumes. The adaptive AI program was then able to assume the author's style, turn of phrase and idioms, to develop the individual characters and even invent new ones.

In the meantime, more and more AI systems are being used in a wide variety of application areas because they consistently achieve better results than conventional IT approaches. Well-known, practical uses include the pattern recognition of objects for cameras on autonomous vehicles (e.g. Google, Tesla, and Mobileye), voice recognition systems such as Alexa (Amazon), Siri (Apple), Bixby (Samsung), Assistant (Google), and Cortana (Microsoft), as well as Google's translation system.

Other areas of application for AI systems are chatbots[4] as a convenient customer interface for hotlines, online sales, and customer services; new protection methods to protect our digital systems from increasingly sophisticated cyberattacks (Leopold et al., 2015); and finally diverse areas of application in the industrial sector, such as improvements in production and maintenance processes, to name just a few. There is hardly an application area today that does not exploit AI solutions.

1.2 AI Made in Austria

AI research has existed in Austria since the 1980s. The Austrian Research Institute for Artificial Intelligence (OFAI) was founded in 1984 by the AI pioneer Prof. Trappl.[5] An expert system was successfully developed by the Alcatel-ELIN research center as early as the 1990s and is a technology which continues to be used by railway operators globally, for highly reliable control systems (Leitner, 2003, p. 268 f.).

A key technological development for today's AI systems was made in 1997 by Sepp Hochreiter, head of the AI Lab at the Linz Institute of Technology (LIT) at the Johannes Kepler University (JKU), and Jürgen Schmidhuber, head of IDSIA, a Swiss research institute for artificial intelligence. They developed the basic AI

[2] KI von Elon-Musk-Startup zerstörte bei Computerspiel-Duell weltbesten Spieler, DerStandard, 12.8.2017, http://derstandard.at/2000062596518/KI-von-Elon-Musk-Startup-zerstoerte-bei-Computerspiele-Duell-weltbesten (last access 19.11.2021)

[3] Computer schreibt sechstes Buch von Game of Thrones, Frankfurter Allgemeine, 30.8.2017, http://www.faz.net/aktuell/wirtschaft/kuenstliche-intelligenz/game-of-thrones-kuenstliche-intelligenz-schreibt-sechstes-buch-15175025.html (last access 19.11.2021)

[4] A new created word from "chat" and "bot" (robot)

[5] www.ofai.at (last access 19.11. 2021)

technology LSTM (long short-term memory) which allowed machine learning systems to effectively process very large amounts of data. Another important figure is Martina Mara, who is leading the Robopsychology Lab at the Linz Institute of Technology, examining an essential aspect of AI.[6]

At the Center for Digital Safety & Security of the AIT Austrian Institute of Technology, several important and international AI initiatives have been established in recent years (Hintermayer, 2020, pp. 96–99): they include the analysis of very large image and audio data for public security and protection of critical infrastructures[7]; the analysis of very large text files in the investigation of cybercrime[8]; traffic analysis of very large amounts of data on the Internet and in telecom networks[9]; analysis of large heterogeneous data structures in historical documents; effective speech-based user interfaces for document management systems[10]; self-learning anomaly detection in IT systems to detect sophisticated cyberattacks at an early stage[11] (Leopold et al., 2015); new analysis methods in text, image, and audio content on the Internet to combat fake shops[12] and discrimination[13]; and the detection of fake news and disinformation (see below). Finally, the center has established a research focus on the explainability and verifiability of AI systems, especially for safety-critical areas, as a key topic for the further development of AI.

In addition to OFAI, JKU, LIT, and AIT, there are other research institutions and companies of international repute in the field of artificial intelligence, including the Know-Center at the Graz University of Technology; the Institute for Basic Information Processing at the Graz University of Technology, under the direction of W. Maass; the Institutes of the Vienna University of Technology, including the Database and Artificial Intelligence Group,[14] and Sabine Köszegi,[15] who deals with

[6]https://www.jku.at/en/lit-robopsychology-lab/about-us/team/martina-mara (last access 19.11. 2021)

[7]KIRAS project FLORIDA—Flexible, semi-automated video forensics system for the analysis of mass video data after terrorist attacks, https://www.kiras.at/en/financed-proposals/detail/florida and EU project VICTORIA—*Video analysis for Investigation of Criminal and Terrorist Activities*, https://www.victoria-project.eu/ (last access 19.11. 2021)

[8]EU H2020 project COPKIT, https://copkit.eu/ (last access 19.11. 2021)

[9]WWTF project Big-DAMA—*Big Data Analytics for network traffic Monitoring and Analysis*, https://bigdama.ait.ac.at/ (last access 19.11. 2021)

[10]Research project "Natural Language Search", 6.3.2018, https://www.pressebox.de/pressemitteilung/ser-gruppe/Forschungsprojekt-Natural-Language-Search-erfolgreich-abgeschlossen/boxid/896124 (last access 19.11.2021)

[11]KIRAS project *CAIS Cyber Attack Information System*, https://www.kiras.at/en/financed-proposals/detail/cais-cyber-attack-information-system (last access 19.11. 2021)

[12]KIRAS project SINBAD—Security and prevention of fake-shop fraud with measures of digital forensics, https://www.kiras.at/en/financed-proposals/detail/sinbad (last access 19.11.2021).

[13]FFG research project "Primming", https://projekte.ffg.at/projekt/3280774 (last access 19.11.2021)

[14]https://www.dbai.tuwien.ac.at/ (last access 19.11.2021)

[15]Chair of the Austrian "Rat für Robotik" and member of the Member of the High-Level Expert Group on AI of the European Commission

gender issues related to AI; and many others. An overview of all AI actors in Austria can be found in EnliteAI (2020), which lists 193 organizations.

With respect to national programs, Austria's Industry 4.0 platform[16] has initiated projects on AI impact assessment for workforces in industrial use cases.[17] In 2018, the "Artificial Intelligence Mission Austria 2030" initiated a comprehensive discourse on the development of a national AI strategy through the broad involvement of many stakeholders (AIM, 2018). Finally, in 2021 a dedicated focus program on "AI for Green" was initiated[18] to enhance research into the use of AI in areas such as the environment, climate, and circular economy.

1.3 Artificial Intelligence Needs Powerful Hardware

The current hype surrounding the impressive performance of AI algorithms goes hand in hand with the availability of extremely powerful and inexpensive processors and storage capacity. Effective AI algorithms process huge amounts of data. Typically, AI learning processes require several million parameters to be calculated within a short period of time. This can only be achieved effectively using the parallel computing technologies available today.

The computer gaming industry, in particular, which needs to process high-resolution graphics, has made a significant contribution to the current success of AI technology. Due to the global popularity of computer games, special hardware has been developed for game consoles, which enables impressive graphic effects on the screens. These high-performance graphics cards with their own processors—so-called graphics processing units (GPUs)—are characterized by their very efficient processing of parallel arithmetic operations. The success of game consoles has led them to be manufactured at volume, in turn leading to low unit prices. Consequently, powerful computing power is now economically available and can be used in a wide variety of AI applications.

The development of powerful processors is advancing rapidly, and many manufacturers are developing new special AI hardware[19]: Google's Tensor Processing Unit (TPU), IBM's TrueNorth neuromorphic chip, and AMD with ATI which it acquired in 2006. In 2017, Intel bought Nervana and Mobileye, and Huawei has also entered this international race with the Mate 10 Pro and its in-house Kirin 970 processor. New start-ups are constantly entering the market in this high-end processor sector, with some of the most prominent names including KnuEdge, Eyeriss, krtkl,

[16] https://plattformindustrie40.at/ (last access 19.11.2021)

[17] https://plattformindustrie40.at/kuenstliche-intelligenz-in-der-arbeitswelt-plattform-industrie-4-0-stellt-menschen-in-den-mittelpunkt/ (last access 19.11.2021)

[18] June 17th, 2021, presentation of the funding programme "AI for Green", https://www.imagine-ikt.at/ai-for-green/ (last access 19.11.2021)

[19] On today's processor technology, billions of transistors are placed on a single square centimeter.

Graphcore, BrainChip, TeraDeep, Wave Computing, Horizon Robotics, NeuRAM3, P-Neuro, and SpiNNaker.

1.4 Forms of Artificial Intelligence: From Rule-Based Systems to Neural Networks

The first "intelligent systems" were created as early as the 1980s, using rule-based expert systems with databases, logical decision trees, and mathematical functions. With the availability of large amounts of data, mathematical concepts and heuristic rules such as statistical evaluations, random samples, and exclusion procedures became important approaches for performing data analyses and achieving "intelligent system behavior."

A clear disadvantage of rule-based systems is that the rule base grows very quickly for larger problems, thus becoming increasingly difficult to maintain. However, the availability of powerful new processors which can support a very large number of parallel processing steps allows new data processing processes such as computer-based neural networks (NNs) to be calculated efficiently. Computer-based NNs attempt to simulate the human brain, carrying out data processing in a manner equivalent to that performed by the neural networks of our brain, i.e., information is mapped through connection structures between individual neurons, with different weightings and rules for activating individual neurons. Such systems are also called "deep learning networks."

1.5 Machine Learning

AI systems are based on various computer science approaches and incorporate a wide variety of mathematical models and algorithms. However, computer-based NNs in particular demonstrate a special ability to find solutions to problems in very specific and limited areas of application. The learning processes used by neural networks constitute a fascinating new data processing method which is summarized briefly below.

1.5.1 Supervised Learning, Training Data, and Ground Truth

Appropriate training data is required to train the parameters of the neuron connections in computer-based NN. Learning takes place by changing the weighting of connections, with information "stored" in millions of connections between neurons. The connections change when new data is taken into account. Thus, a model of the real world is no longer created using a mathematical formula or an analytical description, as was previously the case with conventional software programs, but by training data and feedback mechanisms. In this way an "experience" of the system is realized. This is entirely in line with the human system of learning. As the concrete result, together with the appropriate training data, is made available to the technical system, this process is called "supervised learning."

In AI systems based on NN, machine learning takes place by calculating the parameters of the neural networks on the basis of training data. Large amounts of training data are required to obtain reliable results for real applications, while the meaning of the data for training the NN must be known for the particular application. This means that experts are required to assess and label the existing data accordingly, determining a so-called ground truth in the training data. In summary, we need to consider three basic problem areas for the successful use of AI systems:

Firstly, large amounts of data are a fundamental requirement for training the algorithms. This explains the approach that has long been taken by Google, of collecting data in a variety of application areas. Using Google car with extensive sensor technology for autonomous driving, Google Translate for translating text into various languages, as well as Google's speech recognition systems, very large numbers of data sets are constantly being collected in order to continuously improve the accuracy of their AI systems.

Secondly, specific domain knowledge is required to evaluate the existing data records for the application in question and to annotate them accordingly. In many areas, datasets described and evaluated by experts for AI research and application development have been created by the global research community, such as the MAWI dataset for Internet traffic, or the audio datasets for noise recognition from Google in which over two million short audio files, annotated by humans, are available for AI research. For many new application areas, it is now necessary to develop appropriate training data and to assess and evaluate it using human expertise.

Thirdly, after experts with domain knowledge determine the "ground truth" in training data, and subsequent "feature extraction"—thus determining which parameters are important, and in what way—"modeling" is the third important area of AI technology development. In addition to traditional machine learning approaches such as linear models, which are simple mathematical functions, a wide variety of NN architectures are used, including neural networks with feedback effects, or so-called re-current NN. Finally, the calculations of the stimulation functions of the neurons can be carried out according to a wide variety of mathematical functions.

1.5.2 Unsupervised Learning

In an "unsupervised learning" scenario there is no need for "ground truth" in the training data. Here the system only learns patterns in the available data. The result of unsupervised learning approaches is essentially always a form of “data clustering”.[20]

Using unsupervised learning approaches, an AI system can determine a system’s “normal state” in order to identify any subsequent deviations, i.e., anomalies. An important challenge for such approaches is ensuring that potential anomalies are not already included in the training data. This is a requirement that is often difficult to meet.

1.5.3 Reinforcement Learning

After supervised and unsupervised learning approaches, “reinforcement learning” is a third approach to AI algorithms. Here, in contrast to supervised learning, a previously defined result is not learned through training data, but instead the AI system is trained through permanent feedback of “reward and punishment”. For example, “reaching for a hot stove top” will be painful and the system will be given negative feedback. Thus, in principle, there is no right or wrong answer, but there is a goal, and the AI system learns from ongoing experience in attempts to achieve this goal.

Deep reinforcement learning is about fine-tuning a NN pre-trained with specific training data which is continuously refined with real data during operation.

2 Five AI Challenges

As summarized in the previous section, new AI approaches have triggered a new hype in AI technology due to the development of information processing techniques for large amounts of data and the availability of inexpensive, high-performance hardware. Indeed, AI systems often achieve much better results than conventional IT-based problem-solving approaches.

As a result, to date the focus in AI development has been on the positive aspects in the various application areas. It is important to note that many AI applications are based on simply trying out different NNs: i.e., importing the existing data into available AI applications and then observing which results are achieved with each particular NN approach. If a result is satisfactory, it is usually assumed that the application will also work in real use cases.

However, in order to maintain effective and reliable AI systems, and develop AI technology responsibly and ensure its controllability, five fundamental AI system development challenges must be considered:

[20] E.g., the *k*-means algorithm or the hierarchical cluster method

- *Modelability*—how do we derive models of the real world from data, and how do we create models without errors and bias, i.e., how do we establish an effective "ground truth"?
- *Verifiability*—how do we check the technology?
- *Explainability*—how can we understand the decision-making process of the technical systems?
- *Ethics*—how do we guarantee compliance with ethical principles and values?
- *Accountability*—who is responsible for the decisions made by the system?

These issues must be considered for the entire life cycle of AI systems. The following sections discuss these challenges in more detail.

2.1 Modelability

An AI system is essentially an approach to create a model of the real world using sufficient training data, i.e., an AI machine learns an image of the world which the training data ensures as accurate as possible. This requires a sufficiently large volume of data on the object being viewed in the real world. Expert knowledge is then required to analyze and describe the data in order to derive the best possible model.

2.1.1 Large Amount of Training Data and Ground Truth

As described above, the basis for a functioning AI system is a sufficient quantity of suitable training data, as well as expert knowledge. The training data must be evaluated and annotated by experts so that the machine can recognize the patterns according to which the data should be analyzed. Determining such "ground truth" in the existing training data is one of the essential basic requirements for correct AI functionality.

The available training data, as well as the corresponding expert knowledge for assessing the data, are therefore key to well-functioning AI systems. It is vital that the data do not contain any misinterpretations or prejudices, so-called bias, as this leads to incorrect models being learned as a representation of the real world. The problem of possible misinterpretations of the data is illustrated below.

2.1.2 Overfitting and Superstitions

One problem area with AI systems is the selection of the parameters taken into consideration when assessing the data. If the model chosen is too complicated, and too many parameters must be considered, there is a risk of so-called overfitting in which the machine is no longer able to recognize a pattern within the data. The AI algorithm learns the problem by heart, but without solving the actual problem. We are also familiar with this fundamental problem through attempts to make appropriate predictions from very large amounts of data, such as in economic systems, or with climate and weather forecasts (Silver, 2012).

Another potential problem is that certain features in the training data are recognized by the system as special properties, but they have nothing to do with the actual question. Burrhus F. Skinner, a pioneer in behavioral research and programmed learning, discovered in 1948 that pigeons are superstitious. In an experiment he found that random connections between events, e.g., food provided and random posture, are learned as target patterns. These phenomena are also potentially present in AI algorithms. An example illustrates this problem: researchers trained an AI application to recognize wolves in different images. When testing the system, however, it turned out that the AI system believed it recognized “wolves” even in pictures of dogs. Through more detailed investigations, the researchers found that the machine had not learned the features of a face, but instead had selected and trained the image background as a determining parameter in the training data during the training process (Ribeiro et al., 2016).

2.1.3 Built-in Backdoors in AI Systems

Since the function of AI systems is determined by training data, certain mechanisms can also be trained to remain undetected, so-called backdoor functions. For example, special facial images can be trained in facial recognition systems in order to manipulate the result of a passport check.

2.1.4 Summary

In order to be able to build safe, well-functioning AI systems, we need a large amount of training data, assessed by experts, in order to derive the correct parameters for modeling the real world—a “ground truth.” The training data must not train any undesired effects (bias) or possible fallacies.

It is important to note that it is very difficult to identify hidden patterns after a NN has been trained, as the pattern is only available in the training data and is no longer visible or traceable in the actual AI system.

This brings us to the vital issue of the source of the training data, who verifies it, and how the AI system training processes were conducted. Open-source and certified training data are important approaches to promoting trust in AI systems. In addition, testing and verification of AI systems is an area that needs more attention.

2.2 *Verifiability*

2.2.1 AI Needs New Test Methods

As with any technical system, the issue of whether the behavior of AI systems is sufficiently correct is a fundamental one. A well-known example is an accident involving a Tesla vehicle in which an AI-controlled camera system in the autonomous car was unable to distinguish a white vehicle on the road from the white clouds in the sky, so that the emergency braking system did not react as required.[21]

A fundamental problem in the development of technical systems is the question of the verifiability of the specified functions. Approaches for testing and certifying systems in conventional IT system development have been established over decades. Essentially, test methods are based on engineers knowing and understanding the system design as well as the development processes. This means that system behavior and decision-making can be understood and explained by experts. Consequently, test regulations and test procedures can be designed to check system functions for correctness and also to verify them.

However, the growing complexity of AI systems—from rule-based systems, to statistical functions and mathematics, to neural networks—makes it increasingly difficult for experts to retrace the decision-making process. As AI systems based on neural networks are so-called black boxes, it is difficult to determine the parameters the machine used to arrive at a certain decision.

Fundamentally, we are confronted with the question of how AI systems are tested and correct functioning is verified. Problems such as overfitting and backdoors in training data must be taken into account, as well as ensuring reliable system behavior.

As a model of the real world is formed from training data, the question of the robustness of the model in real operation is essential, as the following examples illustrate.

2.2.2 Deceiving AI Systems by Manipulating the Environment

Many research projects show how AI algorithms can easily be impaired by simple physical changes in the environment. Although such changes are easily noticeable to humans, artificial machines struggle with them. AI-based camera systems can easily be fooled by the targeted application of so-called Robust Physical Perturbation (RP2) attacks, such as graffiti on traffic signs (Evtimov et al., 2017) or printed images on cardboard boxes or T-shirts (Eykholt et al., 2018; Samuels, 2017).

[21] Selbstfahrender Tesla übersieht weißen Lkw vor Wolkenhimmel, Welt, 1.7.2016, https://www.welt.de/wirtschaft/article156727084/Selbstfahrender-Tesla-uebersieht-weissen-Lkw-vor-Wolkenhimmel.html (last access 19.11.2021)

AI systems can also be deceived by changing the digital data, e.g., with only slight changes to a digital image that are invisible to humans. For example, an AI-based camera recognized an ostrich in an image in which some pixels had been slightly changed, instead of a car (Gershgorn, 2016).

2.2.3 Summary

With the use of new AI systems, we are confronted with a paradigm shift in the nature of testing methods. It is important to note that the development of test and certification methods for AI systems is still a very young scientific discipline (Zhang et al., 2020) and research activities in this area need to be increased significantly.

The challenge with AI systems is that it is not easy to determine the parameters in the input data which led to misinterpretations. In order to engender trust in AI systems, it is necessary to understand not only the test procedures but also how and why AI systems make certain decisions, allowing system behavior to be explained, as discussed in the following section.

2.3 *Explainability*

It is not possible to understand the process of decision-making in AI systems to the same extent as with classic IT systems, as the functions were learned through training data, rather than simple if-then-else algorithms. Therefore, new methods are needed to understand or explain AI systems. Explainability means that the results of the machine can be interpreted by humans. It is important to note that explainability should not be equated with transparency: explanatory models can be created without understanding the functioning of an algorithm.

2.3.1 AI Explanatory Methods

Ex-post explanations of AI systems can quantify how strongly individual factors influence a decision, thus creating an understanding of the decision-making. One typical method is the LIME (local interpretable model-agnostic explanations) approach, in which a single prediction of a black box algorithm is imitated with the help of a second, comprehensible, transparent model. The transparent algorithm should reproduce the result of the AI system as closely as possible, with the interpretable model then used to identify those features having the greatest impact on the outcome.

Special explanatory methods for NN are being developed and include the layer-wise relevance propagation (LRP) method. This examines which input parameters, e.g., image pixels, contribute to the activation of a neuron. These are then displayed in a form of "thermal image" for each layer.

2.3.2 AI in Safety-Critical Systems

If AI is to be used in safety-critical systems, such as in aircraft, train and vehicle control systems, autonomous robot systems, control systems for power plants, etc.—where incorrect system behavior poses a significant threat of economic loss or human injury—it is vital that the correct system function can always be ensured. Conventional deterministic technical systems can be comprehensively tested, with their behaviors traced and analyzed by experts, to ensure safe operation. This is vital for safety-critical systems.

As this is not possible for AI systems, we need new approaches for effectively testing AI systems, as well as understanding and explaining decision-making processes. Approaches for such analysis systems are forms of black box test procedures to assess the behavior of AI systems (Aichernig et al., 2019; Bartocci et al., 2020; Fellner et al., 2019); "robustness monitoring," in which test results are used as feedback to automatically assess how "good" or "bad" the behavior of an AI system is (Jaksic et al., 2018, Ničković et al., 2018); or online monitoring analysis methods for control systems that take place at runtime to identify the faulty system component when error conditions are detected (Bartocci et al., 2018, 2019; Ferrère et al., 2019; Ničković et al,. 2020).

Several EU research projects are developing test procedures for AI systems in safety-critical controls, including Enable-S3,[22] AutoDrive,[23] and Valu3S.[24] An overview of various test procedures for autonomous cyber-physical systems (CPS) can be found on the AIT website of the EU project Enable-S3.[25]

2.3.3 Summary

New methods of testing AI systems, as well as achieving understanding and explainability, and preventing misuse are required before AI systems can be used in security and safety-critical areas. This requires more research initiatives.

2.4 Ethics and Moral

When it comes to the question of correct decision-making by machines, it is not only the safety aspect of machines that is relevant, but also whether decisions by machines comply with our ethical principles and rules.

[22] https://www.enable-s3.eu/ (last access 19.11.2021)

[23] https://autodrive-project.eu/ (last access 19.11.2021)

[24] https://valu3s.eu/ (last access 19.11.2021)

[25] https://vvpatterns.ait.ac.at/ (last access 19.11.2021)

A moral action is an action which is considered as right or fair by everybody. Thus, we are talking about a normative system, defined by principles and values, which intends to describe correct behavior that is considered objectively valid by everybody.

2.4.1 AI Systems Can Discriminate

As AI systems essentially derive their function from the training data, the availability and selection of training data is the basis for ensuring ethically correct decision-making. The following examples show how quickly we learn to trust a function, while being inadvertently subject to unethical decision-making.

When the available data from Google Translate is used as training data for a language application, we get results that fundamentally contradict our social gender objectives, as essentially all of the world's languages contain an inherent gender bias (Schindler, 2018).

The situation is even more striking for AI-based facial recognition systems (Lohr, 2018): over the last few years, the accuracy of face recognition systems has been continuously improved. AI already achieves a high level of accuracy, with an error rate of only 1% for white male faces. On closer inspection, however, this error rate rises to 7% for female faces and to 12% for dark-skinned male faces. The error rate in dark-skinned female faces is even more dramatic, at 35%. When these algorithms are applied to images of famous women available on the Internet, in most cases the AI systems recognize a "man" (Buolamwini, 2020). One can imagine how, without human supervision, it is easily possible to generate fundamentally incorrect decisions for individuals with such AI machines.

2.4.2 Ethical Norms Are Defined by Culture and Societies

Since different cultures develop different values and even rules for etiquette, morality, truth, honor, conventions, social expected behavior, etc. it is a dedicated challenge to define ethical decision-making by technical systems for different cultures. This is best exemplified by the well-known discussion concerning the decision-making process for autonomous driving when the mortality of different people must be judged by the AI system. Should the AI system protect elderly people instead of children or consider who has violated traffic rules, etc.? For sure, different regions in the world will define a different ethical understanding for solving this problem statement. AI solutions have to consider this when applied globally.

2.4.3 EU Guidelines for the Design of AI Systems

The European Commission considers this problem as a key challenge. It is developing guidelines on ethics in artificial intelligence with a group set up in 2018, involving an interdisciplinary team of experts and stakeholders. Based on ethical principles such as respect for human autonomy, damage prevention, fairness, and explicability, the EU High-Level Expert Group on Artificial Intelligence (2019, p. 14) has defined seven key requirements for the design and use of technology:

- Human agency and oversight
- Technical robustness and safety, including resilience to cybersecurity attacks
- Privacy and data governance
- Transparency, including traceability and explainability
- Diversity, nondiscrimination, and fairness, including the avoidance of unfair bias
- Societal and environmental well-being
- Accountability, including auditability, minimization and reporting of negative impact, trade-offs, and redress

The aim of the EU guidelines is to promote trustworthy AI which is lawful during the AI system's entire life cycle, and thus complies with all applicable laws and regulations, guarantees ethical principles and values, and is technically robust and reliable. Consequently, the technical system design, as well as the policies and business conditions for the use of the technology, must involve societal discourse.

2.4.4 Summary

A particular challenge for this new technology is the potential to make unethical decisions. Humans should always be the final arbiter in legally relevant decisions so that machines do not make incorrect assumptions and unethical decisions.

2.5 Responsibility

Due to the lack of explainability caused by the black box problem, potential bias, and ethical problems, artificial intelligence represents a huge challenge for legislators and administrators. As AI has such a wide range of possible uses, almost every aspect of our legal system is impacted.

2.5.1 Summary

In order to achieve the goal of sound AI technology, a clear and predictable legal set of rules which addresses the technological challenges is required, as described by the European Commission (2020) report on safety and liability implications of AI.

There is still a lot to do to bring the technical developments in AI technologies in line with the recommendations, guidelines, legal requirements, and regulations. The use of AI systems should always be accompanied by comprehensive impact and risk assessments and designed within an overall architecture so that no automated decisions can be made without human control where these decisions will have a legal impact.

3 Social Threat Potential from AI

3.1 Democratization of Technology

Education at schools and universities, and the availability of open-source software and libraries, has made it very easy to use AI systems without high costs and investments. We can therefore say that this powerful technology has in some way become democratized, allowing it to solve problems in many areas of application. But as with the introduction of every new technology, we experience not only the intended advantages, but are also confronted with potential disruptive impacts and misuse. AI systems allow companies to assess their customers via data analysis, aligning their product and service offerings not only to create an improved customer offering, but also to exclude customers and even discriminate against them. This problem has been addressed in the Austrian preis.wert[26] and Primming[27] research projects in order to develop methods and tools to apply countermeasures.

Even more threatening is that, with AI, the problem of the manipulation of public opinions is taking on a new meaning in our media world. Through targeted, personalized disinformation campaigns, citizens can be influenced and manipulated in their behavior, even including their voting behavior. This raises the question of whether the new technology can be controlled, and whether AI machines might cause damage to people or even entire societies.

3.2 Manipulation of Media

3.2.1 Fake News and Deep Fakes

The Internet is a new media infrastructure that is unique in human history. It is a media system in which every consumer can become a producer. The vast volume of websites, blogs, and social media channels have multiplied our information channels and generated an exponentially increasing volume of media content. Thus, the Internet is an unprecedentedly effective platform for communication, information

[26]Project funded by the Austrian programme Netidee, "*preis.wert*", https://www.netidee.at/preiswert (last access 19.11.2021)

[27]Identification of price discrimination Horizont, 4.2.2020, https://www.horizont.at/digital/news/aufdeckung-von-preisdiskriminierung-ait-und-oeiat-verstaerken-forschungskooperation-73253 (last access 19.11.2021)

exchange, and innovation, and the global "flat world" predicted by Friedman (2007) has become reality.

At the same time, however, the Internet is also a new and unprecedented platform for extensive, falsified communication and disinformation. Although the spread of false information, so-called fake news, has always been part of human history, we have reached a new dimension in terms of scope and distribution efficiency. "Fake news" is the dissemination of false information with the aim of influencing public opinion, groups, or individuals for political or economic interests. Information, i.e., text, images, audio, and video, is manipulated in order not only to communicate untruths, as measured by objective standards, but through subtle suppression, concealment, or the presentation of facts in altered contexts, for the purpose of deceiving and thus manipulating the media consumer, as described by the EU High-Level Expert Group on Fake News and Online Disinformation (2018).

However, it should also be noted that a clear business model is being implemented in the Internet. As every reader of a message, a website, or a blog, and every "like" represents advertising value for influencers or bloggers, there is a high level of motivation to present news as attractively as possible in order to increase business value. It is obvious that this practice does not always respect the truth.

AI technology has also made it possible to create new fake media content using existing data on the Internet. So-called deep fakes create new images, audio content, and videos that look deceptively real in order to defraud companies[28] or even to manipulate political opinion in a country.[29]

AI systems also represent an effective way of determining political attitudes from the available data in our online social media or other digital traces. AI is therefore also a potential tool for misuse in autocratic and dictatorial systems.

3.2.2 The Fact Check: A Necessary Tool Support

New methods and tools are needed to verify news and information. Media companies and authorities are already making considerable efforts to check media content for its truthfulness (European Commission, 2019). With the increasing volume of media, however, we are confronted with the problem that the current approaches to fact-checking do not scale (Graves & Cherubini, 2016). Current procedures take too long and are too expensive to be able to effectively combat the increasing amount of fake news on the Internet.

[28] Deepfake: Betrüger erleichterte Firma dank gefälschter Stimme um 220.000 Euro, DerStandard, 5.9.2019, https://www.derstandard.at/story/2000108225039/deepfake-betrueger-erleichterte-firma-dank-gefaelschter-stimme-um-220-000 (last access 19.11.2021)

[29] Pelosi videos manipulated to make her appear drunk are being shared on social media and YouTube: https://www.youtube.com/watch?v=sDOo5nDJwgA (last access 19.11.2021).

Thus, it is vital that there is increased awareness among the population about the value of correct news, as well as digital users capable of analyzing messages and information. This requires a new digital literacy in our society. Furthermore, we must develop a new understanding for high-quality journalism, in which consumers trust media brands and information platforms as they offer content verification and quality assurance. We need new tools and services to support such objectives, and these in turn will effectively process the very large amounts of information in multimedia data and provide cost-effective support for private and professional users.

AI methods are important approaches to developing effective analysis tools. In Austria, a new initiative was launched in 2020 within the framework of the national security research program KIRAS: the project "Defalsif-AI—Detection of false information using artificial intelligence"[30] is based on comprehensive cooperation between research organizations, media companies, and Austrian authorities.

4 Limits of AI and Diversity of Life

4.1 Singularity: Can AI Surpass Humanity?

Of course, machines are in many ways better and faster than humans and should be used to manage major economic, ecological, and social issues wherever needed. There will never be a limit to the human thirst for knowledge, which is a cause for hope, especially in view of the ever-increasing number of problems faced by our global and digitally networked society.

However, a lack of insight and understanding make us prone to delusions of grandeur. As if in a childlike reaction of defiance, we want to create an artificial human being. In his book (Kurzweil, 2005), Ray Kurzweil describes how human intelligence is replaced by AI in the year 2045. But in order to judge this development we have to critically review the principles of AI technology. What are the fundamental differences between the human mind and the analytical ability of machines? Below we discuss a few basic issues that make the big difference between these two worlds from a philosophical point of view.

4.2 AI Needs a Lot More Intelligence

Despite the promising potential of AI and the current hype associated with this technology, we should keep reminding ourselves that we are only building machines. In this context, despite the euphoria surrounding AI services, Moravec's

[30] https://science.apa.at/project/defalsifai-en/ (last access 19.11.2021)

paradox is an important insight that should be called to mind. Seemingly complicated processes are easier to "calculate" using AI than simple processes like those we use in our everyday lives. Every child moving through a playground manages a higher level of complexity than any AI system. This is also reflected in the fact that, ultimately, today's AI systems still only work effectively in specific areas of application. We are far from creating an all-encompassing super intelligence that surpasses humans.

Being able to analyze even larger volumes of data faster does not create a form of intelligence that can be compared with human intelligence. This is one of the biggest fallacies surrounding big data and AI, and one to which we unprotestingly succumb.

4.3 Life Is Nonlinear

Flexibility, imagination, and creativity are important human characteristics which we need to master our complex lives. We usually refer to them as inspiration, inventiveness, and the ability to find creative solutions. Our life is not a linear development but constantly impacted by disruptive effects, in turn influenced by nonlinearities, errors, and coincidences. These mechanisms will not be imitable by machines for some time.

Offers that correspond to our buying behavior or our preferences may seem intelligent and helpful at first glance, but are only half of our real life. Our life is not a sterile place in which we work through facts—it's fun, exciting, challenging, and sometimes apparently pointless. And most important of all, chance plays a very large, even determining role in all of our lives. True innovation comes from creativity, chance, and the combination of apparently incompatible things (Leopold, 2017, p. 35 f.). In principle, one could say that potentially, AI kills creativity.

4.4 Life Is Not Just About Solving Problems

Life is not just about problem solving. Innovation researchers have long understood that technology and innovation are always the result of constant exchange with other technologies, with the environment, and with the technology user (Leopold, 2017, p. 99). Precht writes: "Life is not a template-based problem-solving process. When we grasp the world, we establish relationships between things. We don't think in terms of concepts, but in terms of relations. All thinking is relational and is therefore relative. Thoughts can never be properly fixed, they are fluid and lie between things" (Precht, 2020, p. 140). "Anyone who believes that life and problem-solving are the same thing cuts down the dimension of life with every problem solved, until in the end a state of carelessness is reached, which can no longer be called life" (Precht, 2020, p. 132).

Artificial intelligence, on the other hand, specifically highlights problems and brings them to the attention of the observer (p. 138). This shows that we are dealing with a tool and not with a form of super intelligence. Human intelligence is what we use when we have not been given instructions on how to solve problems. However, here logic and calculation only play a limited role—human intelligence relies on emotionality and intuition, spontaneity, and association. This also includes empathy for the situation under the influence of existing values (Precht, 2020, p. 25).

4.5 The Data World of AI Is Not Life

There is widespread discussion and fundamental fear that AI will replace humans in many areas. On closer inspection, however, it is clear that AI is just a tool that will help us with various tasks and cannot replace us as humans in any way. AI shows us that we have to reflect on the true values of "being human." Precht (2020) describes fundamental aspects of human behavior which will probably always distinguish us from machines:

- Artificial intelligence is not based on values, has no trust, and knows no truth, freedom, friendship, respect, or willingness to help (pp. 31–37).
- AI knows no community and also no social culture of recognition and appreciation (pp. 31–37). But human behavior and decision-making only manifest themselves through human social contacts that have been experienced.
- AI has no time awareness. As human beings, however, we experience our life through constant reflection on the past and permanently plan our future and permanently align our striving and actions according to subjectively meaningful coordination between our experiences and desired goals.
- AI does not develop a will. To Precht it is incomprehensible that a super intelligence would develop a self-image and a will only by analyzing data and training behavioral patterns (p. 125).

Our abilities to calculate, see, and hear are much worse than those of computers, but we humans orientate ourselves in often infinitely more complex contexts. The limited data world of artificial intelligence only perceives what it is supposed to perceive. But we live according to values, feelings, attitudes, and moral abilities that have developed in our societies during the evolution of mankind because they have proven themselves. The world of AI lacks imagination and emotion, subjectivity and feeling, and is much smaller than human intelligence.

4.6 AI and Morals

Concerning the ethical issues associated with AI, Precht emphasizes a clear message (Precht, 2020, p. 156):

> What makes people human is not due to any a priori stipulations or programming. It is the special way in which we communicate with our environment. We do not just react to it, but we construct it, starting from ourselves, as our world. . . . The use of morality cannot be standardized like the size of dowels and screws. A morality without subjective attitudes, reduced to general considerations and evaluations of life, remains formal and meaningless.

Thus Precht clearly states that we cannot develop a morality suitable for robots. Only through constant discourse on the use of machines to support our lives can we define appropriate policies, regulations, business conditions, and mechanisms to make technology safe, reliable, and as error-free as possible.

5 Conclusions

5.1 Education and Emotional Intelligence to Master the Technology

As described above, human intelligence is what we use when we have not received instructions on how to solve problems. Real problem solving requires intuition, association, empathy, and the application of values and culture. For many problems, processing of instructions and logic does not help in solving them. This is where the important difference between education and training becomes impressively clear.

The question arises as to whether our current education systems also adequately consider these new developments and the challenges they bring. In many disciplines, today's education systems still give extensive factual knowledge a higher priority. Empathy, creativity, and problem-solving skills in the face of incomplete and contradicting facts, and a holistic view of complex systems are often given too little attention.

It is therefore important to fundamentally rethink our education system. In order to live with robots and AI in the future, and to master the technology, we must reflect on our core human competencies of emotional intelligence and true education and replace the trivial learning of facts with soft skills. We cannot compete with AI and robots; instead we must master them.

5.2 Responsibility for the Development of Technology

As AI becomes increasingly popular, we must discuss methodologies for system development and system operation, standards, and certification of technical systems to ensure a system behavior which meets our ethical principles, regulations, and laws. If we do not wish algorithms to eventually rule us, we must retain ultimate responsibility for what the machines do. The value of life is not determined by purely economic criteria, nor can it be determined according to mathematical formulas. Therefore, AI must not be allowed to become a new religion!

In his 1961 comedy *The Physicists*, Friedrich Dürrenmatt discussed the fundamental question of human responsibility for technological development that brings with it the danger of threatening the world. Technological mastery is an absolute must in all of our endeavors—the essence of being human revolves around freedom, will, emotion, and creativity.

If our society does not face up to these emerging dangers, we will have a potentially uncontrollable technical system that will have an impact in many areas of our lives, and criminal actors will have new and effective tools for their objectives. However, we have the option of establishing the dominance of the "good" use of AI. It is essential to set up research initiatives to recognize and understand the new threats associated with AI technologies and to develop suitable countermeasures. We must find mechanisms to increase the robustness of our society as well as to ensure the sustainable control of AI systems.

5.3 AI Needs Standardization

We should note that technological advances in AI are ahead of governance. Although a number of standardization efforts are ongoing, standardization bodies, regulators, service providers, and operators, as well as public stakeholders, must consider this dynamic development and scale up standardization activities to support regulatory efforts. A report drafted by the European Observatory for ICT Standardisation (EUOS) provides an overview of existing standardization activities, reviews, guidelines, and published white papers for different AI subjects (Lindsay, 2021).

5.4 A Broader Approach to AI Research

In order to arrive at appropriate AI systems, we need a comprehensive research program which, in addition to the actual core issues, deals with important issues such as the security of AI methods, testing and verification, annotation tools and methods for training data, and sets of rules in order to comply with legal and ethical requirements. Leopold et al. (2020) describe a comprehensive AI research program.

To rule out the improper use of AI, it is imperative to understand the new technology better, to define standards and implement policies and regulations for its socially acceptable use, as well as to develop skills and tools so that authorities, companies, and every single citizen have the opportunity to participate in society's positive development as a responsible media and technology consumer (Brundage et al., 2018).

However, there remains a big difference between whether a machine can be trained to perceive emotional moods or whether this ability is inherent. Only humans have a body-soul constellation, as well as cognitive abilities to shape the world, and experience events in the form of feelings. Only grief, joy, love, empathy, social competence, and willingness to help create that unique tension that sets humans apart from technical systems.

Thus, the fact remains that even the most sophisticated machine will probably never be able to compete with humans in their multi-dimensionality of cognition, emotion, and physicality and in their sensual perception of the world!!

References

Aichernig, B. K., Bauerstätter, P., Jöbstl, E., Kann, S., Korosec, R., Krenn, W., Mateis, C., Schlick, R., & Schumi, R. (2019). Learning and statistical model checking of system response times. *Software Quality Journal, 27*(2), 757–795.

AIM. (2018). Artificial Intelligence Mission Austria (AIM AT) 2030—Die Zukunft der Künstlichen Intelligenz in Österreich gestalten. Bundesministerium für Verkehr, Innovation und Technologie (BMVIT) und Bundesministerium für Digitalisierung und Wirtschaftsstandort (BMDW), Wien. https://www.bmk.gv.at/themen/innovation/publikationen/ikt/ai/aimat.html. Last access 19 Nov 2021.

Bartocci, E., Deshmukh, J., Gigler, F., Mateis, C., Ničković, D, & Qin, X. (2020, September 20–25). Mining shape expressions from positive examples. In *International conference on embedded software (EMSOFT), Hamburg, Germany*.

Bartocci, E., Ferrère, T., Manjunath, N., & Ničković, D. (2018, April). Localizing faults in Simulink/Stateflow models with STL. In *Proceedings of the 21st international conference on hybrid systems: Computation and control HSCC '18, part of CPS Week* (pp. 197–206). https://doi.org/10.1145/3178126.3178131. Last access 19 Nov 2021.

Bartocci, E., Manjunath, N., Mariani, L., Mateis, C., & Nickovic, D. (2019) Automatic failure explanation in CPS models. In *SEFM 2019* (pp. 69–86).

Brundage, M., Avin, S., Clark, J., Toner, H., Eckersley, P., Garfinkel, B., Dafoe, A., Scharre, P., Zeitzoff, T., Filar, B., Anderson, H., Roff, H., Allen, G. C., Steinhardt, J., Flynn, C., ÓhÉigeartaigh, S., Beard, S., Belfield, H., Farquhar, S., Lyle, C., Crootof, R., Evans, O., Page, M., Bryson, J. J., Yampolskiy, R., & Amodei, D. (2018). *The malicious use of artificial intelligence: Forecasting, prevention, and mitigation*. Future of Humanity Institute, University of Oxford, Centre for the Study of Existential Risk. University of Cambridge, Center for a New American Security, Electronic Frontier Foundation, OpenAI, February 2018.

Buolamwini, J. (2020, January 18–20) The coded gaze: Bias in AI, DLD conference "What are you adding?", Munich. https://www.dld-conference.com/, and https://www.youtube.com/watch?v=rjesnx_Pp5w&list=PLxaUSBUUlSviSAoJ9Xuoet0s2rDOSHcey&index=29. Last access 19 Nov 2021.

EnliteAI. (2020, October). AI landscape Austria—The Austrian AI landscape: An overview of the entire ecosystem covering startups, companies, research institutions as well as their geographic distribution and growth. https://www.enlite.ai/works/ailandscapeaustria. Last access 19 Nov 2021.

EU High-Level Expert Group on Artificial Intelligence. (2019, November 8). Ethics guidelines for trustworthy AI, European Commission. ISBN 978-92-76-11998-2; https://op.europa.eu/en/publication-detail/-/publication/d3988569-0434-11ea-8c1f-01aa75ed71a1. Last access 19 Nov 2021.

EU High-Level Expert Group on Fake News and Online Disinformation. (2018, March 12). Final report of the high level expert group on Fake News and Online Disinformation. European Commission. ISBN 978-92-79-80419-9. https://digital-strategy.ec.europa.eu/en/library/final-report-high-level-expert-group-fake-news-and-online-disinformation. Last access 19 Nov 2021.

European Commission. (2019, October 29). Code of Practice on Disinformation one year on: online platforms submit self-assessment reports. https://ec.europa.eu/commission/presscorner/detail/en/statement_19_6166. Last access 19 Nov 2021.

European Commission. (2020, February 19). Commission report on safety and liability implications of AI, the internet of things and robotics, Brussels, COM(2020) 64 final. https://ec.europa.eu/info/sites/default/files/report-safety-liability-artificial-intelligence-feb2020_en_1.pdf. Last access 19 Nov 2021.

Evtimov, I., Eykholt, K., Fernandes, E., & Li, B. (2017, December 30). Physical adversarial examples against deep neural networks. BAIR Berkeley Artificial Intelligence Research. https://bair.berkeley.edu/blog/2017/12/30/yolo-attack/. Last access 19 Nov 2021.

Eykholt, K., Evtimov, I., Fernandes, E., Li. B., Rahmati, A., Xiao, Ch., Prakash, A., Kohno T., & Song, D. (2018, April 10). Robust physical-world attacks on deep learning visual classification. In *CVPR 2018 conference*.

Fellner, A., Krenn, W., Schlick, R., Tarrach, T., & Weissenbacher, G. (2019). Model-based, mutation-driven test-case generation via heuristic-guided branching search. *ACM Transactions on Embedded Computing Systems, 18*(1), 4:1–4:28.

Ferrère, T., Nickovic, D., Donzé, A. Ito, H., & Kapinski, J. (2019). Interface-aware signal temporal logic. In *HSCC 2019* (pp. 57–66).

Friedman, T.L. (2007). *The world is flat: The globalized world in the twenty-first century: A brief history of the globalized world in the twenty-first century*. Penguin. ISBN 9780141034898.

Gershgorn, D. (2016, March 30). Fooling the machine—The Byzantine science of deceiving artificial intelligence. *Popular Science*. https://www.popsci.com/byzantine-science-deceiving-artificial-intelligence. Last access 19 Nov 2021.

Graves, L., & Cherubini, F. (2016). The rise of fact-checking sites in Europe. *Reuters Institute for the Study of Journalism*. https://reutersinstitute.politics.ox.ac.uk/our-research/rise-fact-checking-sites-europe. Last access 19 Nov 2021.

Hintermayer, N. (2020). Innovative Sicherheit. *Forbes*, nr. 6-20, pp. 96–99, https://lnkd.in/e22Rf5H. Last access 19 Nov 2021.

Jaksic, S., Bartocci, E., Grosu, R., Nguyen, T., & Nickovic, D. (2018). Quantitative monitoring of STL with edit distance. *Formal Methods in System Design, 53*(1), 83–112.

Kurzweil, R. (2005). *Singularity is near*. VIKING, published by the Penguin Group, Erstausgabe. ISBN 0-670-03384-7.

Leitner, K. -H. (2003). *Von der Idee zum Markt: die 50 besten Innovationen Österreichs*. Böhlau, p. 268 ff.

Leopold, H. (2017). *Social communication for corporate innovation management*. PhD, Lancaster University, School of Computing and Communications, Lancaster, UK

Leopold, H., Bleier, T., & Skopik, F. (2015). *Cyber Attack Information System—Erfahrungen und Erkenntnisse aus der IKT-Sicherheitsforschung*. Springer. ISBN 978-3-662-44306-4.

Leopold, H., Krenn, W., King, R., & Mateis, C. (2020, March 17). Artificial intelligence landscape—An introduction in technology fields & research areas. AIT Technical Report. https://www.researchgate.net/publication/332471378_Artificial_Intelligence_Landscape_-_An_Introduction_in_Technology_Fields_Research_Areas. Last access 19 Nov 2021.

Lindsay, F. (2021). Landscape of AI standards—Report of TWG AI, The European Observatory for ICT Standardisation (EUOS), Lindsay Frost (Ed.), EU Horizon 2020 project StandICT.eu. https://doi.org/10.5281/zenodo.4775836., https://standict.eu/. Last access 19 Nov 2021.

Lohr, S. (2018, February 9). Facial recognition is accurate, if you're a white guy. *New York Times*. https://www.nytimes.com/2018/02/09/technology/facial-recognition-race-artificial-intelligence.html. Last access 19 Nov 2021.

Manhart, K. (2018, January 17). Eine kleine Geschichte der Künstlichen Intelligenz. *Computerwoche*. https://www.computerwoche.de/a/eine-kleine-geschichte-der-kuenstlichen-intelligenz,3330537,2. Last access 19 Nov 2021.

Nations, D. (2020). What is a CAPTCHA Code?, updated on February 27, 2020. https://www.lifewire.com/what-is-a-captcha-test-2483166. Last access 19 Nov 2021.

Ničković, D., Lebeltel, O., Maler, O., Ferrère, T., & Ulus, D. (2018). AMT 2.0: Qualitative and quantitative trace analysis with extended signal temporal logic. In *Tools and algorithms for the construction and analyses of systems (TACAS) proceedings, part II, 2018* (pp. 303–319).

Ničković, D., Lebeltel, O., Maler, O., Ferrère, T., & Ulus, D. (2020, August 3). AMT 2.0: Qualitative and quantitative trace analysis with extended signal temporal logic. *International Journal on Software Tools for Technology Transfer, 22*, 741–758. https://doi.org/10.1007/s10009-020-00582-z. Last access 19 Nov 2021.

Precht, R. D. (2020). *Künstliche Intelligenz und der Sinn des Lebens*. Wilhelm Goldmann Verlag.

Reichl, P., Frauenberger, C., & Funk, M. (2020). Homo Digitalis—Wiener Kreis zur Digitalphilosophischen Anthropologie. Last update January 2020. www.homodigitalis.at. Last access 19 Nov 2021.

Ribeiro, M. T., Singh, S., & Guestrin, C. (2016). "Why should i trust you?": Explaining the predictions of any classifier. In *ACM, KDD 2016 San Francisco, CA, USA*.

Samuels, M. (2017). Hacking risk for computer vision systems in autonomous cars. *SecurityIntelligence*, 10.8.2017. https://securityintelligence.com/news/hacking-risk-for-computer-vision-systems-in-autonomous-cars/. Last access 19 Nov 2021.

Schindler, A. (2018, November). AI and the need of a responsible and inclusive innovation policy. In *OVE - Digitalisierung im Überblick: GIT-Newsletter 2015-2018, Newsletter Social Media, "Beherrschen wir die künstliche Intelligenz*" (pp. 146–149). Band 97 der OVE Schriftenreihe. ISBN 978-3-903249-07-3.

Silver, D., Schrittwieser, J., Simonyan, K., Antonoglou, I., Huang, A., Guez, A., Hubert, T., Baker, L., Lai, M., Bolton, A., Chen, Y., Lillicrap, T., Hui, F., Sifre, L., van den Driessche, G., Graepel, T., & Hassabis, D. (2017). Mastering the game of Go without human knowledge. *Nature, 550*, 354–359. https://doi.org/10.1038/nature24270. Last access 19 Nov 2021

Silver, N. (2012, September 27). *The signal and the noise: Why so many predictions fail—But some don't*. Penguin.

Turing, A. M. (1950, October). I. Computing machinery and intelligence. *Mind—A Quarterly Review of Psychology and Philosophy, LIX*(236), 433–460. https://doi.org/10.1093/mind/LIX.236.433. Last access 19 Nov 2021.

Zhang, J. M., Harman, M., Ma, L., & Liu, Y. (2020) Machine learning testing: Survey, landscapes and horizons. *IEEE Transactions on Software Engineering*. https://doi.org/10.1109/TSE.2019.2962027.

Helmut Leopold has more than 30 years of experience in the IT and communication technology industry. He is currently the Head of Center for Digital Safety & Security at the AIT Austrian Institute of Technology. In this role he leads 200 research professionals in a digital research program covering areas such as artificial intelligence and data science, cybersecurity, post-quantum encryption, 6G, photonics, and electronic-based highly reliable system development. Having led international efforts such as the Broadband Services Forum (BSF), San Francisco, Mr. Leopold is now leading several national initiatives and expert groups as well as initiatives supporting international organizations such as the IAEA and the United Nations Office of Counter-Terrorism. Additionally, he is the key organizer of the International Digital Security Forum (IDSF).

Before joining AIT, Mr. Leopold worked at Alcatel and at Telekom Austria, the country's largest network operator. As acting CTO at Telekom Austria from 1998 to 2008, he was responsible for the digital transformation of Telekom Austria into a modern multimedia and broadband company. He was a key driver for the introduction of broadband Internet and interactive television in Austria and is today recognized as a digitalization expert of national and international repute. Helmut Leopold was born in 1963 in Hohenems, Vorarlberg. He holds a degree (Ing.) from the technical college HTL for electronics and telecommunications in Rankweil, Vorarlberg; a masters' degree (Dipl.-Ing.) in computer science from the Vienna University of Technology; and a PhD in computer science from the University of Lancaster, UK.

Technology Serves People: Democratising Analytics and AI in the BMW Production System

Matthias Schindler and Frederik Schmihing

Abstract Individualisation and an increasing number of variants characterise the production of premium vehicles in particular. This implies an increase in complexity in manufacturing. The principles of lean production form the basis for the continuous improvement of processes. Further optimisation of production can be achieved with data analytics and artificial intelligence (AI) methods.

These innovations raise the issue of corporate social responsibility, CSR. This article describes the BMW Group's approach to corporate social responsibility and the sustainable development of digitalisation in production. In addition to the technical aspects of data analytics and AI, the organisational implications are highlighted. For the BMW Group, the claim that 'technology serves people' means that production employees must understand the quality figures in their area and be able to independently carry out a root cause analysis in the event of an error. AI systems must be designed intuitively so that employees can tailor them to their specific application in self-service.

The BMW Group places people—in the production system, especially the direct production employees—at the centre. Data analytics and AI must contribute to making work in the BMW production system more pleasant and even more attractive. The goal is a strength-based division of labour between humans and IT systems.

Keywords CSR · Data analytics · Artificial intelligence · Production · Quality

1 Digitalisation and Production: A Complex and Dynamic Environment

The premium segment of the automotive industry is characterised by intense competition. Production has to be correspondingly cost-effective—the aim is to produce premium quality under maximum efficiency. The foundation for the day-to-day

---

M. Schindler (✉) · F. Schmihing
BMW Group – Innovation, Digitalisation, Data Analytics, Munich, Germany
e-mail: matthias.schindler@bmw.de; frederik.schmihing@bmw.de

R. Schmidpeter, R. Altenburger (eds.), *Responsible Artificial Intelligence*, CSR, Sustainability, Ethics & Governance, https://doi.org/10.1007/978-3-031-09245-9_7

increase in efficiency in production systems is formed by the design principles of lean production (Dombrowski et al., 2015). Continuous improvement aims to systematically examine processes and design them in a way that any form of waste is minimised or ideally eliminated (Brunner, 2017). In order to be able to react as well as possible to the development of demand in the global sales markets (Zipse, 2018), the BMW Group pursues the approach to run many different derivatives on the same production line (Majohr, 2008). The current trend of hybridisation and electrification results in an increase in the number of variants in the powertrain. As a result, the integration approach means a considerable increase in complexity for the BMW production system.

Especially for premium products, quality is non-negotiable—this has so far implied high expenses for quality inspections and, in some cases, rework to repair any defects (Jochem, 2010). Since ensuring consistently high quality is the permanent goal of any production, daily routines are used in manufacturing to ideally design the production processes and optimise quality so that as little rework as possible is required (Deckert et al., 2015). In the premium car market in particular, derivatisation and an extensive range of individualisation options play a key role; the current BMW 3 Series Sedan, for example, is available in more than one billion variants (Grüneisl, 2018). The complexity induced by the immense number of variants and integration takes established methods of process improvement to their limits. In particular, the analysis of the causes of errors becomes significantly more laborious due to this complexity (Töpfer & Günther, 2007). In addition, the logic of quality inspections in long control loops that has been applied up to now is proving to be inefficient and cost-intensive (Ehrlenspiel et al., 2014).

On the one hand, data analytics and artificial intelligence are seen as central enablers to manage complexity in production systems. On the other hand, these technologies are partially discussed critically in society and science. The focus is on the extent to which these technologies are accessible and operable for people outside of IT and what impact they have on jobs (Zimmermann, 2017). Visual analytics enables interaction with large amounts of data such as quality data in production (Ware, 2004). Artificial intelligence allows, for example, the robust categorisation of components and is thus useful for shortening quality control loops.

In the context of these IT innovations, corporate social responsibility (CSR) has the essential task of ensuring the reliable use of digital technologies and the sustainable implementation of innovative applications of data analytics and artificial intelligence for all stakeholders (Knaut, 2017). In this context, the circle of stakeholders includes at least customers, production employees, management, software developers and society worldwide.

For customers, ensuring premium quality is at the highest priority. Every customer expects a perfect product, built and delivered in exactly the individualised configuration he ordered. Employees must benefit from innovations in everyday production. For them, the focus is on comprehensibility and usability of the new IT solutions. Innovations find long-term application in the production system when the production employees understand the underlying procedures and can classify the benefits that technology brings them and the effort they have to invest in the realisation of a new application. In order to establish data analytics and AI deeply in production, management must also understand the basic principles of the impact correlations and become familiar with the necessary prerequisites. At the same time, macroeconomic effects with regard to the number and design of jobs are the focus of interest for society. Thus, the effects on the attractiveness of jobs in production are also of concern. Another pillar of social responsibility describes the democratisation of innovations and algorithms. Technology must offer people added value; to this end, the broadest possible circle of users should understand it, be able to use it and benefit from it.

The BMW Group combines innovative technologies, emotional products and individual customer care to create a unique overall experience. The claim to innovation leadership also applies to the production system. In this article, the focus is on the implications of the innovations in data analytics and artificial intelligence for quality work in the BMW production system in order to shed light on CSR within this system boundary. This paper is therefore oriented towards the following core questions regarding corporate social responsibility: It will be discussed how corporate responsibility is changing through the use of data analytics and AI. For this purpose, it will be presented in which form the BMW Group deals with the consequences and possibilities of AI and how the potential risks are dealt with. It also shows what AI means for the company's (global) value chain and strategy and how this is changing the company's social responsibility. In the data analytics and AI sector, cooperation is essential and the question arises how to deal with different approaches to responsibility and sustainability. Finally, this article explores the question of what demands data analytics and artificial intelligence place on managers at all levels in production.

2 Status Quo

2.1 Quality Work in Production: A Critical Review

In quality work, the principles of lean production manifest themselves in the continuous improvement process. Here, the visualisation of quality key figures holds an essential role. On so-called process boards or Andon boards (Oeltjenbruns, 2000), highly aggregated key figures in accordance with key performance indicators, KPIs, are usually presented (Richter & Rico-Castillo, 2015). On the one hand, the KPIs are calculated and aggregated up to plant level over a given period of time. On the other hand, the production employees, together with the responsible foremen and supervisors, deal with these KPIs cyclically in a structured routine in order to identify negative effects, find causes and eliminate the causes of errors.

Although this error correction process is highly effective in production, the time required to identify the cause can be very long, as individual components pass through hundreds of production and logistics stations, some of which are not directly visible to the employees. Individual effects and correlations in possible error patterns are difficult to determine, and in this regard the aggregation of production-specific key figures only adds little value. Key figures are indispensable for monitoring production processes, but in many cases this aggregation only allows a trend to be identified, which does not immediately reveal the specific cause of the error.

The high degree of individualisation described above, along with the logic of integrating different drive forms and derivatives on the same line, implies a high degree of complexity in the BMW production system. For the production employees, this is accompanied by a high risk of errors. As a result, manual and automatic checks are used for quality inspection. In the case of manual inspections, production employees are assigned to inspect relevant parts according to the so-called four-eyes principle (Regber & Zimmermann, 2009). In this way, they do not serve the direct creation of value, but check whether other employees have worked correctly. In addition, automatic camera portals with conventional image processing are used (Demant et al., 2011). In these camera portals, there is one target photo for each attribute. To inspect the installed object, a photo of the corresponding component is taken and compared with the specified target photo on a pixel-by-pixel basis. Typically, this is done using the grey value of each pixel or a selection of pixels. If a small number of pixels deviate from the target by more than a tolerable margin, an error entry is generated for that component. In this case, a foreman attempts to locate the car on the production floor and checks the component. If this categorisation of the camera check is incorrect, there is a 'false positive'—a pseudo error (Deuse et al., 2017). The foreman deletes all pseudo or sham defects from the quality system. All actual defects ('true positives') must be reworked before the vehicle can leave the factory.

Camera portals are usually set so sensitively that they do not miss any actual errors ('false negatives'); correspondingly, false errors occur frequently. For the pixel-based image comparison as described to work, very constant environmental conditions are required. Each photo of a component must always look the same—day and night, in summer and in winter. Camera portals are therefore constructed as follows: The portals are integrated into a box or booth that serves to shade the ambient light. Inside are numerous cameras and lighting units (Demant et al., 2011). Each camera is aligned with a feature in such a way that it captures it in an almost format-filling manner. The illumination (white or red light) is synchronised with the corresponding camera. Red light sources are applied to minimise reflections, as these impede recognition. To ensure that all photos look similar and can be compared with the target photo, the vehicles are usually stopped in production to take the photos in the camera portal. This is contrary to the imperative of production flow from lean production.

Due to the space required and the interruption of the production flow, the number of camera portals in production is limited—especially in plants with very limited space such as the BMW Group's core plant in Munich. This results in a small number of such portals per production line for inspection, which means long quality control loops. This is accompanied by rework, in the course of which components sometimes have to be detached in order to replace a defective component. In addition, there is the disadvantage that with this logic of automatic quality inspections, parts are initially installed incorrectly and have to be reworked afterwards. In practice, fluctuating environmental conditions lead to insufficiencies in defect detection. A high rate of pseudo defects implies an immense effort for the foremen, as they have to check all defect entries and correct them if necessary.

2.2 *Quality Work: Quo Vadis?*

Visual Analytics: Enabler for Value-Added Data Analysis

The relevance of the targeted use and analysis of data has experienced significant growth in recent years and will continue to do so, and has established itself under the term big data or data analytics (Albertson, 2018). This has led to the development of numerous and entirely new business models, and topics such as data security and transparency have become central issues in society and industry, even though customer data is not used in the context of production. One way for the BMW Group to contribute to this in production is to bring data analysis as close as possible to the point of value creation. This has the distinct advantage that the production specialists can carry out their data analysis independently and derive appropriate recommendations for action.

The primary problem, however, often manifests itself in a lack of know-how in the field of data analytics. Production employees are considered absolute domain experts within the complex processes of highly flexible automotive production. One option here is to train these employees accordingly and qualify them for statistics and data analysis. Given the fact that a person can only learn limited subjects as a specialist, the BMW Group is pursuing a different approach: In addition to conveying a basic understanding of data analysis, the aim is to reduce the complexity of a data analytics application in such a way that it is as low as possible at the point of value creation, thus also reducing the training effort. Visual analytics in particular has dedicated itself to precisely this approach of so-called self-service analytics. This scientifically relatively young field is becoming increasingly relevant in research and practice. Visual analytics is the science that facilitates analytical thinking through interactive visual interfaces (Keim et al., 2008). The central aspects here are the appropriate visualisation of data and a correspondingly intuitive interaction by the user.

In the concrete application in production, the BMW Group focuses on quality work in addition to applications from maintenance. Due to the aforementioned complexity of a highly flexible production system, ensuring premium quality becomes a central challenge. In addition to error prevention through lean production methods, error detection and cause identification are of immense importance. This is especially valid in labour-intensive assembly, where a new customised vehicle is produced every minute for instance.

Figure 1 shows an example of a possible layout of an interactive dashboard. The simple presentation and interaction can significantly reduce the time required for the employee to find the cause. While the upper figure demonstrates the initial state,

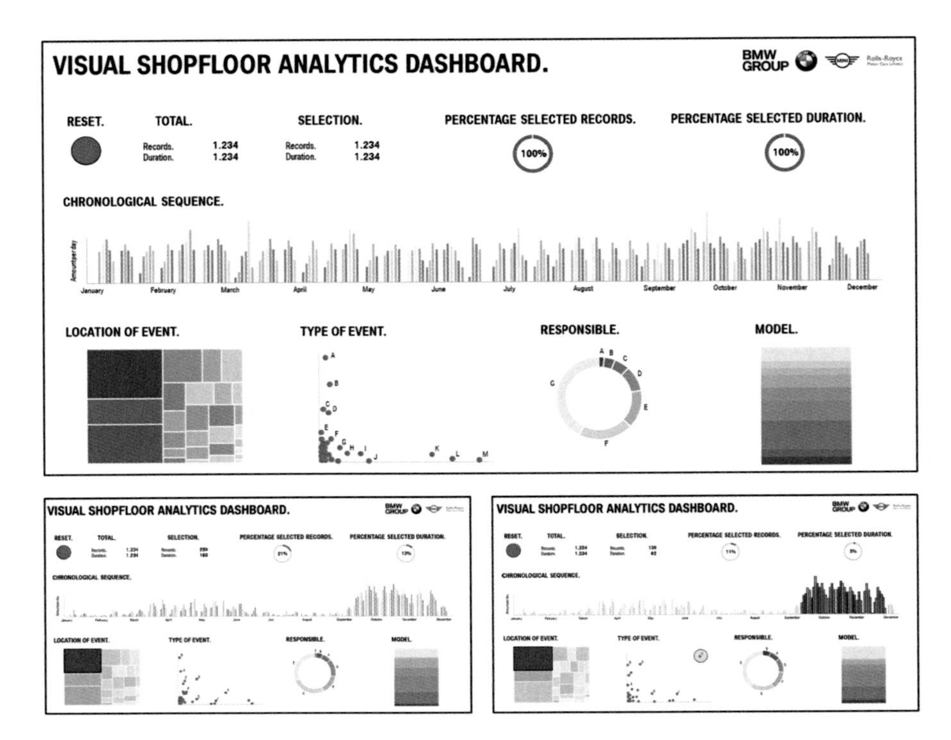

Fig. 1 Interactive dashboard for data analysis in self-service (fictitious numbers)

the figures from bottom left to bottom right show the respective iteration steps of the interaction. The production worker can click through the different views step by step and intuitively select the most critical areas and thus identify a root cause.

Various dimensions play a central role in quality work—especially in the case of assembly. For example, it is of interest which vehicle or model it concerns, when it stands where in the production line, which features a possible defect has or also where or on which component a defect has occurred. By interacting with the quality data, the relevant area for the production employee can be filtered within seconds. For example, the focus can be placed on a newly launched model, a special assembly or even a specific time window in order to efficiently analyse the development of the defect dimensions.

Artificial Intelligence: Deep Learning-Based Object Detection

Artificial intelligence offers an innovative technology for image processing. So-called deep learning (Goodfellow et al., 2016) relies on artificial neural networks to perform tasks such as classification, object detection or segmentation on photos (Yao et al., 2012). When using neural networks, a distinction must be made between the training phase and the productive phase—'inference' (Park et al., 2018).

A record of photos is taken during the training. Subsequently, all photos have to be labelled. The purpose of labelling is to provide the available photos with meta-information in the sense of a digital tag (Russell et al., 2008). Subsequently, the weights in the neural network are automatically modified over thousands of calculation steps so that the network detects the training data as ideally as possible. This creation of the adapted network is supported by frameworks such as Tensorflow, Digits or others (Marburg & Bigham, 2016). A large number of frameworks and architectures for neural networks are now available free of charge as open source. A very high level of dynamics can be observed, as international researchers and companies participate in the further development and publish their results online (object detection).

To prepare an AI for the live operation, the network is tested on unknown data. If this test proves a sufficiently robust result, the network can be used for productive operation. When taking the training photos, the data set must be designed in such a way that the variation of positioning, orientation, focusing, illumination, colour, reflection, etc. covers a spectrum that is as realistic as possible. These parameters therefore should be varied during data acquisition in such a way that they encompass the variations expected in later live operation. Depending on the difficulty of the recognition task, at least hundreds of photos per feature type are required as a basis for creating a robust neural network.

In practice, such neural networks prove to be particularly powerful: The robustness with regard to the conditions mentioned above is significantly higher than that of camera portals that rely on conventional image processing. The number of pseudo errors can be reduced to zero in numerous applications. For production, such robust solutions offer the advantage that they provide a precise quality statement and therefore do not tie up foremen. It also offers the option of automating control activities that today can only be performed manually. Such AI solutions allow an almost arbitrary variance of the parameters for photos.

Among others, the following visual inspections were realised in the production system of the BMW Group using AI:

- Object detection of the type plates: An AI checks if the type plate that has been assembled at the rear of the vehicle (e.g. '540i') matches the specific customer order 100%; see Fig. 2.
- Real-time tear detection in the press shop: With the help of a neural network formed metal parts are examined within milliseconds to find micro tears. It must be absolutely reliable to differentiate between cracks and non-critical features such as oil drops, metal splinters or dust particles (Fig. 2).
- Steering empty containers: In a logistics application, an AI decides whether empty containers consist of a single large case or several small boxes that have to be lashed down before transport.
- Checking the vehicle interior: In production, it must be ensured that certain objects such as first-aid kits or warning triangles are present in the vehicle. Here, an AI supports the corresponding inspection.
- Detection of small objects such as screws, caps or plugs.

Fig. 2 AI in the BMW production system—checking the type plates (left) (BMW Group, 2019b) and tear detection in the press shop (BMW Group, 2019a)

3 CSR in Visual Analytics and Artificial Intelligence

For the BMW Group, socially responsible action and sustainability play an essential role—in the product itself as well as in all corporate processes and the entire value chain. This chapter therefore uses the guiding questions outlined at the beginning to work out how corporate social responsibility is represented in the BMW production system. Data analytics is primarily applied in the BMW production system to optimise quality work and improve maintenance in such a way that unplanned machine downtimes are avoided. The main area of application in quality is the reduction of rework. In the following, an overview is given of the targeted measures that the BMW Group has initiated in the context of digital innovations to optimise quality work. The structure of the chapter is based on the guiding questions on CSR.

3.1 How Does the Use of Data Analytics and AI Change Corporate Responsibility?

The benefits of artificial intelligence in production manifest themselves, among other things, in the efficient analysis of (semi-)structured mass data and robust image processing. Thus, for example, quality inspections can be automated at a remarkably stable level. The immediate added value for the production system lies in a minimum rate of apparent defects with simultaneous detection of all components that are not in perfect order. Here, AI systems such as neural networks prove to be superior—compared to conventional software for image processing as well as humans. For direct production employees, this technological progress through AI means on the one hand that they can be relieved of reluctant control activities that previously had to be carried out manually. On the other hand, the elimination of pseudo defects results in more time for the foremen—they no longer have to search for products to inspect and delete defect entries from databases. This means they can concentrate on their actual tasks and, for example, support the direct workers on the assembly line. Overall, more accurate quality auditing can create an even higher awareness of quality, as false defects are no longer a distraction. A high false defect rate, for example, can typically result in negligence in final inspection, as a level of frustration begins to set in for quality specialists. One facet of responsible handling of technical innovations is the active integration of users. In the BMW Group's specific AI applications, feedback from users in production is obtained on a cyclical basis. This feedback is exclusively positive, as production employees are no longer entrusted with monotonous control tasks, but can instead take on value-added tasks. These include, for example, direct assembly activities, process improvement or support in the integration of new vehicles. In addition, the employees involved report increased motivation.

Careful selection of the application is elementary for the trustworthy use of artificial intelligence systems. It is crucial to select the use case in a way that it provides direct added value to the employees, which becomes immediately transparent to them. The primary maxim the BMW Group here is 'technology serves people'—AI must be easy to use and must not, under any circumstances, create a technical hurdle. AI needs to facilitate processes so that employees expand their toolbox to include the AI's method of checking quality. For AI to be applicable to production workers or other users whose field of employment is outside of mathematics or computer science, users must learn and understand the process and basic workings of AI. Any AI application must be designed to be simple. In the concrete application of object recognition, this means, on the one hand, the users from production must understand the training process for modelling a neural network—including the requirements that result from this for training photos and labels. On the other hand, an IT tool is needed that allows the employees to perform the labelling in a way that is intuitive for them. It is of eminent importance that every employee is enabled to use AI themselves, so that they can build and test the specific AI for their particular use case and decide for themselves whether to use it. The BMW Group has therefore designed its own labelling tool that is available to every employee. The software development process was deliberately designed to be agile. In the course of this, pilot users from production were permanently integrated into software development. With the help of their short-cycle feedback, a system was designed that is particularly easy to use. In this respect, the self-service idea for production employees is paramount. The logic is to consciously not turn production employees into data scientists through months of training, but to design the IT systems in such a way that those employees can operate them immediately without any training effort. In this manner, the technology supports all direct production employees, whereby they can set up their systems themselves. This approach of democratisation was omnipresent in the development of the IT system 'BMW Labelling Tool' that enables the training of an artificial intelligence. The entire technical complexity of optimising the neural network and mathematical operations is encapsulated in the IT system so that the interface to the user is kept particularly simple. In this way, truly arbitrary end users can initiate their specific AI for their manufacturing task. They only need the training photos and can annotate them with the help of the labelling tool.

Every technological innovation that is capable of automating processes must also be discussed in terms of the implications for jobs. The applications presented show that AI opens up new possibilities for automation. It is the responsibility of companies to use such solutions profitably and sustainably. This means that they must select use cases in such a way that AI adds process-related value and supports employees. In the BMW production system, technology serves people prevails. All potential application scenarios are categorised on the basis of this basic idea. In concrete terms, AI is used where it supports humans—either because it is more robust, has a lower error rate or relieves them of fatiguing, boring tasks. AI can thus contribute to realising a division of labour that is ideal for humans, with IT systems or machines taking over the repetitive, monotonous, physically demanding tasks while humans work on creative, value-creating processes. The approach is that only people see products through the eyes of the customer. Therefore, people will always work in processes that involve subtlety and finesse.

In addition to the availability of cost-efficient storage and multi-processor computers, the performance of AI frameworks and the large number of pre-trained neural networks can be seen as key enablers of why AI is now being widely used. The idea of open source takes a central position here: A large number of powerful frameworks for creating artificial intelligence can be found freely available on the Internet (Github, 2020). Competition between researchers and companies produces more powerful developments almost every week (object detection). For an IT specialist, the application of an AI is possible very quickly with the help of these ready-made solutions. The basic technology behind this is 'transfer learning', which is characterised by the fact that the user does not have to train a neural network from scratch, but rather adapts a pre-trained network with the help of a framework (Tan et al., 2018). This generally requires significantly less training data.

For the BMW Group, acting sustainably includes applying AI systems responsibly in production. It also means that it not only obtains, tests, uses or further develops open-source software components in order to provide BMW employees with optimal access to AI, but also that it publishes its own enhancements and makes them available to the global community: The BMW Group has published parts of the AI algorithms it has developed and uses in production on the online platform for open-source Github (Hemmerle, 2019). This makes it possible for software developers around the world to obtain and use these algorithms, which were developed by a team from the Innovation Department of Production and the Innovation Lab of the BMW Group IT. The content includes so-called programming interfaces, which serve to accelerate the training and inference of AI models, as well as the BMW Labelling Tool, which is particularly intuitive to use (Kamradt et al., 2020). The BMW Group has specifically published these software modules online in order to make them accessible to a broad mass of users and thus contribute to the open-source community and the democratisation of AI. At the same time, the large group of testers can highlight potential for improvement and contribute impulses for the further development of the algorithms. Independent of the software development at BMW, third parties can use this code and extend it individually. So, on the one hand, the aim is to give something back to the developer community and take responsibility. On the other hand, the BMW Group would like to use this step to encourage other companies to act just as responsibly and make innovative software publicly available. The BMW Group's vision is to simplify AI algorithms to the point where users without IT expertise are able to build their own AI—with just one click—in the same way that office software is installed today. If developers create and publish interfaces to other powerful AI frameworks, the application of AI will be simplified and accelerated even further in the near future.

3.2 How Does the BMW Group Deal with the Consequences and Possibilities of AI? How Are the Potential Risks Dealt with, and What Are the Possible Solutions?

Basically, the AI user no longer writes software, but the AI optimises itself. This has the advantage that neural networks are, for example, adaptable and work robustly under a wide range of conditions. The optimisation of a neural network during the training process usually proceeds over several hundred thousand computational steps. As a result, it is difficult or impossible to understand why an AI system behaves in a certain way. Therefore, the testing of an AI before productive use is particularly relevant. The BMW Group applies AI in the production system for applications of widely varying criticality. The more critical the planned application is, the more detailed the test must be before operation. For all quality aspects, the company has established its own standard of the evaluation index, 'BI' for short. This ranks quality-related features between BI 8 (excellent) and BI 1 (unacceptable), with three levels of criticality. In the training process, the AI tries to adapt as well as possible to the training data set. Given these circumstances, it is important to prepare an independent test data set to evaluate the performance of the AI on this unseen data. Before a productive application, it also conducts a so-called hot test to determine the performance under real conditions. A central tool in all these tests is the so-called confusion matrix (Patro & Patra, 2014). An essential requirement for the application of AI systems in the production systems is that no test slippage (false negatives) should occur. Secondly, a minimum number of pseudo errors are aimed for.

Only if this hot test proves successful will the AI be applied. Otherwise, additional training data must be acquired and the modelling phase must be run through again, or neural networks must be discarded completely for the specific application. In order to take the specifics of each production area into account, the BMW Group has defined a proficiency test in each instance, within the framework of which the neural network is checked. In addition to the structured approval and the application-adequate definition of test procedures, a cyclical check of AI solutions must be carried out in order to recognise possible negative trends. In parallel to these technical and organisational measures, the BMW Group is also working intensively on how to design innovative solutions to secure AI algorithms. In order to control the behaviour of AI systems as far as possible, it deliberately refrains from using the so-called reinforcement learning (Li, 2018) (in this method, a neural network is further modified in parallel with its operation). Once a trained neural network has been applied live in the BMW production system, it is accordingly not possible to change it autonomously. Consequently, every adaptation requires the active contribution of an employee—be it the integration of an additional feature or the training of a new product. Accordingly, the process of training and adapting AI systems is properly integrated into production planning. This ensures that the quality systems are up to date and stable for the start of production of a new product.

In general, artificial intelligence implies black-box systems to a certain extent: It is normally difficult to predict the precise behaviour of an AI during the training process or to transparently reconstruct the decision paths. The technical term of non-deterministic behaviour describes in this context that the same training data set can result in different neural networks under the same boundary conditions (Nagarajan et al., 2018). To make AI applicable to critical processes in particular, the BMW Group is also working on the explainability of AI systems. The goal is to expand the range of applications to include problems that have so far been handled exclusively by humans due to the degree of risk or an evaluation index that is too critical.

Regardless of the test procedures and the comprehensibility of AI, there are technical limits that play a significant role. Today, quality specialists in production are interested in an AI application that, for example, accurately performed the detection of the smallest anomalies such as damage on surfaces. Responsible action in this context also includes serious expectation management. An extremely sophisticated AI application entails corresponding risks: On the one hand, the technical feasibility is only given if an immense data set can be aggregated. This results in effort to examine and collect photos from existing systems, to take additional photos and to intentionally simulate errors. Subsequently, the labelling will be equally expansive. On the other hand, a particularly voluminous data set does not guarantee a 100% robust application. Currently, there is no theoretical framework for assessing the chances of success in such cases. Central to this is the competence to assess which potential use cases are feasible as well as the transparency towards all stakeholders to communicate this.

3.3 What Does AI Mean for the Company's (Global) Value Creation and Strategy and How Does It Change the Company's Social Responsibility?

The implementations of data analytics and artificial intelligence in the BMW production system show that these innovations provide significant added value in the optimisation of quality work. Some of the applications implemented are directly transferable to other areas in addition to the technologies introduced in the automotive industry. For one thing, there are similar challenges in data visualisation across value chains and industries. For another, the AI solutions are just as suitable in areas such as after-sales, incoming goods screening at suppliers as they are for leasing or rental returns. Data analytics and AI are therefore applicable along the entire value chain—on the manufacturer's side as well as with all suppliers. In particular, the meta-benefits such as the facilitation of quality work by means of visual analytics and the exceptionally robust quality auditing by means of deep learning can be applied in an identical manner regardless of the manufacturing stage or the degree of refinement of a product. Furthermore, the methods of artificial intelligence are suitable for more advanced applications in image processing such as semantic segmentation and the generation of artificial photos. In particular, the latter approach of synthesising photos and thus training data can be considered particularly promising for initiating AI systems more quickly and with less effort in the future. This advantage manifests itself in the fact that AI will soon also be used for planned applications that could not be realised until now due to the lack of options for recording training data. In particular, critical product features and small series should be mentioned here, where it is not possible to generate a sufficiently large number of training images or to simulate them using actual hardware setups. In addition, the automated analysis of mass data with different levels of structure offers a large field of application for AI. Thus, AI is predestined for modern questions of advanced analytics such as the analysis of texts, chatbots and pattern recognition—be it anomalies in transactions or conspicuous features in machine states.

Global value creation must also be analysed in the context of business intelligence and AI training software. All relevant and powerful frameworks for the creation of AI models come from the USA and China. Currently, there is no sufficiently competitive counterpart from Europe. Although a framework alone does not guarantee application-adequate AI, there is still a risk of dependency here. On the one hand, it is therefore relevant for the BMW Group to build up know-how on how to arrive at good AI systems in the specific context. On the other hand, all critical components of the BMW production system were analysed in a structured way in order to make a make-or-buy decision. The BMW Group considers it fundamental to have sovereignty over its data and to maintain this throughout all stages of the process of training, testing and applying AI. This was the reason why the labelling tool was designed and programmed internally. From a strategic point of view, it is crucial for the BMW Group to exclusively internally inspect and evaluate error images—simulated as well as real errors.

For the visual analytics applications, commercially available software packages were compared. Within this scope, a purchased solution from the global market leaders is used, whereby the implementation expertise is always in-house. This means that employees in the central innovation department create application-specific dashboards. To do this, they coordinate the design with the users on the shop floor and continuously develop it further. This also ensures that the application can be run and operated in the BMW Group's IT resources, so that the underlying quality data never leave the company. By means of a central data hub, all internal company data is available to employees for visualisation, subject to appropriate authorisation.

In order to shape research on data analytics and AI in a purposeful way, the BMW Group is an active partner in local and international research projects. For instance, the exemplary production process of milling is being used to investigate how data exchange between companies can be organised in partnership (Buhl, 2019). Furthermore, as a shareholder of the German Research Center for Artificial Intelligence, DFKI, the company actively shapes AI development in this area. In addition to opting for an internal intranet-of-things platform for production and the cross-company exchange of data and software modules, the BMW Group has clearly committed to initiating a European digital ecosystem and is a founding member of the Gaia-X (Koch et al., 2020) or Catena-X initiative.

3.4 Which Cooperation Is Required and How Are the Different Approaches to Responsibility and Sustainability Dealt with?

The multitude of powerful frameworks for AI and almost all modern network architectures are found as open source. On the one hand, this means great opportunities and low entry barriers for users in any industry; on the other hand, as of today, it requires specialists who are able to set up these solutions in such a way that they can be used by end users. Setting up hardware and software, including installing drivers and debugging, is currently non-trivial in most cases. In addition to the open-source offerings, IT providers have developed commercial AI products, which they preferably make available via the cloud. The large American software enterprises represent the market leaders here. These products are typically linked to a business model in which the respective cloud occupies a central position. Since these professional cloud applications can be used instantaneously as ready-to-use products, they are predestined for quick feasibility studies or a rough statement regarding the confidence or robustness achievable with a given data set, for example. As an end user, the costs per application are of elementary importance when deciding on the productive use of such commercial offers. In addition, the strategically relevant problem of lock-in must be taken into account—as soon as a dependency on a solution provider arises for productive systems, this can have an unfavourable effect on the price development. At the same time, there is no economic option to change providers due to high transaction costs. Furthermore, information protection plays an essential role. Regardless of the specific IT implementation or solution, intellectual property must be protected from access by third parties. This is particularly relevant if insights into manufacturing processes are granted or, for example, the quality achieved is recorded as part of the training or productive phase of an AI solution. In addition, the neural network, i.e. the AI model itself, must be encapsulated so that it cannot be published outside the company. As outlined above, the idea of open innovation is particularly important in AI, as frameworks and architectures for AI are available as open source. When publishing software, it is highly relevant to adequately protect intellectual property. Even in such a collaboration with a world-wide developer community, it is important to encapsulate company-specific, internal information and protect it from external access. For example, the neural networks themselves, which are used live in production, are not to be published.

In the BMW production system, it is not so crucial to select a specific IT product or the world's best framework for AI, but rather the information technology system integration. Here in particular, it is highly relevant that the integrator team has in-depth expertise in the areas of production, IT, artificial intelligence, software development and operation. The BMW Group takes appropriate care when selecting partners for the conception, realisation and series support of AI in the BMW production system. At the heart of these partnerships is the idea of designing AI in the production system end-to-end. The close cooperation with worldwide IT integrators has proven to be exceptionally successful, in which AI applications for production were strategically prioritised, designed and implemented. In practice, the end-to-end idea means that AI systems must be easy to operate for a broad mass of users, they must be robustly integrated into the landscape of existing IT systems via interfaces and they must be operated. Only in an atmosphere of mutual trust between customer and IT partner can solutions for quality work in production be realised.

In addition to software, the qualification of manufacturing employees, management and IT and planning staff of today and tomorrow will play an important role. The BMW Group is therefore actively engaged in strategic collaborations with research institutions worldwide in order to directly incorporate the latest findings on AI into business solutions. In addition, the BMW Group invests in the promotion of university partners such as the Technical University of Munich (Pighi, 2020), where the BMW Group is involved in lectures, for example. As another pillar of research and development, the BMW Group is active in funded research projects and supports numerous industry doctoral students in the field of data analytics and artificial intelligence.

3.5 What Challenges Do Data Analytics and Artificial Intelligence Pose for Managers at All Levels in Production?

It is central for managers to understand the basic principles of data analytics. This includes an overview of the capabilities, the rough technical functioning and the premises. To inform and train management across the board accordingly, the BMW Group has therefore designed an Analytics Education and Experience Centre and set it up inside the Research and Innovation Center in Munich. In structured half-day workshops, managers learn about the technical possibilities, the basic mathematical and information technology principles and explicitly the limits of the currently available analytics technologies. For production managers, it is crucial to find out in each case how the available digital technologies can be applied to their production area. All technologies along the value chain are considered—press shop, body shop, paint shop, assembly, logistics, engine and component production, purchasing and the local IT itself. This helps to identify useful fields of application and, at the same time, takes a holistic look at the implications of innovative technical solutions.

Other relevant contents taught in this centre are data accessibility, IT system integration and data quality. All IT solutions for analytics require the availability of the necessary data. In practice, however, this seemingly banal premise proves to be a challenge. Technical and organisational hurdles have to be overcome in order to obtain data in the originally recorded form. Another hurdle is typically to set up a live interface to existing IT systems in production so that up-to-date data is permanently available in real time. From an organisational perspective, the person responsible from production must release the data and agree to a connection. In this context, it is the task of IT to set up and operate the corresponding interface. System integration plays an important role here, because in so-called brownfield plants (Bracht et al., 2018), which have already existed for several product generations and are constantly being restructured, the IT landscape also has a certain history. This also applies to the corresponding data. With the help of visual analytics and concise dashboards, there is an option to also optimise this data quality based on visualisation (Liu et al., 2018). In order to structurally enable the essential topic of data availability, the BMW Group has made the strategic decision to establish a data hub for production data. The decision is accompanied by an investment in the corresponding platform as a central access point to data. Responsible handling of data also entails the precise definition of roles and rights. In addition, the data must be transformed so that it does not contain any data protection-relevant content. The Analytics Centre is scheduled to run for 18 months. The company's goal is to inform managers across departments in a timely manner.

In the field of artificial intelligence, the company's first step in the production system is to intentionally focus on object recognition in photos. This is also because a significant proportion of the entries in the quality data are due to spurious errors. This is where AI and visual analytics mesh seamlessly: Real-time dashboards empower quality experts to filter and evaluate the entries stored in the quality database. As a result of the pseudo errors found, a first goal of AI is to increase robustness and reduce bogus errors in photo-based quality inspections. In addition, the strategic goal of reducing rework requires the shortening of quality control loops—components must no longer be incorrectly installed, inspected and then reworked. In the future, only the correct components should be assembled. This is where AI-based object detection makes a decisive contribution.

For managers, it is of eminent importance to understand what AI systems are capable of doing, for example, by means of object detection in quality work. It is equally important to estimate the expected effort for a potential use case in order to plan resources, seriously calculate business cases and prioritise applications. For this purpose, the production system has developed its own method that production specialists and managers can use to analyse the feasibility of a potential AI application and estimate the effort involved. This tool is being continuously developed and expanded to include other areas such as the classification or segmentation of images.

In order to delight the top management level of the advantages of AI from the very beginning and to start the project in the department and IT, the MVP (minimum viable product (Duc & Abrahamsson, 2016)) idea was omnipresent: In the course of an agile development, simple prototypes were designed, realised, tested and demonstrated again and again in short cycles in order to sustainably establish the innovative AI solutions. In order to organise development and global roll-out efficiently, the AI innovations are organised in a chessboard logic. In this concept, exactly one manufacturing location takes the lead for a specific use case in terms of the innovation leader. This use case is pre-developed, validated and standardised in an agile team consisting of the business unit, which provides the product owner, and the development team from the Innovation Lab. This is followed by the worldwide roll-out to other locations. The stringent definition of a location for each application avoids redundant activities. Clear and transparent communication ensures that all activities remain synchronised across locations. When selecting the development location for each innovation, process maturity is decisive for the BMW Group. The basic principle in the BMW production system worldwide is 'lean before digital'—an innovation must serve to further improve lean processes; it must not lead to process turbulence or capacity bottlenecks in value creation. In addition, the timelines of upcoming product integrations and the expected financial benefits play a crucial role in the prioritisation and allocation process.

For the BMW Group, responsible handling of AI systems also means setting up and applying all systems in such a way that data protection for employees is ensured at all times. On the one hand, the selection of a location in production is always done in consultation with the social partners before the deployment of an application, i.e. going live. On the other hand, only those data that are recorded are actually needed and analysed. In concrete terms, this means that in assembly with comparatively long cycle times, only individual photos are recorded per vehicle—no permanent video stream. In addition, algorithms are being developed to ensure anonymisation. Algorithms are moreover being developed to ensure anonymisation. For example, in cooperation with the Innovation Department and the IT Innovation Lab, a powerful software package was designed and developed to crop people who may have been photographed out of photos or frames in real time and to make the corresponding pixels unrecognisable, while preserving the rest of the photo. For this purpose, an AI is once again used. The logic is absolutely identical to all other AI applications: The relevant objects that shall be anonymised in the course of this application are labelled (in the simplest case, the rectangular bounding boxes are used as labels; an even more sophisticated anonymisation along contours can be realised via semantic segmentation, which works pixel-precisely). In addition to the type of object recognition, this anonymisation solution offers the option of adapting the granularity and intensity of the anonymisation. In the BMW production system, this software tool is used intensively to anonymise persons in photos and thus ensure data protection. However, this application is designed in an open way—so every possible feature can be learnt in order to realise any use cases in which special objects are to be anonymised or censored. Since the BMW Group is fundamentally committed to the democratisation of AI, this AI solution for anonymisation was also published as open source in the form of ready-to-use programming interfaces, so that developers worldwide can download, test and further develop this software free of charge.

With the self-service approach introduced for all AI software and solutions in the BMW production system, the management puts the responsibility totally in the hands of the direct employees. This implies a high level of trust in terms of selection and robust implementation, including functional testing and operation. If the first AI application runs stably at the development site, the roll-out follows the motto 'copy with pride'. Digital innovations are applied as a supplement to the principles of lean production when employees experience appreciation for their previous process improvements and they feel the added value of the novel technologies for their daily tasks. Management must exemplify that digitalisation is not an objective in itself—an inefficient or wasteful process cannot ever be digitalised. The BMW Group's executives must trust the production staff to realise these innovations on their own responsibility.

4 Conclusion

Digital innovations such as visual analytics and artificial intelligence expand the set of tools available for optimising the BMW production system. The basis of process improvement on the shop floor is the lean design of manufacturing. Digitisation must at no time be seen as an end in itself or a universal solution. The innovations presented require a basic technical understanding on the side of the users and then offer considerable added value in quality work for the production employees and the company. Visual analytics brings about the paradigm shift—away from extensive printouts of highly aggregated key figures, towards clear dashboards that allow every production employee to focus on exactly those topics in real time and to analyse them in a data-based manner that are currently relevant for their area of responsibility. Saving paper does not only satisfy the aspect of sustainability. This approach of data analytics in self-service enables process improvements in complex production systems such as today's automotive production in the first place. The prerequisites are the (information) technical availability of and organisational access to data.

Artificial intelligence manifests itself in much more attractive production jobs. For example, methods such as deep-learning-based object detection are proving to be particularly robust for inspecting quality features. The fact that such AI systems function stably under practically any environmental conditions mean that solutions can be made more lightweight on the hardware side and thus more sustainable. In addition, AI shortens the quality control loops. Employees benefit directly from optimised work content—they do not have to check a number of discrete features on vehicles and enter the result in a database or even recheck the results of conventional image processing systems. Instead, they can henceforth concentrate on value-added manufacturing steps and process improvements. Since AI systems require an effort for the recording and labelling of the data in the course of training, the initial effort must be seriously estimated and an economic efficiency analysis must be carried out.

Strategically, the five core values are crucial for the BMW Group's culture—openness, appreciation, trust, transparency and responsibility. The core values apply to all spheres—including, of course, the innovative applications of visual analytics and artificial intelligence in the BMW production system. This concluding chapter summarises the concrete meaning of corporate social responsibility against the backdrop of these five core values.

The BMW Group dedicates itself extensively to informing its employees. Especially when it comes to the topics of data analytics and AI, it is crucial to instruct production employees, management and planning with regard to the advantages, technical principles and effects associated with these technologies. Openness means the open presentation of benefits and costs creates a realistic idea of performance. In the concrete applications in the BMW production system, this has induced significant demand from employees for these innovative solutions.

In the BMW production system, 'lean before digital' is valid. On the one hand, only a lean production system can provide the basis for digitalisation. On the other hand, new technologies will be implemented in the long term if production employees feel appreciated and their previous process improvements are rewarded according to the principles of lean production. The aspect of appreciation towards the employees also includes the user-friendly design and a particularly intuitive design of the systems.

To engage employees, it has proven essential to demonstrate the benefits of visual analytics and AI in large-scale communication campaigns via videos and live demonstrations. Furthermore, managers must also become seriously acquainted with the performance of the innovations. In addition, it is important to clearly communicate the necessary technical premises as well as current limits of the technology. Such honest expectation management creates trust in innovative technologies. In the BMW production system, the communication of the first successful AI realisations has triggered an immense demand from production employees for corresponding, innovative applications.

In all new applications, the human being occupies the role of integrator. For the BMW Group, the motto 'technology serves people' means that AI or technical systems never overrule people. For data analytics, this means that an algorithm supports the visualisation and makes suggestions. The final decision is made by the human being. For AI systems, this means that the AI does not take on a life of its own. In the BMW production system, reinforcement learning does not take place in such a way that the AI would develop autonomously in an uncontrolled manner. In this way, trust in the technology is maintained. New features are always integrated via the structured planning process, learnt and checked by humans. There will be no complete autonomy of machines—humans will remain the final decision-making authority.

With visual analytics, production employees are given the opportunity to analyse mass data themselves in a simple form. In the same way, managers have the opportunity to examine key figures and get to the root of possible causes of errors. An essential prerequisite for this form of visual analytics in self-service is the transparency of the corresponding data. This transparency with regard to the processes and especially the quality requires careful handling of the data.

The field of artificial intelligence in particular is experiencing enormous momentum. New algorithms and architectures are available as open source virtually free of charge. For the BMW Group, as a user of open-source software, responsible behaviour also includes making a contribution to this worldwide community of developers. The BMW Group is committed to sustainability and publishes its developed programming interfaces so that popular frameworks can be used even more easily. In this way, the BMW Group would like to encourage other companies to make similar contributions and contribute to the democratisation of AI. In addition, this movement can motivate developers from science and industry to endow their algorithms with programming interfaces that are comparably convenient.

For the BMW Group, responsible use of AI means defining problem-adequate test procedures to safeguard systems for live operation in production, which are characterised by their not completely deterministic behaviour. When selecting partners, their use of AI in particular plays a major role. Thus, it is crucial how they develop and secure artificial intelligence. Even more important is how partners offer AI and for what purposes they use it. The BMW Group is aware of its social responsibility. It is therefore actively involved in research projects and initiatives of German and European relevance. In addition, it enters into targeted cooperation with universities and research institutions.

Analytics and artificial intelligence only work together with humans. Therefore, the appealing design of IT systems and joint process integration are given the highest relevance. In practice, the agile refinement of information technology solutions in close coordination between production, IT and a select group of first-time users proves to be useful.

Technology serves people. Only when production employees see the added value for their processes and the positive contribution to their individual jobs will these digital innovations find their way into sustainable realisation.

References

Albertson, M. (2018). Software, not hardware, will catapult big data into a $103B business by 2027. Edited by SiliconANGLE Media Inc. Online available https://siliconangle.com/2018/03/09/big-data-market-hit-103b-2027-services-key-say-analysts-bigdatasv/, zuletzt geprüft am 30.06.2020.

BMW Group. (2019a). Fast, efficient, reliable: Artificial Intelligence in BMW Group Production. Online available https://www.press.bmwgroup.com/global/article/detail/T0298650EN/fast-efficient-reliable:-artificial-intelligence-in-bmw-group-production?language=en, last checked 24.08.2021.

BMW Group. (2019b). Artificial Intelligence Identifies Model Designations at BMW Group Plant Dingolfing. Video. Online available https://www.youtube.com/watch?v=V6MmMYjXScI, last checked on 24.08.2021.

Bracht, U., Geckler, D., & Wenzel, S. (2018). *Digitale Fabrik. Methoden und Praxisbeispiele. 2., aktualisierte und erweiterte Auflage*. Pringer Vieweg (VDI-Buch).

Brunner, F. J. (2017). *Japanische Erfolgskonzepte. KAIZEN, KVP, Lean Production Management, Total Productive Maintenance, Shopfloor Management, Toyota Production System, GD3—Lean Development* (4., überarbeitete Auflage). Hanser (Praxisreihe Qualitätswissen).

Buhl, U. (2019, June 14). *Forschung für die digitale Zukunft. 81 VHB-Jahrestagung—Session Digitale Transformation*. Verband der Hochschullehrer für Betriebswirtschaft e.V. (VHB).

Deckert, F., Krenkel, P., & Mielke, T. (2015). Null-Fehler-Prinzip. In U. Dombrowski & T. Mielke (Eds.), *Ganzheitliche Produktionssysteme. Aktueller Stand und zukünftige Entwicklungen* (pp. 81–95). Springer (VDI-Buch).

Demant, C., Streicher-Abel, B., & Springhoff, A. (2011). *Industrielle Bildverarbeitung. Wie optische Qualitätskontrolle wirklich funktioniert* (3rd ed.). Springer.

Deuse, J., Schmitt, J., Stolpe, M., Wiegand, M., & Morik, K. (2017). Qualitätsprognosen zur Engpassentlastung in der Injektorfertigung unter Einsatz von Data Mining. In N. Gronau (Ed.), *Industrial Internet of Things in der Arbeits- und Betriebsorganisation* (pp. 47–61). GITO Verlag (Schriftenreihe der Wissenschaftlichen Gesellschaft für Arbeits- und Betriebsorganisation (WGAB) e.V).

Dombrowski, U., Krenkel, P., & Mielke, T. (2015). Struktur Ganzheitlicher Produktionssysteme. In U. Dombrowski & T. Mielke (Eds.), *Ganzheitliche Produktionssysteme. Aktueller Stand und zukünftige Entwicklungen* (pp. 26–31). Springer Vieweg (VDI-Buch).

Duc, A. N., & Abrahamsson, P. (2016, May 24–27). Minimum viable product or multiple facet product? The Role OF MVP in software startups. In H. Sharp & T. Hall (Eds.), *Agile processes in software engineering and extreme programming. 17th International Conference, XP 2016. International Conference XP 2016. Edinburgh, UK* (pp. 118–130). Springer Open.

Ehrlenspiel, K., Kiewert, A., Lindemann, U., & Mörtl, M. (2014). Einflüsse auf die Herstellkosten und Maßnahmen zur Kostensenkung. In K. Ehrlenspiel, A. Kiewert, U. Lindemann & M. Mörtl (Eds.), *Kostengünstig Entwickeln und Konstruieren* (pp. 165–415). Springer.

Github, Inc. (Ed.) (2020). #Deep learning. Microsoft. Online verfügbar unter https://github.com/topics/deep-learning, zuletzt aktualisiert am 13.06.2020.

Goodfellow, I., Bengio, Y., & Courville, A. (2016). *Deep learning*. MIT Press.

Grüneisl, M. (2018). Smart Data Analytics und Künstliche Intelligenz. Produktionssystem der BMW Group: Wirksame Anwendungen helfen, Komplexität besser zu beherrschen. In Euroforum (Ed.), *Die Zukunft der Industrie. Digital Smart Vernetzt. Handelsblatt Journal* (pp. 15–16). Handelsblatt Media Group.

Hemmerle, A. (2019). BMW Group shares AI algorithms used in production. Edited by BMW Group. Online available https://www.press.bmwgroup.com/global/article/detail/T0303588EN/bmw-group-shares-ai-algorithms-used-in-production?language=en, last checked on 24.08.2021.

Jochem, R. (2010). Was versteht man unter Wirtschaftlichkeit von Qualität? In R. Jochem (Ed.), *Was kostet Qualität?* (pp. 27–54). Carl Hanser Verlag GmbH & Co. KG.

Kamradt, M., El Achkar, C., Saller, E., Alkazzi, J. -M., Nassif, J., Sleiman, J., et al. (2020). BMW Innovation Lab. Repositories. BMW Group. Online verfügbar unter https://github.com/BMW-InnovationLab/, last updated on 07.07.2020, zuletzt geprüft am 10.07.2020.

Keim, D., Andrienko, G., Fekete, J. -D., Görg, C., Kohlhammer, J., & Melançon, G. (2008). Visual analytics: Definition, process, and challenges. In A. Kerren, J. T. Stasko, J. -D. Fekete, & C. North (Eds.), *Information visualization. Human-centered issues and perspectives* (pp. 154–175). Springer (Lecture notes in computer science, 4950).

Knaut, A. (2017). Corporate Social Responsibility verpasst die Digitalisierung. In A. Hildebrandt & W. Landhäußer (Eds.), *CSR und Digitalisierung. Der digitale Wandel als Chance und Herausforderung für Wirtschaft und Gesellschaft* (pp. 51–59). Springer Gabler (Management-Reihe corporate social responsibility).

Koch, M., Hanke, T., & Kerkmann, C. (2020). *"Moonshot" Gaia-X—Die wichtigsten Fragen und Antworten zur europäischen Cloud*. Edited by Handelsblatt Media Group. Online available https://www.handelsblatt.com/politik/deutschland/datenprojekt-moonshot-gaia-x-die-wichtigsten-fragen-und-antworten-zur-europaeischen-cloud/25888416.html?ticket=ST-4149249-VvyGbqdLdUeYfSIDc5gJ-ap5, last updated on 04.06.2020, last checked on 07.07.2020.

Li, Y. (2018). Deep reinforcement learning. Online verfügbar unter https://arxiv.org/pdf/1810.06339.pdf, last checked on 30.03.2020.

Liu, S., Andrienko, G., Wu, Y., Cao, N., Jiang, L., Shi, C., Wang, Y.-S., & Hong, S. (2018). Steering data quality with Visual Analytics: The complexity challenge. *Visual Informatics, 2*(4), 191–197. https://doi.org/10.1016/j.visinf.2018.12.001

Majohr, D. (2008). Optimierung von Vorbehandlungsanlagen in der Automobilindustrie. Universität Rostock. Online available http://rosdok.uni-rostock.de/resolve/id/rosdok_disshab_0000000103.

Marburg, A., & Bigham, K. (2016, September 19–23). Deep learning for benthic fauna identification. In: IEEE (Ed.), *OCEANS 2016. Monterey, CA, USA. Institute of Electrical and Electronics Engineers; Marine Technology Society; Oceanic Engineering Society* (pp. 1–5). IEEE.

Nagarajan, P., Warnell, G., & Stone, P. (2018, July 14). The impact of nondeterminism on reproducibility in deep reinforcement learning. In *2nd Reproducibility in machine learning workshop at ICML 2018. Stockholm, Sweden*. Online verfügbar unter https://arxiv.org/abs/1809.05676.

Object Detection. Online available https://paperswithcode.com/task/object-detection, last checked on 20.06.2020.

Oeltjenbruns, H. (2000). Organisation der Produktion nach dem Vorbild Toyotas. Analyse, Vorteile und detaillierte Voraussetzungen sowie die Vorgehensweise zur erfolgreichen Einführung am Beispiel eines globalen Automobilkonzerns. Zugl.: Clausthal, Techn. Univ., Diss. Shaker (Innovationen der Fabrikplanung und -organisation, 3).

Park, E., Yoo, S., & Vajda, P. (2018, September 8–14). Value-aware quantization for training and inference of neural networks. In: V. Ferrari, M. Hebert, C. Sminchisescu, & Y. Weiss (Eds.), *Computer vision—ECCV 2018. 15th European conference proceedings, part I. The European conference on computer vision (ECCV). Munich* (pp. 580–595). Springer (Image processing, computer vision, pattern recognition, and graphics, 11205).

Patro, V. M., & Patra, M. R. (2014). Augmenting weighted average with confusion matrix to enhance classification accuracy. *Transactions on Machine Learning and Artificial Intelligence—TMLAI, 2*(4), 77–91. https://doi.org/10.14738/tmlai.24.328.

Pighi, M. (2020). *BMW Group fördert Digitalisierung von Studiengängen an der TU München mit einer Mio. Euro*. Edited by BMW Group. Online available https://www.press.bmwgroup.com/deutschland/article/detail/T0307747DE/bmw-group-foerdert-digitalisierung-von-studiengaengen-an-der-tu-muenchen-mit-einer-mio-euro/, last checked on 10.07.2020.

Regber, H., & Zimmermann, K. (2009). *Change-Management in der Produktion. Prozesse effizient verbessern im Team* (1. Aufl.). mi-Wirtschaftsbuch.

Richter, T., & Rico-Castillo, M. (2015). Standardisierung. In U. Dombrowski (Ed.), *Lean Development. Aktueller Stand und zukünftige Entwicklungen* (pp. 46–59). Springer (VDI-Buch).

Russell, B. C., Torralba, A., Murphy, K. P., & Freeman, W. T. (2008). LabelMe: A database and web-based tool for image annotation. *International Journal of Computer Vision, 77*(1–3), 157–173. https://doi.org/10.1007/s11263-007-0090-8

Tan, C., Sun, F., Kong, T., Zhang, W., Yang, C., & Liu, C. (2018, October 4–7). A survey on deep transfer learning. In V. Kůrková, Y. Manolopoulos, B. Hammer, L. Iliadis, & I. Maglogiannis (Eds.), *Artificial neural networks and machine learning—ICANN 2018. 27th international conference on artificial neural networks, proceedings, part III. ICANN 2018. Rhodes, Greece* (pp. 270–279). Springer (Theoretical computer science and general issues, 11141).

Töpfer, A., & Günther, S. (2007). Steigerung des Unternehmenswertes durch Null-Fehler-Qualität als strategisches Ziel: Überblick und Einordnung der Beiträge. In A. Töpfer (Ed.), *Six Sigma. Konzeption und Erfolgsbeispiele für praktizierte Null-Fehler-Qualität. 4th, aktualisierte und erw. Aufl* (pp. 3–40). Springer.

Ware, C. (2004). *Information visualization. perception for design* (2nd ed.). Elsevier/Morgan Kaufman (Morgan Kaufmann series in interactive technologies, 22.

Yao, J., Fidler, S., & Urtasun, R. (2012, June 16–21). Describing the scene as a whole: Joint object detection, scene classification and semantic segmentation. In: IEEE Computer Society (Ed.), *IEEE conference on computer vision and pattern recognition (CVPR), 2012. CVPR 2012. Providence, Rhode Island* (pp. 702–709). IEEE.

Zimmermann, K. (2017). Digitalisierung der Produktion durch Industrie 4.0 und ihr Einfluss auf das Arbeiten von morgen. In B. Spieß & N. Fabisch (Eds.), *CSR und neue Arbeitswelten. Perspektivwechsel in Zeiten von Nachhaltigkeit, Digitalisierung und Industrie 4.0, Bd. 84* (pp. 53–72). Springer Gabler (Management-Reihe corporate social responsibility).

Zipse, O. (2018, March 13). Zeitenwende in der Produktion: Wie Systemintegration zum entscheidenden Wettbewerbsfaktor wird. Münchener Management Kolloquium. TCW Transfer-Centrum; TU München.

Matthias Schindler has been 'Head of AI Innovation' in the BMW production system since 2018. Since 2017, he has been in charge of the innovation cluster Data Analytics in the department 'Innovation, Digitalisation, Data Analytics'. In addition to artificial intelligence, the focus lies on data analysis in quality work and predictive maintenance. Until 2016, he worked in technical planning at the BMW Group: In the 'Industry 4.0' department, he led the innovation field 3D digitisation for integration planning in brownfield factories. There he wrote his PhD thesis, embedded in a research project on cyber-physical production systems. Schindler completed his Master of Science in Mechanical Engineering and Management at the Technical University of Munich (TUM) in 2013.

Frederik Schmihing has been working in the BMW Group's 'Innovation, Digitalisation, Data Analytics' department since 2017. After completing a dual degree in industrial engineering, he wrote his master's thesis in this department as part of his Master of Business Administration and Engineering at the Rosenheim University of Applied Sciences. Based on this, which aimed to localise critical areas in the BMW production system, a doctoral project was initiated in collaboration with the Technical University of Berlin. The aim of this project is to develop a concept that enables production employees to carry out data analysis for quality improvement directly at the point of value creation with the help of visual analytics.

Sustainability and Artificial Intelligence in the Context of a Corporate Startup Program

Frank Barz, Hans Elstner, and Benedict Ilg

1 TechBoost, a Startup Program Designed to Drive Sustainability Through Innovation in an B2B Environment

"If we do not work with startups we will have no future" is the main goal behind this 2017 developed startup program of Deutsche Telekom.

Today, it supports over 800 startups in the area of big data, cloud, and artificial intelligence. All startups get free cloud vouchers between 15 k€ and 100 k€ for 1 year on Deutsche Telekom's own public cloud infrastructure, the Open Telekom Cloud, using 100% renewable energy.

During this time, a partnership is developed where Deutsche Telekom and startups together support sustainable project at Deutsche Telekom's business customers.

The main business drivers are the CO_2 reduction, supply chain transparency, and ESG reporting, although the scope grows daily from collaboration tools, over OCR software, to event platforms enabling customer to conduct positive impact on environment and society.

The success model behind this program is matchmaking between corporates and startups to reach a competitive advantage and support a mindset change in the way projects are conducted to reach the sustainable development goals of corporates.

F. Barz (✉)
Telekom TechBoost, Essen, Germany
e-mail: Frank.Barz@telekom.de

H. Elstner
rooom.com, Jena, Germany

B. Ilg
Flip, Stuttgart, Germany

R. Schmidpeter, R. Altenburger (eds.), *Responsible Artificial Intelligence*, CSR, Sustainability, Ethics & Governance, https://doi.org/10.1007/978-3-031-09245-9_8

On the one side, an own KI platform (TechBoost Connect) is used to enable customers to bring in their business requirements for sustainable solutions, and on the other side, startups with the required competencies are matched, and the TechBoost team supports the entire project from selection of the "best fit" to the successful implementation of every single project. A team of coaches and project managers ensure that both sides (startups and corporates) will benefit from this sustainable project using a KPI-driven approach. Sustainability is driven by innovation, and therefore, it is already "best practice" that startups are bundling their competencies to generate the best solution for the customer's need on sustainability. Customers already have acknowledged that they can only solve their scope two to three sustainable goals with corporation and collaboration with startups.

The challenge of today's sustainable innovation is that the success stories on sustainable project are not shared within the economy and different industries. Using the platform free of charge, the business customer and the startup have to agree to share their success or failure story into the TechBoost community.

Sustainable startups solve mostly problems of climate change, poverty, and inequity. How big data and artificial intelligence are used to reach the sustainable development goals of Deutsche Telekom's business customers is explained in two examples:

Flip, a collaboration app, mostly used by non-desktop workers in the logistic, production, health, and retail segment
rooom, the enterprise metaverse solution

2 Flip App: Sustainability in Collaboration Using a Messenger App

Eighty percent of the world's population works in deskless areas. Yet only 1% of all software solutions are developed to meet the needs of the deskless workforce. Especially in internal communications, a profound communication gap has developed over the years. Flip, the employee app empowering deskless work, provides a remedy. With Flip, for the very first time, deskless workforce in retail, healthcare, and production can take part in the internal communication process of their companies simply and intuitively.

Perfectly tailored to the needs of deskless workforce, Flip combines top-down and bottom-up communication tools to offer added value for all users: The digital bulletin board, the so-called newsfeed, displays all relevant information for the user. The single and group chats allow employees to communicate across departments and locations. Other helpful functions are a survey tool, task management, and an automatic translation function in 26 languages.

In addition, there is the possibility of integrating the existing HR and IT infrastructure via intelligent interfaces, and the employee app also adapts visually to the customer's corporate design. The employee app was developed together with work councils, as the issue of data security is a top priority in many companies. In contrast to private messengers, this means that, for example, the user's private mobile phone number is not readout.

Apart from all the functional benefits of the app, Flip empowers the inclusion of deskless workforce. This allows companies to achieve considerable savings potential and fundamental competitive advantages. Deskless workers have decades of accumulated knowledge. Making this knowledge available and giving all employees a voice in the process can become a key driver for the company's success. In addition, companies can positively influence their sustainability in various areas using the employee app.

Flip ensures a sustainable company-wide communication concept. By bundling different systems in one app, the entire company is relieved of the complexity of internal communications. The transfer of information is accelerated, communication channels are shortened, approval processes are streamlined, and working time is freed up for productive activities. In many large companies, the transfer of information is associated with many stumbling blocks and complex approval processes. With Flip, it is possible to establish a direct communication channel between the management level and the workforce, which can be decisive for the company's success in the long term. At the same time, the use of the employee app enables direct feedback from customers to be collected and passed on to the head office without delay. In addition to internal communication, this can also have a lasting positive effect on the customer experience. Ultimately, a sustainable communication concept is a key indicator for the sustainable growth of the company.

Through this sustainable communication concept, further factors can be influenced that have a direct impact on the eco-balance of the company. By digitizing all communication, Flip enables huge savings in terms of paper costs.

Especially in production, retail, and healthcare, notices are printed out several times a day to be posted on the bulletin boards. Other information such as the shift plans, meal plans, or holiday requests is mostly printed on paper, too. Most of the printouts will not be recycled and therefore harm the company's eco-balance. How high the actual savings potential and the influence on sustainability can be is shown in the example calculation based on a retail company with 1000 employees and 80% deskless workforce. This company could save €15,000 per month in paper and printing costs and save about 30 trees from deforestation by using an employee app.

Another advantage of having a digital communication channel that enables cross-location and cross-country exchange is the savings in travel expenses. Business trips have a strong impact on the corporate sustainability balance. Still, business travel is one of the biggest contributors to emissions left by the business world. Often companies are not even aware of how big the environmental footprint left by unnecessary traveling is. Especially when cross-location communication in the company cannot yet be managed with the help of a digital communication channel, employees often cover several hundred kilometers by car and air to ensure a cross-functional communication exchange. Ensuring a simple and reliable way of cross-location communication for all employees, including deskless workforce, travel expenses can be reduced drastically and leave a positive ecological footprint.

In this way, companies can benefit from potential savings on the one hand and make a positive contribution to the sustainability of the entire business world on the other.

However, sustainability can also be considered in the context of employee retention and satisfaction. Companies are currently in a highly competitive war for talents. Especially for the next generation, the sustainability of the employer is often a decisive criterion. A tool like Flip thus not only ensures that communication takes place in a way that is native to the new generations but also fulfills the desire for sustainability at the same time. Moreover, while many companies strive for sustainable growth in success, they do not consider that sustainability in employee retention is a core factor for this. Hiring and firing often involve high costs and a tremendous amount of work for both employer and employees. The goal of companies should be to retain good employees in the long term and thus sustainably.

In addition to the many sustainability effects that result for companies that use Flip, the company itself attaches great importance to leaving a sustainable ecological footprint. The company uses the employee app for overall communication to benefit from a sustainable communication concept within the company. Business trips that are not relevant are avoided, and, if a trip is still necessary, the most sustainable method of travel is chosen. Flip also relies on sustainably grown coffee, locally produced office furniture, and a balanced and sustainable working environment. One big part of the sustainable working environment is the paperless office policy. With the help of different digital platforms, it is thus possible to do without printing out information and forms. At the same time, the use of these digital platforms influences the way of working. This makes location-independent mobile working possible. The possibility of working remotely from any location with the help of these platforms reduces commuting and the associated known disadvantages for the environment. All these factors are intended to create a general sustainable working environment.

2.1 How Can the Flip App Drive Sustainability with Digitization and Artificial Intelligence

Increasing digitization also means that many resources, especially energy, are used more frequently. This change in the world of work should nevertheless always take place in consideration of the sustainability line. Therefore, the technical background of the applications that are used is an essential factor. One way to protect resources is to cut back to the essentials. This means, for example, that all of a company's communication takes place exclusively via one technical stable channel. However, this channel must be understandable for all employee groups. To ensure this, Flip relies on simplicity. The entire technical structure of the app is designed to create a simple and intuitive user experience. The users of the app are mostly not very tech-savvy, have never written a professional email, and have not conducted a video conference. That is why technically highly complex systems are often not understandable for them.

To provide a simple solution that is understandable for everyone in the company, simplicity has to start with the development of the app. Through a special backend and frontend development, the app can be downloaded within seconds.

This makes the app much easier to access for all workforce groups. On the basis of this advanced technology, it is thus possible not only to switch from analog to digital communication and to support the sustainability of the entire company but also to enable a bit of flexibility and self-determination for deskless workforce. Strong algorithms and a solid technical base of apps and tools are fundamental criteria to ensure that the use of these tools empowers a sustainable new way of working.

2.2 *What Kind of Ethical Principles Has Flip Adapted into Their Software Development*

With regard to business ethics, the cultural fit of all employees represents the base of the underlying moral values, on which important topics such as sustainability or data protection are built. Therefore, cultural fit plays a decisive role, starting with the recruitment process and extending to the daily work of employees, and is part of sustainable HR management.

It is important to Flip GmbH that employees are the most important resource not only in the Flip app but also in the company itself. To make this possible, Flip's corporate values represent the core of its sustainable human resources management and cultural fit. In this context, the values "Bold," "Authentic," "Responsible," and "Focused" are the guiding principles that allow every employee to grow and work together toward the company's vision.

IT professionals at Flip can grow sustainably with Flip's success and to develop personally and professionally. Because a job should no longer just mean working but should also represent an environment, where participating directly and sustainably in the company's success is possible, a motivated and functioning frontend and backend team is essential to develop an employee app that meets Flip's and customers' standards. In addition, an open feedback culture allows the IT specialists to learn from each other, be bold in trying out new approaches, and encourage a cross-departmental exchange of knowledge.

Internal communication activities should be as efficient as possible and not cause unnecessary extra work. In addition, internal information should always remain internal. That's why data protection plays an important role—also when it comes to choosing the right communication tool. In general, the security of sensitive data is becoming increasingly important; many companies still communicate with private messengers in business. However, these do not meet the data security requirements and represent a potential security risk when communicating with sensitive data.

Therefore, Flip data protection is an essential part of their business ethics. The app handles all data GDPR compliant and offers the highest security standard for sensitive data. Thus, a secure way for internal company communication is created. To always ensure data security, Flip's IT employees are trained to develop a stable and lasting system that can grow sustainably and adapt to each customer's needs.

In addition, Flip is a BYOD—bring your own device—service. The app is designed to enable employees to be reached securely via their private smartphone. Employers thus do not have to worry about internal information being compromised. This guarantees that the app has the highest security standard in the long term.

2.3 How Does the Partnership with a Corporate Supports the Sustainability Strategy of Flip

With Telekom TechBoost, Flip has found a reliable partner who supports the company with a suitable network and profound expertise. The collaboration allows Flip to participate in major Telekom projects and events such as Digital X, make new contacts, and expand its network.

Through Telekom TechBoost's profound expertise, Flip is supported widely in customer inquiries and stands by as a partner in decision-making. This also includes strategic support—whether on marketing strategy or sales channels—where Flip always has the right contact person for every request.

Apart from the expertise of the Telekom TechBoost team, the partnership focuses on sustainability. Flip aims for long-term cooperation with Telekom TechBoost, which is strengthened by an appropriate network and strengthens the joint success in the long term. In addition, Telekom TechBoost also focuses on sustainability in its actions. An important part of its long-term sustainability strategy is the aspiration to green IT and to make digitization sustainable with a green cloud. This plan ensures transparency in sustainability and links up with Flip's commitment to making its app fully sustainable. The joint path with Telekom TechBoost enables Flip to grow sustainably with a reliable and long-term partner.

2.4 Future Developments at Flip App

In the past, high absence rates and considerable turnover, in addition to the fact that there was no channel to the workforce, were a major problem for the management of big companies. Nevertheless, the answer to this problem was obvious: Deskless workforce was not heard, was unable to speak up, and did not have the right tools to communicate. This often resulted in an enormous loss of potential.

Today's managers know that non-desk workers are crucial to a company's competitiveness and that they need to be involved in the continuous improvement process. Deskless workers are often the first point of contact for customers and determine the efficiency of production. So, they are involved in every step of the value chain, as well as always being part of the customer experience. Therefore, it is essential to include deskless employees in the internal communication process. Many partners and business angels of Flip that mostly operated in the management of global companies are therefore familiar with the difficulty of reaching deskless employees. Thus, they understand the need for a unified communications solution that provides equal access to information for all employees. The potential of it is enormous. With a mobile, digital solution, workforce is always informed; it offers practical help for everyday work, provides a comprehensive communication tool, increases efficiency, and at the same time supports the sustainability of the company. Moreover, only by continuously involving all employees is it possible to build a corporate culture which supports a sustainable employee engagement.

This conveys a consistent image to customers, applicants, and other interested parties and reduces employee turnover. Thus, it has a direct impact on the costs incurred, and HR costs are reduced. It also ensures that internal knowledge is retained. With a digital solution, knowledge transfer can even be made easier and more comprehensive—regardless of location, workplace, and department.

Considering that 80% of the world's population are deskless employees, it becomes clear how far-reaching a change in communication with these employees can be.

A big part of the employee app provider's vision is changing communication, but it goes far beyond that. Communication is a key factor for successful digitization, the chance for every employee to have a voice.

The future of work may have changed for the office workforce, but it is still the same for the deskless workforce. This is precisely why the company's vision is to empower every deskless employee by providing a communication solution that meets the needs of this target group.

It's obvious that the shift from bulletin board to digital employee app is a transformational one. For some employees, it can seem threatening at first—whether due to a lack of technical know-how or language barriers. By listening to the users, staying in close contact with customers, and being able to meet individual requests, most of these fears can be eliminated. Developing the app together with customers helps to understand how fast the evolvement of communication actually is in the companies.

Just like sustainability, communication is an international problem. With a full 80% of the world's population working operationally, it is more than time to address the issue and provide a solution.

3 rooom.com: How the Metaverse Is Driving Sustainability with Digitization and AI

rooom is a globally operating startup headquartered in Jena, an epicenter of innovation, and has several offices in Germany and the United States. The company is known for its Experience Platform which is built for creating, managing, and promoting engaging 3D, AR, and VR experiences. The software is web-based, runs on all desktop and mobile devices, and has become an essential tool for creating corporate brand experiences in the metaverse.

The platform company was founded in 2016. Until now, it was possible to well-known customers and partners, such as the Berlin Tradefair, PWC, and Deutsche Telekom. With a big set of features, it is possible for all customers to bring people together through the digital products such as rooomEvents[1], rooomProducts[2], rooomSpaces[3], and more.

With all these customers and partners, rooom had the possibility to continuously expand and refine its solutions. Further the platform offers intelligent solutions for the data-reduced representation of three-dimensional objects and spaces, the reduction of data, and the long-term use of these. We are now giving an insight into the platform, how it uses modern technology and brings it in line with sustainable and ethical principles and thereby makes an important contribution to sustainable digitalization. Before that, however, we go into the general corporate culture and where these principles are lived.

Digitization is an important factor for sustainability, and rooom is guided by this. As a digital company, rooom relies as far as possible on the use of digital alternatives to paper. As a partner of the "Schutzgemeinschaft Deutscher Wald," the startup also support the planting of new trees in Thuringia. In addition, rooom AG employees have the option of working in a home office and are provided with e-scooters or company bicycles, which can help reduce emissions from commuting. Employees stay connected in different ways through all our offices. As a board member of the sustainability platform "aware," rooom is doing everything we can to work together with other companies toward a more sustainable and digital society. Above all, however, the rooom platform itself offers great potential for a more sustainable business world. In terms of resource conservation (travel cost, paper, and time savings as well as reduced returns in online retail incl. reduced use of

[1]A flexible and comprehensive solution that allows customers to plan and implement all their internal and external events simply quickly on the web.

[2]Product manufacturers and sellers can present their products online in 3D. Customers can experience the 3D models without an app directly via the browser from all angles in 3D. It allows to change colors and surface textures through a configurator and can be implemented into different webshops. Models can be created from existing 3D data, from photos or videos, or by using the rooom 3D scan app.

[3]Companies can present their services in interactive online 3D showrooms, including 360-degree tours based on scans. Customers can explore rooms, halls, or other scenes using a mobile or desktop device and can collaborate through avatars and voice chat.

materials) and the use of hybrid events, we offer forward-looking products internationally.

3.1 How the rooom Software Supports Sustainable Principles

To clarify the ways in which rooom's platform drives sustainability, it is important to briefly review the meaning. Sustainability is based on three pillars; we should keep in mind as persons but also as an international company with a growing impact on economy and society: Environmental, Economic and Social factors. First our actions are evaluated according to their impact on the ecological dimension of our lives. In this context, it is important to take responsibility for future generations and to keep an eye on the effects on our nature and all living beings in the world. It is important for rooom as a company to keep this horizon of responsibility in mind. Therefore, we are developing solutions which save resources and lower return rates in e-commerce. In our own company environment, we are implementing a culture of working remotely without a lot of traveling and staying in touch virtually, as mentioned before. As we can learn from the substitution rule[4], exhaustible resources used for fuels, for example, should be offset by an equal amount of functionally similar resources. As a company, we can contribute to this by engaging in projects and supporting associations dedicated to protecting our resources. Much more relevant for us, however, is the creation of technical innovations or the development of effective solutions. These can be found, for example, in the algorithms which are used for the minimization of data volumes. The visualization of our 3D models is not facilitated with a lot of computing power. Research has enabled us to develop technologies that allow us to prepare and calculate 3D models in the data center in a very efficient and energy-saving way. In everyday operation, we use only a few servers to deliver 3D experiences. And these servers are hosted in the Open Telekom Cloud, whose data centers are powered exclusively by green energy. In addition, we have found clever solutions to perform most of the computations of 3D renderings directly on the user's device without significantly increasing the power consumption on the end devices. And since most mobile devices have very low power consumption anyway, we can dramatically reduce the overall power consumption. Our AI-supported 3D model creation continues to contribute to the improvement of efficiency in consumption. Different objects can be recognized and tagged with keywords. The software recognizes similar objects quickly and needs only a fraction of the time of a conventional 3D modeling process. Images, scan data, or 3D data can be converted into 3D models in a very short time, which can then be used effectively

[4] Substitution rule. The consumption of exhaustible resources should be in proportion to the expansion of functionally equivalent resources or technical innovations such as the optimization of electricity consumption. [Detzer 99-101 ; Regan 2004].

due to the aforementioned small amount of data. In terms of the optimization rule[5], we try very hard to keep the resource intensity low. Our solutions do not require the purchase of additional hardware. We deliver our experience on common end devices that are already available to the greatest part of the public.

Another significant opportunity to improve sustainability can be found in online retailing. rooomProducts provides various applications that enable this sustainable optimization of online trading. A detailed 3D model provides key data for the customer. The 3D models are delivered by the special 3D viewer. As described before, this works with small data. The information that can be transported by these solutions enables the customer to make a reliable decision.

For technical devices, product specifications can be recognized well. The projection of an object in augmented reality allows to show size ratios, components, and colors of products directly in the customer's home, without having to order different variants. The integrated configurator can also reduce order quantities. A proper selection can be made even before the goods are shipped. Savings in packaging material can be achieved here. Instead of a large package with 1000 grams of CO_2 equivalent, a smaller one can be chosen, because less variants of a product must be ordered. A single folding box corresponds to only 20 grams of CO_2 equivalents (www.umweltbundesamt.de).

In addition, 3D visualizations for individual customers can reduce returns by up to 20% for individual products. This is also due to the detailed presentation. Further wrong orders and possible destruction of returned objects can also be avoided using 3D viewers. This also saves enormous emissions during transport. It is becoming clear that the field of e-commerce and online retail is moving more toward sustainability in many areas and solutions that support this idea are being integrated more and more frequently. This is also reflected in the high request for rooom's digital solutions.

The latest developments in digitalization toward the metaverse[6] now bring new possibilities. Comprehensive measures can also be taken here to provide companies with sustainable solutions for their digital strategies. These are being pushed further and further by additional rooom products.

3.2 *Sustainability and Responsibility in the Metaverse*

"Sustainable development is a development that meets the need of the present without compromising the ability of future generation to meet their own needs"—

[5] Optimization rule. The productivity of resources should be optimized so that resource intensity can be minimized. This can be achieved by using existing resources such as hardware and targeted recycling.

[6] Metaverse: A metaverse is a digital space where virtual, extended and physical reality merge. Access requires a digital device in the form of a computer, smartphone, virtual reality glasses or augmented reality glasses.

Detzer, K.A u.a. 1999. The quote doesn't refer directly to development as we know as a tech company; nevertheless, it is a guidance for us. With building parts of the metaverse and a multitude of digital solutions, we create long-lasting solutions which are perfectly fitting the needs of the current time, can exist for a long time, and are adjustable for changed needs. Further, we are developing solutions that save resources, which benefits the next generation. Exemplary for this are rooomEvents and rooomSpaces, which enable digital meetings, interactive work, and collaboration. Here, we also rely on our technology, which was described earlier. Since everything is based on one platform, access to data on the web is guaranteed, and any content is available and customizable at any time. Everything that's produced for a virtual event, workshop, showrooms, or web shops can be reutilized in the needed context. Nothing is just lost or a one-time thing. This saves important energy and time and conserves content. This approach should be essential for everything that is created in the metaverse.

3.3 Virtual Events in the Metaverse

Particularly relevant in the years of the pandemic and for future developments for the metaverse is the hosting of hybrid and virtual events. Virtual and hybrid event solutions can contribute to climate protection. Virtual meetings are part of the future and not just a replacement for physical trade shows and business meetings. In terms of sustainability, it's easy to prove that you can save a lot of CO_2 when attending a virtual event, which speaks in a special way for the new technologies.

A classic event does not have a good energy balance, which becomes clear when looking at the following figures. According to atmosfair, a provider of CO_2 offsetting, 70% of an event's main emissions are caused by travel to and from the event, 15% by accommodation, and 10% by catering. Accordingly, only about 5% of the emissions are attributable to energy consumption, infrastructure, grounds, and the like. In the case of a virtual event, travel to and from the event, accommodation, and catering are not included at all. This means that 95% of the main emissions of a "normal" event do not even occur with a digitally organized event on rooomEvents. When the IFA 2020 was held hybrid, it was possible to reduce enormous emissions caused by traveling. Instead of 260,000 visitors from all over the world, only 3000 were there in person. For an on-site 2-day conference with 150 participants, for example, atmosfair calculates total emissions of 29,604 kg CO_2. To absorb this amount of CO_2, a beech tree must grow for about 25 years. The same 2-day conference with 150 participants conducted digitally thus saves about 27,000 kg CO_2.

There is indeed still some of technical equipment needed to engage in a virtual event. But the fact that digital events reduce emissions through less travel is abundantly clear. However, all participants in the office or home office still have to find their way to the event, also digitally. This is done using their computers, screens, speakers, webcams, and other equipment. A study by Öko-Institut e.V. found that the CO_2 footprint of a person with intensive digital use in a year is around 1009 kg. According to quarks.de, a round-trip flight between Munich and Berlin produces around 245 kg of CO_2 emissions per person. For the distance Berlin-New York, the figure is around 2500 kg.

Further, no special hardware is needed to access the metaverse with a rooom-based solution, neither at the customer hosting an event nor at the end users. Extra purchases are therefore not necessary, which is a big plus for the environmental balance.

In addition to meetings and events that can take place in the metaverse in the future, it also offers opportunities to sustainably digitize, preserve, and exhibit cultural assets. We have already described how energy can be saved by digitizing content. If the opportunity is taken to make exhibitions accessible from any location without having to present valuable exhibits in a logistically complex way in temporary exhibitions, additional emissions can be saved.

As mentioned before, avoiding the need for special technology, rooom focuses on very small data volumes when creating 3D content. This has the positive side effect that less computing power is required, which in turn reduces energy consumption. In addition, participation in conferences or industry events usually takes place during working hours, in other words, during a period when people would probably have been sitting at their PCs anyway. This option makes a huge difference and allows valuable natural resources to be saved in the long term.

4 Outlook

Most startup ecosystems are not well connected with each other's to find the "best fit" to solve the sustainable developments of business customers. Today, we find public and corporate incubators and mixed ecosystems. Today's resources are used to solve the challenges of on a single business or geographic perspective. What would happen if all "success" and also "failure" stories are shared within economies and projects are evaluated by the entire community? It has been mostly acknowledged that the reach of sustainability goals can only achieved with collaboration, know how sharing in an automated way (with the help of sustainable matchmaking platforms).

Startup innovation to achieve positive impact for environment and society are mostly supported by artificial intelligence, and the collaboration between startups and corporates will not only speed up the developments (matchmaking) but also drive sustainable change within and outside organization.

To support sustainable projects using artificial intelligence AI, Deutsche Telekom has set up a guideline because we see that AI can have a positive or negative impact. During this project, we work together with all parties to have a common view on our guiding principles on artificial intelligence. These guidelines can be found in https://www.telekom.com/en/company/digital-responsibility/details/artificial-intelligence-ai-guideline-524366.

Frank Barz —technology pioneer, future strategist, and consultant for digital transformation across disciplines—is currently director at Deutsche Telekom. In his role, he leads the TechBoost startup program which matches some 850 startup companies with business partners of Deutsche Telekom. Tapping into his pioneering spirit and partnering with startup companies for novel technologies, he develops and implements a wide range of innovative, digital, and sustainable solutions for customers, providing support all along the change process. Prior to this role, he worked as digital consultant for T-Systems, responsible for the areas energy, e-mobility, and smart city.

Hans Elstner, CEO and Founder of rooom.com, the German metaverse.

Benedict Ilg, CEO and Founder of Flip, the employee app.

Exploring AI with Purpose

Benno Blumoser

Never get complacent: Developing AI solutions doesn't just take expertise. It also means fostering an intrapreneurial work culture while keeping in mind the greater good our work serves. That's what we do at the Siemens AI Lab.

By Benno Blumoser, Head of Siemens AI Lab

The more opportunities technology creates, the more important it is to ensure we apply it to the right purpose. Our generation is by no means the first to discuss responsible use of technology. But given the emergence of artificial intelligence and our immense expectations for it to shape the world, our generation probably faces more pressure than any before when it comes to matching the most powerful technology of its time with the most urgent challenges. It's no longer the time to deploy technology for technology's sake or just because it's cool. We have to consider the consequences of its deployment. Today, we don't just need any technology, but technology with a purpose. Specifically, we need AI with a purpose that serves the greater good. In this chapter, we want to describe this challenge and the actions we're taking at Siemens to assume this leadership role.

Perhaps it's helpful to begin by reminding ourselves of what AI can do and its two-sided nature. Essentially, AI applications fall broadly into three categories. The first is what we call a **Transparent World**. This means using AI to understand and interpret the world around us in a new way with unprecedented levels of transparency. This especially involves supervised machine learning as one of the driving technologies behind many AI applications. It allows us to detect patterns in data sets like pictures, videos, or time series data, as the relevant algorithms are trained with historic, labeled data. As a result, corresponding patterns can be detected on new, unlabeled data sets. In agriculture, this can help in recognizing visual patterns on a field surface; and as a result, you can optimize cultivation methods by sowing the

B. Blumoser (✉)
Siemens AI Lab, Munich, Germany
e-mail: bernd.blumoser@siemens.com

R. Schmidpeter, R. Altenburger (eds.), *Responsible Artificial Intelligence*, CSR, Sustainability, Ethics & Governance, https://doi.org/10.1007/978-3-031-09245-9_9

type of crop that best suits any given surface. By the same token, social media companies can apply these algorithms to the track record of your Facebook interactions, revealing, for instance, an inclination toward a certain product. The second category can be labeled **Human Augmentation**, whereby using AI systems makes us more efficient in tasks we want to perform—whether it's providing the right information snippet at the right point in time, taking over tedious process steps that would then free us to do higher-level work, or even supporting us in our creative endeavors. And the third is **Machine Autonomy**. Here, AI technologies make a range of applications possible, from self-driving vehicles to automating infrastructure systems and sub-systems, for example, optimally parameterizing devices in a low-voltage grid, scheduling manufacturing tasks, or operating gas turbines in an energy-efficient way.

Beyond improving specific processes, applying these capabilities can make a real difference. Crucially, it contributes to solving global problems greatly affecting us. For instance, fully automated train operation increases existing rail capacity by up to 30%[1] while reducing energy consumption by 30%. At the same time, passenger comfort is enhanced, especially when autonomous trains are combined with other means of self-driving transportation, such as streetcars. As a result, these multimodal systems make public transport more attractive to passengers and more competitive vis-à-vis other forms of transportation. What's more, they embody zero-harm systems for consumers and especially the environment.

But let's be clear: Artificial intelligence of the three types mentioned can be used for purposes we may not approve of so easily. Moreover, in some instances, it's not easy to determine the trade-off between the positive and negative implications of a given AI solution. It might even strongly depend on varying cultural and political value systems when it comes to assessing the respective ambiguity and finally deciding the most suitable approach.

For instance, there are AI technologies used for infrastructure surveillance that increase safety and efficiency. They detect when pipelines are threatening to burst or when power lines are about to short-circuit because trees have grown too far into them or identify obstacles on rail infrastructure impeding safe operation that need to be removed. At the same time, these technologies can also be applied to people in public areas and detect the ways they move when, from where, and where to. In the European legal system, for instance, these applications can only be implemented once the highest standards of data privacy rights are fulfilled. Even some applications using AI to identify people's faces in public areas are prohibited, as stipulated by the EU Artificial Intelligence Act proposed in April 2021. Yet in other parts of the world, a different value judgment is being made.

The same is true for AI systems with the capability of generating new design artifacts through so-called generative adversarial networks. Computer-generated generative designs can be tremendously useful in industrial design and engineering

[1] https://www.railjournal.com/signalling/db-and-siemens-demonstrate-automated-s-bahn-train-in-hamburg/

processes. AI generates a wealth of proposals experts can choose from. It speeds up the design and engineering process. As with all accelerated processes, it also creates opportunities to scale these process steps, e.g., by applying them much more often, exploring the breadth of options more intensely, and developing a higher level of design quality. On the other hand, these technologies can be misused to generate fake photos and videos, or manipulate public opinion, especially when people aren't trained to recognize and understand these forms of betrayal of their trust. The danger this generates in democratic systems and the risk of causing social instability are widely discussed. More importantly, it underlines that what matters is not merely the technology, but its very application and purpose. The latter two must be submitted to open-minded and value-based discussion.

As it's difficult to decide whether a new AI solution is also an "AI with purpose," how do we decide which AI is which? Which has a real purpose? In the following, we'd like to explain the criteria for addressing this question systematically. For us, any AI application will have to tick three important boxes.

First, starting with the obvious point for any profit-seeking enterprise at the peak of the pyramid, it must guarantee **scalable return on investment.** We have a great deal of responsibility not just toward the owners but also to our employees as well as the general public to run profitable businesses. Siemens secures 293,000 direct jobs globally and many more indirect jobs; countless families and local economies depend on our success. Of course, any AI application we pursue will have to contribute to this goal.

For sure, there's an enormous economic potential. According to a study by McKinsey[2], values of 3.5 to 5.8 trillion US dollars will be unlocked through artificial intelligence. Yet this "unlocking" isn't as trivial as it initially appeared during the first hype cycles when highly profitable, albeit at times ethically questionable, internet business cases, such as Google Ads, were expected to be easily transferable to the industrial B2B world. Many of these business cases could only reach this level of profitability because the decisive factor came to the internet giants for free: user personal data. Still today, they can be exploited without any fair value sharing mechanisms, as long as we don't consider the "free use" of search engines and social media platforms to be adequate compensation for every user's contribution.

Interestingly, this hardly differs from previous business models, in which operating resources had been recklessly exploited at a time when corresponding regulatory framework existed to ensure a fair share for every contribution. Think, for instance, of the extraction of raw materials with disastrous consequences for the environment and local tribes or offshoring jobs in textile and other industries to developing countries with low social standards at the cost of workers' health and safety.

In industrial AI, we're not living in "blind" gold rush times. Today, there's no single AI-based business model where big money is being made at the expense of

[2] https://www.mckinsey.com/featured-insights/artificial-intelligence/notes-from-the-ai-frontier-applications-and-value-of-deep-learning

somebody or something else. Rather, we're witnessing a whole range of AI solutions as part of various ecosystems that allow each contributor a stable return on their investment. At Siemens, we have several levers for ensuring a commercially sustainable approach. One key lever is the "MI Core," a technical framework for adopting and scaling use cases: By providing best practices as re-usable assets to anyone in the company with a similar challenge, we've experienced a significant acceleration in our development cycles. Another lever is an appropriate "data strategy" across Siemens' various organization units, which means cleaning, processing, and cross-linking relevant data, informing, and accelerating business cases in usually disparate fields.

Second, it must be **designed responsibly.** Here, it helps to turn to the work of regulators. Critical for us are the seven requirements defined by the European Commission for securing responsible design in AI. Focusing on lawfulness, ethics, and technical robustness, these guidelines ensure not only that AI enables human agency but that also proper oversight and accountability are practiced. Moreover, apart from demanding product safety and data privacy as well as transparency and non-discrimination, they require AI applications to promote social and environmental wellbeing. Although many of these principles aren't new, they are based on existing legal principles that can be simply applied to AI solutions that have to be reinvented and sharpened, as AI poses questions and options that haven't been explored before. Nevertheless, as this field has been widely discussed during the past years on corporate, public, and especially political bodies, we are forced to confront the trade-offs we must solve.

Principles, such as "explainability" of algorithms, are of course highly reasonable, but never absolute. They always have to be implemented in a pragmatic way. The EU's current approach of offering variability according to levels of criticality has been recognized as a very good compromise because it facilitates innovation and mitigates risks at the same time. Moreover, technology provides another important means for ensuring better trade-off solutions concerning so-called trustworthy AI. For example, with technical solutions, such as "AI-on-the-edge" or "federated learning," we can fulfill all necessary privacy requirements while benefiting from AI. One example is "occupancy detection," which relays the mere number of people in a space without transmitting the actual images the count is based on. Here, strict European regulations could have resulted in technical USPs applicable also in other parts of the world.

Third, which is related to the last item on the EU's requirement list, any AI application must have a **meaningful intention.** As elucidated above, given the vast breadth opportunities, we have to decide which purposes are relevant to us and which to do away with. Deciding for a meaningful AI application can't merely be decided by factors, such as existing portfolio, customer, investor, employee expectations, or even the overall intention and a company vision. We should also apply external criteria to shape these goals.

At the moment, there's probably no better synonym for "meaningful" than the 17 Sustainability Development Goals introduced by the United Nations in 2015, which set out numerous targets and indicators for monitoring the progress of each target. According to a Perspective article in *Nature*, 79%[3] of these targets could be positively impacted by artificial intelligence. While the areas related to energy and resource efficiency, smart infrastructure, health, and wellbeing are very central to our activities at Siemens, we never lose sight of the entire set of goals and frequently review our impact on these.

Though these lines aren't per se contradictory, they nevertheless need to be well aligned with each other to form an effective synergy. And on a practical level, there are numerous examples of favorable alignments. For instance, think of the optimized use of resources in industrial production where AI-powered predictive maintenance solutions help avoid breakdowns and waste along the production phases and save service costs by monitoring machinery operations with ease; or the aforementioned example of autonomous trains increasing capacity and helping operators transport more people and goods to provide city mobility systems with "zero harm" to people and the planet; or monitoring technologies for environmental purposes.

So how do our products measure up against these standards? Our technologies are purposeful because they support people in leading better lives and in very different ways. Let me give three examples.

As mentioned above, **generative design**, as developed by Siemens Technology with the respective business units, supports product design with AI capabilities, for example, by intelligently utilizing all sorts of information on a component or machine. In this case, whenever it's relevant for the current design process, a digital companion supplies all this information, for instance, on predecessor models, performance data, or alternative circuit board designs.

Another example is a **patient's digital twin**, as currently envisioned by our colleagues at Siemens Healthineers[4]. Toward this long-term vision, they developed the first parts that can be implemented today. For instance, the digital twin of an organ is greatly beneficial, e.g., in liver cancer treatment. Here, multiple sources of medical as well as physical data, such as a 3D scan of the liver, are used to create a bio-physiological model of a patient's liver, including the cancerous region. Based on this model, they can simulate the potential effects of different therapies for the patient, predict their success, and thereby single out the best individual therapy. This improves the overall effectiveness of therapies and significantly enhances the treatment of patients as individuals.

Finally, as a third example of purposeful AI use, worth mentioning is stress monitoring on trains. Here, AI can help detect stressful situations on trains to improve everybody's safety. This pilot system I'm referring to is being developed by Siemens Mobility. It uses AI locally to analyze visual sensor data from the train's interior to detect critical moments and raise the alarm when a dangerous situation occurs. This not only ensures higher safety levels but also protects an individual's

---

[3] https://www.nature.com/articles/s41467-019-14108-y.pdf

[4] https://www.siemens-healthineers.com/services/value-partnerships/asset-center/white-papers-articles/value-of-digital-twin-technology

right to privacy by having AI work on sensitive data at the edge of the train's network system, meaning the data doesn't leave this circumscribed environment.

As these examples hopefully make clear, at Siemens, it's our goal to be a company that can rightfully claim a leadership role in the sustainable and responsible application of artificial intelligence. Therefore, as a company traditionally focused on hardware, we aim at strongly promoting a culture of digital innovation. A lot is new about this culture, but not everything. To begin with, it must combine the best of what one could call the assets of a big corporation, such as highly efficient processes, its installed base, and its expertise, with the best of startup culture, such as focus on a purposeful vision that's implemented by concrete innovative ideas, entrepreneurial spirit, and passion.

For this synthesis to work successfully, a new type of intrapreneurial employee is needed, who can manage the varying degrees of freedom a digital innovation culture requires. Even in our established corporation, we need people with the mindset of a startup employee. Not only do we want them to be able to manage the varying degrees of freedom but claim it. This, in turn, requires a different leadership style in delegating autonomy to employees, allowing for higher levels of creativity, resilience, and motivation.

Experimenting with holacratic structures as a future format as well as organizing teams by decentralizing responsibilities to all team members without losing transparency and system control is equally important as organically intertwining work formats with mindful moments, such as check-ins or introspective sessions for reflecting on team tensions and shaping purpose statements.

At Siemens, many units, and surprisingly those related to AI, are exploring this area. For instance, at the Munich-based Siemens AI Lab, exploration doesn't only involve identifying and technically validating promising AI use cases but also testing future team mechanisms. So far, this approach has been well demonstrated during corona lockdown phases, when our co-location credo collided quite fundamentally with the pandemic's new rules. During these challenging times, we needed to adapt, and it helped that we could rely on a system that enables team participants to restructure their work with a common purpose in mind.

Let me briefly describe how this played out in detail. Within days of the first lockdown, the team's "communication lead" came up with the idea of launching a series of podcasts in line with her task of propagating AI literacy within our network. The "orientation concept owner" started virtualizing on-site workshop concepts. And the "chief happiness officer"—yes, we have that—started working on the team's resilience to the psychological challenges posed by the pandemic. All this work was self-guided, and no manager gave specific instructions to this effect.

With the Siemens AI Lab being just one of many examples, holacracy is a trending leadership style that can be found in many teams, especially those moving ahead with digital innovation. They cultivate a leadership style that delegates autonomy to team contributors based on common interaction rules, mutual trust, and a purpose statement. All activities are directed toward one goal, without a predefined roadmap.

For Siemens as a technology company, this is best epitomized by "Technology with Purpose," which we talked about in the initial section. It describes how we think and direct our work as well as going beyond the mere necessity of being profitable. Our global initiative "Tech for Sustainability" illustrates this approach. Here, students, researchers, customers, and employees are invited to work on seven real-world sustainability challenges from our business units, where they come up with solutions and explore their practical feasibility in a hackathon. If deemed successful, some of these contributors are invited to work together with our Siemens experts to push these innovations even further.

For example, we invited researchers and universities to find a sustainable track heating system for the Siemens Mobility's Neoval transit system, which is based on a rubber tire vehicle running on a concrete track. When the system is running outdoors on viaducts, cold weather can create black ice, which could cause slippage and vehicle collisions. The aim is to construct the tracks so that they remain unaffected by ice in an energy-efficient manner. Another challenge, which addresses customers, suppliers, and startups on a broader level, aims to drive the "green factory" along the manufacturing value chain: from sustainable product design, sustainable procurement, energy efficiency, to forecasting auxiliary material usage and the remaining life of products and machines. Other ideas concerned nanomaterials for coating, AI for control of energy generation, or adapting roofs for maximizing solar panel use. On the whole, all the participating teams offered a great variety of solutions that we at Siemens couldn't have come up with ourselves had we not reached out. We therefore tremendously appreciate this fruitful collaborative experience. Still, it's clear that as we collaborate, we also need to have a high tolerance for failure. Many projects fail, and even some of those that appear successful might not prevail in the long run from a financial point of view.

Summing up, as far as I can assess our capabilities, resources, and willingness for working together, at Siemens, we are well on track towards claiming such a leadership role that aims for sustainable progress, as well as productive joint efforts in the field of AI.

But we certainly can't and shouldn't be complacent. We need to keep our curiosity level high and never stop learning. This means we need to continue learning how to orchestrate our innovation ecosystems with clear, fair, and motivating rules of collaboration. It also forces us to regularly evaluate which part of our established PLM processes we need to improve on internally and which to open up as external challenges. In the process, we need to keep a cool head above it all by exploring the capabilities of technologies without losing sight of why they're needed in the first place. These are the daily challenges we face while developing AI not only at the AI Lab and at Siemens but also in the greater tech community. To meet today's global challenges, AI offers technological solutions we simply can't do without. That's why we're committed to exploring, developing and using AI with purpose.

Bernd Blumoser is Co-founder and Head of the Siemens AI Lab, which serves as co-creation platform for exploration of Industrial AI.

Relying on his management consulting experience, he held leading positions in Siemens' Open Innovation & Trend Scouting Program and was an essential driver for the development of the Siemens AI strategy. He is convinced that AI must be designed responsibly and "for the good," and is intrigued by the implications of technologies at the intersection with sustainability goals, the future of work and new digital leadership culture.

He holds a diploma in "international cultural and business studies" and an M.A. for "musical education" from the University of Passau. In his private life, Benno is passionate about Latin America, hiking in the Alps and playing music. He's married and has three kids.

Developing Responsible AI Business Model

Sundaraparipurnan Narayanan

1 Setting the Context

In the age of mobile apps, Internet of things, or connected devices, we are in the process of creating more data by 2025, than the data generated cumulatively during 2011–2020. Data is one of the precious resources in today's time and cannot be undermined (Antonio Neri, March 2020).

This thought is becoming critical as new sources of generating data are emerging in multiple ways including image processing, wearables, video streaming, robotic sensors, and so on. These contribute to the ever-evolving digital technology landscape of our world enabling effective adoption of innovative technology.

Digital technology has a significant role in transforming our lives and the generations to come. Artificial intelligence has significantly powered digital technology in enabling better access, better information, and even better governance through multiple channels (social media, digital businesses and services, Internet of things, mobile apps, etc.), thereby evolving to contribute or influence the way we and our communities live. Businesses and economies are building a digital world to enable civilization at large.

Digital world deserves a balance. While we expect most of the civilization on earth to adapt to the digital world over time, it may still not be in equilibrium. The person who holds data that influences the masses has far deeper advantages than the person whose data is held, thereby giving rise to a new economy—"Data and AI Economy."

S. Narayanan (✉)
AI Tech Ethics, Bengaluru, Karnataka, India
e-mail: sundar.narayanan@aitechethics.com

R. Schmidpeter, R. Altenburger (eds.), *Responsible Artificial Intelligence*, CSR, Sustainability, Ethics & Governance, https://doi.org/10.1007/978-3-031-09245-9_10

There may be some who would have evolved with the power of digital access and some who would have not. Power of data ownership and access will always have an imbalance, with economics and politics on the one side and rights and transparency on the other. Data possession and use for economic advantage with AI can impact societies in the form of discrimination (gender, race, or community) and widen the inequality gap.

Recognizing the above, in 2019, OECD adopted artificial intelligence principles that are innovative and trustworthy and that respect rights and democratic values. The five principles in OECD points toward artificial intelligence that (1) enables inclusive growth, sustainable development, and well-being; (2) values human-centered principles, fairness, and necessary safeguards toward the context; (3) requires transparency and explainability disclosure; (4) manages risks through robustness, security, and safety; and (5) fixes responsibility and accountability on organizations that develops them.[1] There are several lenses of view that have emerged over the past couple of years including ethical AI, responsible AI, trustworthy AI, fair AI, robust AI, etc. These lenses, however, converge and overlap in their definitions. Hence, in this chapter, the focus will be toward responsible AI as it stands in the intersection of the third (Obligation to do what is right, fair, and just. Prevent harm) and fourth layers (Build the AI ecosystem to address societal challenges) in the Pyramid of Social AI.[2]

2 Understanding the Current Ecosystem of Responsible AI

Responsibility is the state of being accountable. Responsible AI is about artificial intelligence and autonomous systems being accountable for the decisions, actions, or influence of humans thereof. While looking at responsible AI, it is essential to examine the key ecosystems that are evolving with reference to it. There are three key ecosystems that need consideration, namely, regulatory ecosystem, research ecosystem, and business ecosystem.

2.1 Regulatory Ecosystem

There is an increased awareness and debate on the need for regulations to govern AI systems. Countries across the globe have adopted AI principles or AI strategies, many attempting to ensure that the AI systems are fair, equitable, safe, secure, and accountable and preserve human values among others. Select geographies have

[1] https://www.oecd.org/going-digital/ai/principles/

[2] Figure 1 and Table 1 of research paper – Socially Responsible AI Algorithms: Issues, Purposes, and Challenges, https://arxiv.org/pdf/2101.02032.pdf

proposed regulations that focus on governance and accountability of AI systems, while some have extended such regulations to any autonomous systems (not limiting it to AI).

One of the most prominent and largest such propositions is the European Union Artificial Intelligence Act (Draft).[3] The Artificial Intelligence Act brings among others three key factors, namely, (a) prohibited and high-risk systems, (b) conformity assessment for high-risk systems including a requirement to have a detailed risk management mechanism, and (c) quality process and technical documentation requirements for AI systems. The regulatory attempt is to bring harmonized standards that enable the whole of Europe to evaluate effectiveness of these systems in a consistent manner, thereby entrusting responsibility of compliance on providers of AI systems.

Such an effort establishes the minimum expectations from organizations working on developing and deploying AI systems to adopt measures that consistently attempt to minimize risks and avoid or remove harms that can be caused to humans by these AI systems.

2.2 *Research Ecosystem*

Over the past few years, there has been a constant evolution of the responsible AI research ecosystem. It is modulating toward specific focus areas, based on demographic relevance and criticality. However, at an overall level, they tend to be expanding in six key areas, namely, (1) policy and advocacy work which is focused on building better policy (in respective geography or across geographies);[4,5] (2) awareness and education work that intends to bridge the existing knowledge gap and contribute to capacity building;[6,7] (3) standards and certification aimed at adopting unified approaches of measuring compliance/maturity;[8,9] (4) machine learning- or technology-driven solutions for addressing bias[10,11] (including

[3] Proposed Regulation of the European Parliament and of a Council laying down Harmonized Rules on Artificial Intelligence and Amending certain union legislative acts https://eur-lex.europa.eu/legal-content/EN/TXT/?uri=CELEX%3A52021PC0206

[4] https://www.amnesty.org/en/latest/news/2021/01/ban-dangerous-facial-recognition-technology-that-amplifies-racist-policing/

[5] https://www.brookings.edu/research/the-eu-ai-act-will-have-global-impact-but-a-limited-brussels-effect/

[6] https://ecornell.cornell.edu/certificates/data-science/data-ethics/

[7] https://certnexus.com/certified-ethical-emerging-technologist-ceet/

[8] https://standards.ieee.org/industry-connections/ecpais/

[9] https://www.responsible.ai/certification

[10] https://aif360.mybluemix.net/

[11] https://fairlearn.org/

debiasing) or enabling explainability[12,13] or robustness or computational ethics;[14] (5) operational processes like MLOps[15] aligned to process-driven approaches for mitigating risks of AI; and (6) human-computer interface focused on behavioral research including ethical by design, ethics of attention engineering, data use, etc.[16,17,18]

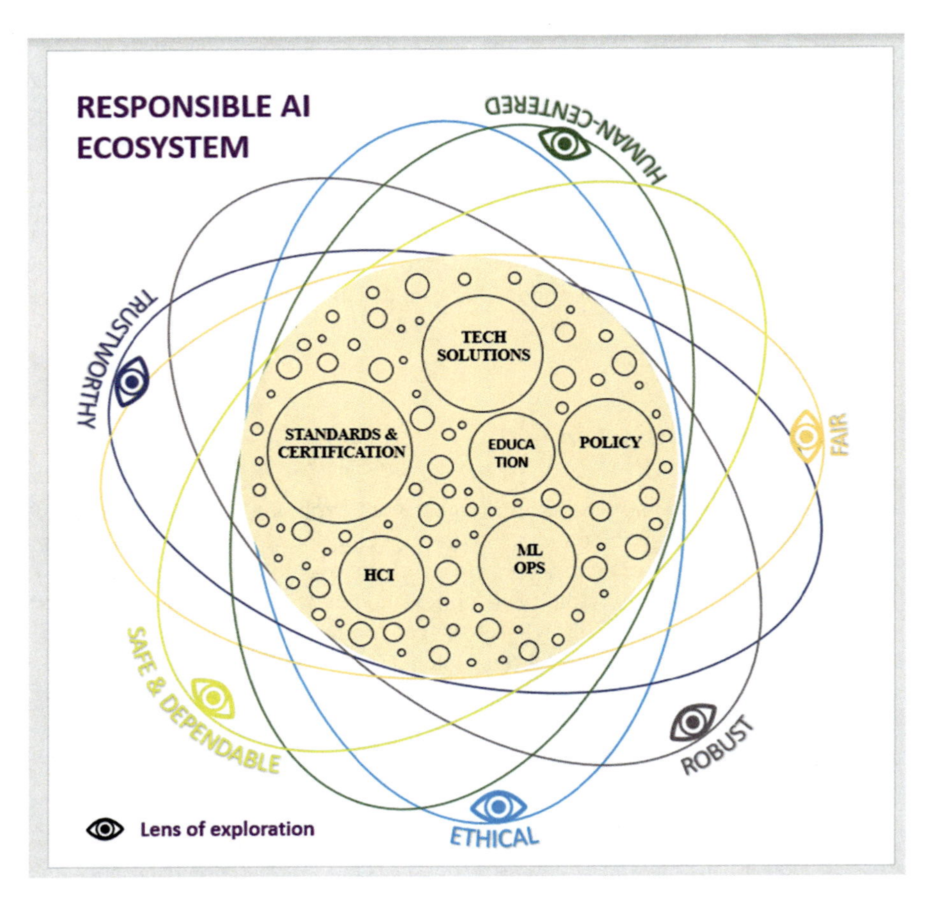

The diagram above is an illustrative reflection of the current state of responsible AI

[12] https://ethical.institute/xai.html

[13] https://github.com/SelfExplainML/PiML-Toolbox

[14] https://www.sciencedirect.com/science/article/pii/S1364661322000456#

[15] https://www.deeplearning.ai/wp-content/uploads/2021/06/MLOps-From-Model-centric-to-Data-centric-AI.pdf

[16] https://www.hcii.cmu.edu/research

[17] https://research.google/research-areas/human-computer-interaction-and-visualization/

[18] https://www.microsoft.com/en-us/research/group/human-computer-interaction/

These six areas have inherent inter-relationships between them. These in some circumstances may not be independent approaches, but well-integrated approaches. The lenses represented are illustrative and are some of the key definitions that have emerged used in the past couple of years explaining select aspects of overall responsibility.

2.3 Business Ecosystem

Adoption of AI principles by commercial organizations is on the rise with many of the Big Tech giants (e.g., Google, Facebook, Accenture, Twitter, Salesforce, etc.) establishing their organizational framework for responsible AI.[19] Further, some of these organizations have started consistent disclosure of their efforts toward responsible AI in practice.[20] These efforts directly relate to the significance of such tech giants in committing to a standard. It's an inherent factor that such efforts exhibited by tech giants are a necessity given a deep reputational impact for failing to take those steps can have a resultant effect on their market capitalization. However, efforts toward such adoption of AI principles are not democratized across the business ecosystems and commercial organizations beyond the selective few referred above and are far from aligning to or adopting responsible AI principles. Further, while there exists an interwoven interface between the research ecosystem and the business ecosystem, such interfaces are sporadic and not consistent in all areas of research.

3 Stages of Responsible AI Maturity

As eluded above, all these ecosystems are in their early stages of maturity. Examining the history over decades of evolution of sustainability will reflect the transition of thought process from exploitation of resources to regulated use. It would emphasize on regulated use to fair price for use, fair price for use to replenish the resources, and replenishment of resources to developing the society where resources have depleted (Business Standard, November 2019).

Drawing an analogy, the approach toward using available natural resources have evolved in five stages: (1) exploitation, (2) consent to extract, (3) approval from regulator for access, (4) contribute to deprived societies that are dependent on such resources or region in which such resources exist, and (5) re-establish ecology by reinstating or taking efforts to bring balance. Responsible AI would also essentially

[19] Artificial Intelligence at Google: Our Principles https://ai.google/principles/

[20] Privacy Progress Update https://about.facebook.com/privacy-progress

follow the same route. It's unlikely we will have a different outcome, as it's a trend of human evolution.

Stage 1—Collect and exploit available data with AI: This is the phase where data is exploited for commercial use without broad-level governance factors.

Stage 2—Seek consent for data use: This is a phase where governments in different regions bring regulations that stipulate conditions for data access and use, including consent from data subject.

Stage 3—Pay for data and responsible data use: This is a phase where the data subject is compensated for the data use. The economics around the price will stabilize over a period, and newer governances and exchanges for price limits will come into existence. It would extend to the stage of leading a fair pay model for use of and access to data.

Stage 4—Contribute to developing societies that are impacted by AI deviations: This is a phase where business's social responsibility emerges to support the society that is deprived of data access or the society that is impacted by AI harms or discrimination.

Stage 5—Responsible use of AI and balanced act of sustainable development: This is a phase where the principles of sustainability are formulated and active social participation in AI businesses is encouraged to develop a sustainable future digital world.

Besides the above, responsible AI will evolve through two critical paradoxes that will add to the journey toward sustainability. They are:

A. Equitable bargain between sustainability and economics: The economic bargaining power against responsible AI and the economic implications of irresponsible AI practices will determine the time it takes for the social voice to emerge toward responsible AI.
B. Sustainability challenges caused by computational processes involving data: The sustainability (environmental) impact caused by data processing power required for computation will also drive dimensionality to emergence of responsible AI thoughts.

Essentially AI shall be used to bridge the existing societal/economic gaps, but not extend them. Data possession and use for economic advantage can impact societies in the form of discrimination (gender or race or community) and widen the inequality gap. For the above reasons, responsible AI needs to be considered from a sustainability perspective. Whereby, it helps in bringing equality and reducing opportunities for discrimination and divide.

Currently, the artificial intelligence is in the first two stages, and while regulations like the General Data Protection Regulation (GDPR)[21] and California Consumer Privacy Act (CCPA)[22] are attempting to bring in need for consent, it will evolve to

[21] General Data Protection Regulation https://gdpr.eu/

[22] California Consumer Privacy Act (CCPA) https://oag.ca.gov/privacy/ccpa

the rest of the stages over the coming years. EU AI draft regulation (Harmonized Standards) is the effort in the right direction to move this context forward. Given this context, it is highly relevant for leaders to consider building the businesses on responsible AI approaches to be valued by stakeholders.

4 Responsible AI Business Model

Responsible AI is about being accountable for the decisions, actions, and consequent human influence toward the decisions and actions that are triggered by AI. To bring parity across society with the adoption of responsible AI, commercial organizations shall commit to put principles into practice. Commercial organizations which are designing, developing, deploying, using, and decommissioning AI and autonomous systems need to find a necessary balance in adopting principles and implementing responsible AI practices. The efforts shall be driven by understanding (a) principles, (b) pillars, (c) business model canvas, and (d) actions toward responsible AI.

4.1 Principles

Principles are guiding propositions that serve as a foundational thought process for business. Primary principles of commercial organizations would be to ensure that business practices built on AI stay accountable, consider the safety and security of humans, and uphold human values. There are ten key principles to consider in this context. They are:

- Defining, building, and maintaining corporate ethics policies and expectations with specific reference to ethics of artificial intelligence
- Upholding standards for corporate governance
- Driving culture of responsibility (toward society) across organization
- Committing to not using technology for purposes that can harm humans
- Ensuring ethics in data collection processing and use
- Ensuring trust and safety for humans
- Upholding transparent communication and timely disclosure
- Maintaining auditability or traceability of decisions across value chain
- Demonstrating accountability for outcomes, actions, or decisions from AI and autonomous systems
- Defining, implementing, and measuring progress toward sustainability goals

4.2 Pillars

Pillars provide essential support to maintaining or sustaining the principles across the business environments. There are four pillars of responsible AI. They are:

- Trustworthy (transparent and reliable)
- Fair and inclusive (non-discriminative)
- Sustainable (environmental)
- Safe and secure (robust)

It's important for organizations to formulate appropriate definitions of these pillars as it relates to their business and align their processes with these pillars purposefully.

4.3 Business Model

Business models are an essential rationale of how a commercial organization creates, delivers, and captures value with its products and/or services. These could be in economic or social context. Responsible AI requires a strong business model that aligns to the principles and integrates the pillars of responsible business and sustainability. Responsible business models enable businesses to defocus on choices that may be irresponsible in this context. There are two key aspects that are necessary for enabling this business model, namely, (I) responsible AI business model canvas, which helps the businesses to define economic or social rationale for delivering value, and (II) responsible AI decision canvas, to guide decision-makers at compelling junctures of business toward principles. Both these aspects are aimed at paving the way for sustainability to responsible business practice.

4.3.1 Responsible AI Business Model Canvas

There are three key essential myths to be debunked prior to exploring the canvas. Firstly, it is necessary to recognize that responsible business model need not be a not-for-profit model. It establishes adding a layer of responsibility and not diminishing opportunity for profits. Secondly, data monetization approaches need not conflict with responsible AI or the underlying business models. Data monetization shall evolve to provide newer dimensions to the existing business models (e.g., revenue share with data subjects). Thirdly, these business models need not have a community-driven approach. In select cases, community-driven approach could be core to the business model, but that is not a necessity in all cases.

Business model canvas proposed by Alexander Osterwalder in his book *Business Model Ontology* contains nine building blocks as segments structured as a one-page canvas. The building blocks are customer segment (segment, persona, and problem/

needs), value proposition (problems/needs solved for persona), channel (avenues linked to persona and segments), customer relationship (ways enabled for customers to engage with solution/product), revenue streams (commercial models tied for solution/products use by persona), key activities (critical actions tied to the overall business), key resources (resources necessary to deliver the actions and provide the value proposition for the customer), key partnerships (relationships that are critical to augment capability to fulfil the activities), and cost structure (cost associated with delivering product/solution, performing key activities, and resource/partnership costs).

Responsible AI business model canvas builds on the business model canvas by characterizing key questions to consider in each building block. These blocks balance the focus on customers, stakeholders, and data subjects. The key questions for each building block are:

Customer Segment

- Does the segment consider an inclusive environment of including under-represented populations?
- Whether adequate accessibility requirements are considered for the customer segment?

Value Proposition

- Whether value propositions are evaluated for the safety and security of people?
- Whether the value propositions exhibit accountability and transparency?

Channels

- Whether the channels adopted are not untowardly inconsistent to the principles (e.g., harmful influence)?

Customer Relationship

- Are ethical aspects of data acquisition and use considered?
- Are design considerations for ethical nudges included in attracting and engaging customers?

Revenue Streams

- Whether the value propositions and revenue streams defined are sustainable?
- Whether the revenue streams exclude income opportunities from solutions that could be harmful or exploitative (exploiting cognitive bias)?

Key Activities

- Whether policy and monitoring activities are aligned to the principles considered?
- Whether an adequate consideration is made for quality and risk management?
- Whether quality considerations cover documentation across lifecycle including tests or validations and the results thereof are considered?

- Whether risk management plans include monitoring risk including key risk indicator monitoring, monitoring of concept drift, and post-market monitoring?
- Whether traceability and disclosure requirements are considered?

Key Resources

- Are adequate people with necessary skills for driving responsible AI considered?
- Whether upskilling considerations and responsible AI alignment training are considered for key resources?
- Whether adequate tools (model pipeline, MLOps, etc.) are considered for maintaining the quality of the model design, development, deployment, and decommissioning process?
- Whether ways to evaluate data quality and annotation quality are considered as part of the process?
- Whether adequate consideration for consistent research to evaluate effectiveness or examine potential harm on an ongoing basis?

Key Partnerships

- Whether civil society organization(s) included as partners for providing feedback on fairness?
- Whether adequate levels of external experts are appropriate for relevant customer segments (psychologists who deal with children)?
- Are adequate measures considered for ensuring compliance by partners?

Cost Structure

- Are costs relating to paying data subjects for gaining access to their data considered?
- Are costs and methods associated with developing synthetic data for models considered?
- Whether cost relating to resources and partnerships for responsible AI is considered?

This said, responsible AI business model canvas is not a fool-proof mechanism; it is a guidebook for structuring value propositions, defining collaborations, and establishing economic metrics. These questions enable better consideration while framing the responsible AI business model.

4.3.2 Responsible AI Decision-Making Canvas

Business decision-making is interwoven with dilemmas. While principles guide in alignment, certain practices may be weighed more than others due to multiple reasons including competing values within the organization, psychological, and cultural forces that influence behavior and disparate value propositions of various stakeholders. However, a structured approach toward responsible AI decision process allows consistent adoption of principles.

Responsible AI decision-making canvas has six progressive layers. These layers not only help in making decisions but also influence others in the decision process, where the influencer is not a decision-maker. The layers are:

Layer 1: Establishing Principles and Conflict—It is important to establish the key principle which is in question from the AI principles adopted by the organization and define the underlying conflicts. Conflicts are best expressed as choices along with their underlying economic or social push/pull factors. Providing reference to the context or used case scenario in question is also necessary.

Layer 2: Defining Criticality and Priority—Defining criticality of decisions with the potential value at risk and priority of decisions considering the timeline along with potential impact are relevant. These help in enhancing the context of the decision-making scenario.

Layer 3: Alternatives and Effectiveness—Assimilating the alternatives that exist in the decision process, beyond the laid down conflicting choices explaining the reasons for non-consideration of those including effectiveness measure as relevant, is another essential element.

Layer 4: Documenting Impact to Organization and People—Collating critical information on the impact including the impact to people (both internal and public at large) and organization enhances the perspective on decision-making. Specifically, if the decision relates to the choice of monitoring metrics or thresholds for a nAI system or its underlying machine learning model, potential safety impact to people with such model is an essential consideration.

Layer 5: Consultation and Framing—With AI used cases extending across spectrums, decision-making process will require consultation with experts including psychologist, anthropologist, and civil society. Identifying and consulting with such specialists adds more clarity to the decision process. Further, it is important to frame the underlying conflict with the information gathered thus far. Needless to mention that the way objectives are framed influences the cognitive decision process.

Layer 6: Structuring Actionable and Communication—Post the conflict being framed for decision, it is essential to look at the actionable necessary for such decision-making. Also determining crisp communications on the decision or for the decision enables in influencing stakeholders toward the responsible AI business.

4.4 Steps Toward Responsible AI Business

Businesses have existed prior to adopting the AI principles, and as they progress toward aligning to the principles, they need to take into consideration eight key steps that enable them to progressively engage and transform the process. They are:

4.4.1 Step 1: Understanding RAI Landscape

Gaining introduction to the RAI ecosystem including the importance of design, data acquisition and use, model choices, and human-AI interaction in the broader scheme of organizational mission.

4.4.2 Step 2: Assessing Current Gaps in AI Lifecycle

Adopting a structured methodology in conducting risk assessment, identifying gaps, and mapping the gaps to scale of priority across the design-development-deployment-decommissioning (DDDD) cycle.

4.4.3 Step 3: Establishing Business Value of RAI

Understanding the key components and drivers of business value. Examining the role of RAI in enhancing business value through such components or drivers.

4.4.4 Step 4: Developing a Framework

Frameworks are built on need assessment considering the current gaps in AI lifecycle. Frameworks are intended to be a broad structure that directionally guide the actions on RAI and align with the culture of the organization. Developing or adopting a framework given the business and context of organization is focused on integrating adopted principles and aligned pillars to put them to practice.

4.4.5 Step 5: Aligning Principles to Framework

Varying principles are relevant in varying circumstances, and some of them have significant importance in some selected environments than others, aligning principles and prioritizing them for making the framework robust.

4.4.6 Step 6: Structuring Actionable Plan for RAI

Mapping the drivers of value to actions regarding RAI to work toward a sustainable RAI model for stakeholders. These actions are formulated into action plans for short term (6 months) and long term (18–24 months) along with specific measures to monitor their progress and measure their success.

4.4.7 Step 7: Integrating Skills for RAI

Identifying, sourcing, and integrating essential skills and capabilities for RAI has utmost importance for organizations, considering the field is currently evolving. The process will also demand developing mechanisms to build capacity and progressively plan for skills and specialization to cater to the needs of RAI.

4.4.8 Step 8: Putting RAI in Practice

Defining ways to put RAI into practice including demonstrating innovative approaches in implementation. These may include integrating RAI as part of the broader compliance-controls-culture approach of influencing organizational stakeholders and driving assurance.

5 Convergence of Social Responsibility

Businesses exist for serving the community albeit with commercial intent. In the current business and regulatory environment, there are no structured measures that collectively look at how a business can be trustworthy and responsible combining all these factors. Customers would want to buy products from companies that are trustworthy. Governments and financial institutions would want to work with companies that are ethical. Younger generation would be looking at working with organizations that exhibit responsibility in society. These are not isolated factors, but collective factors that represent the trustworthiness and responsible behavior of organizations.

Organizations looking at improving trust would invest in mechanisms that extend beyond legal and compliance requirements, catering to the needs of the ecosystem responsibly. Responsible AI is not a choice, but a value that the brand stands for and demonstrates consistently over time.

Sundaraparipurnan Narayanan is an ethics professional with 15 yrs of experience in advising corporations in developing policies, creatingcontent, training people, conducting risk assessments, and assisting in fact-finding reviews. Sundar is an artificial Intelligence (AI) Ethics researcher with a focus on ethical issues and downside risks associated with artificial intelligence systems. Sundar is also a Fellow of ForHumanity, an organization working towards building criteria for audit of AI systems. Co-authored the Risk Management Framework for audit of AI systems as part of ForHumanity. He advises AIGovernance and Risk platforms in implementing AI ethics effectively for their clients. He also consults companiesin conducting algorithmic impact assessments of their machine learning models.

ESG Fingerprint: How Big Data and Artificial Intelligence Can Support Investors, Companies, and Stakeholders?

Pajam Hassan, Frank Passing, and Jorge Marx Goméz

Abstract Current research is investigating the extent to which measurements of corporate sustainability through environmental-social-governance (ESG) controversies have an impact on a company's valuation. Early investors, stakeholders, and companies are already using the ESG data and ratings generated to inform their investments or strategic decisions in companies. On the other hand, it is apparent that measurements are often based on static indicators collected annually. Furthermore, analysis has shown that mainstream ESG ratings lack a consistent ESG framework. Similarly, ESG rating indicators are often predefined and do not provide users with sufficient transparency to integrate them into their daily business processes. This is where this chapter comes in and develops an ESG taxonomy based on historical ESG events from which risk patterns, the so-called ESG fingerprint, are automatically extracted. These help to reduce complexity and enable the design of artificial intelligence-based ESG information systems that map the risk management process across phases.

Keywords ESG · Risk analysis · AI-based information systems · Investment · Due diligence · Sustainable development goals

P. Hassan (✉)
intuitive.ai GmbH, Hamburg, Germany
e-mail: pajam@intuitive-ai.de

F. Passing
intuitive.ai GmbH, Hamburg, Germany

IU International University of Applied Sciences, Bad Honnef, Germany
e-mail: frank@intuitive-ai.de; frank.passing@iu.org

J. M. Goméz
Carl von Ossietzky University of Oldenburg, Oldenburg, Germany
e-mail: jorge.marx.gomez@uni-oldenburg.de

R. Schmidpeter, R. Altenburger (eds.), *Responsible Artificial Intelligence*, CSR, Sustainability, Ethics & Governance, https://doi.org/10.1007/978-3-031-09245-9_11

1 Status Quo

To date, more than 3000 organizations with an investment volume of almost 100 trillion US dollars have signed the Principles for Responsible Investment.[1] The fundamental principles state that attention to environment, social, and governance (ESG) is essential to sustainable investing. To translate these paradigms into the actions of investment banks, companies, and stakeholders, processes, tools, and procedures have been developed in recent years to manage sustainability in terms of ESG in banks and companies. Sustainability has the strategic goal of achieving the Sustainable Development Goals (SDG) in 2030.

Organizations are making commitments to their internal and external stakeholders to consider the environmental and social impacts of their operations. In response to these pressures, many companies have initiated and implemented a variety of sustainability initiatives. Details of these initiatives are increasingly disclosed publicly in corporate sustainability reports or equivalent. However, stakeholders often have difficulty interpreting the meaning of the reported information. A number of ratings, awards, and indices have emerged to highlight companies with exemplary sustainability. Primarily, company reports and websites of the respective companies and industry reports are used for this purpose, which are evaluated according to structured indicators. The indicators collected are translated into ESG ratings and often revised annually. To date, there are no standardized ESG criteria according to which a company is rated.[2]

In recent years, digitization has led to the development and collection of many new, so-called alternative data that lead to further insights for the assessment of ESG in companies. In particular, measuring ESG controversies in alternative data such as news, blogs, forums, and other social media channels is proving to be a practical tool for investment decisions as well as strategic decisions in companies. In addition to traditional knowledge brokers and rating agencies,[3] startups[4] are also developing new technological ways[5] to harness the wealth of available data. Two application domains in particular are becoming established. While one domain focuses on sustainable investment, other providers are focused on sustainability in the supply chain. What they have in common, however, is their focus on a key performance indicator (KPI) system whose limitations are still insufficiently evaluated. For example, Escrig-Olmedo et al. (2010) explain that a large number of ESG rating and information providers cater to the increasing demand for ESG-related information, but nevertheless the evaluation criteria used are incompletely explained in the

[1] https://www.unpri.org/pri/about-the-pri

[2] Escrig-Olmedo, E.; Fernandez-Izquierdo, M.A.; Ferrero-Ferrero, I.; Rivera-Lirio, J.M.; Muñoz-Torres, M.J. Rating the raters: evaluating how ESG rating agencies integrate sustainability principles. Sustainability2019, 11, 915.

[3] For example, Refinitiv, MSCI, S&P, Morningstar.

[4] For example, Arabesque, Sustainalytics, intuitive.AI, EcoVadis.

[5] By its own account often AI.

Fig. 1 Sustainable Development Goals of the United Nations

context of risk management.[6] Similarly, the methods lack cross-process applicability in risk management. From this, it can be concluded that the design of information systems is not sufficiently considered. Furthermore, there is no typology to characterize and classify ESG objects—in the sense of information system research. In addition, AI-based information systems for ESG risk management procedures offer further potential in the operational risk management process.

This is where the present work comes in and examines existing approaches for a suitable ESG risk management framework in the context of information system research. Furthermore, a taxonomy is developed that characterizes ESG-related opportunities and risks and makes them classifiable. The draft ESG fingerprint serves as a framework for designing an ESG risk information system.

2 Introduction ESG Risk Management and Information Systems

In September 2015, the 195 member states of the United Nations (UN) agreed on the 2030 Agenda for Sustainable Development. The most important components of the agenda are the 17 Sustainable Development Goals (Fig. 1).

---

[6]Escrig-Olmedo, E., Muñoz-Torres, M.J. and Fernández-Izquierdo, M.Á. (2010) 'Socially responsible investing: sustainability indices, ESG rating and information provider agencies', Int. J. Sustainable Economy, Vol. 2, No. 4, pp. 442–461.

The 17 overarching goals set globally consistent benchmarks for sustainable development priorities and targets through 2030, serving organizations, investors, and companies alike to have measurable and investable sustainability goals. At the same time, these incorporate ESG criteria to a large extent.

ESG is understood to mean the consideration of criteria from the areas of environment (environment), social (social), and responsible corporate management (governance). These criteria are used to evaluate companies with regard to their progress in the area of sustainability. In recent years, the importance of ESG criteria in investment decisions and sustainable procurement processes has increased significantly, not least as a result of the EU's action plan for financing sustainable growth within the framework of the Capital Union.[7] This is also clearly illustrated by the UN PRI initiative of the United Nations, which has drawn up six principles for responsible investment and intends to implement them.[8] The signatories of the UN PRI manage to date more than 100 trillion US dollars (investments under management) with strongly increasing numbers.[9] They already consider ESG in their investment processes. A large number of studies have examined the connection between ESG criteria, the success of a company or its performance, and a possible reduction in risk. This has led to adjustments and expansion of risk management to include ESG sustainability criteria, particularly in decision-making processes in the areas of investment, strategic planning, and supply chain.

The consideration of risks plays a strategic role for those companies and investors that are increasingly exposed to complex, rapid changes and often find it difficult to anticipate risks. In this context, a risk is to be understood as a possible deviation from planned results, which may result in a decline in value creation.[10] In this thesis, potential risks are considered from the perspective without internal company information. In the context of this work, the term controversy is therefore used to make it clear that a risk is involved, but that it does not necessarily have to be.[11] This is precisely where risk management provides support, which is to be understood as a process of identifying, assessing, mitigating, and monitoring negative events or situations that may significantly affect an organization or a company as a whole (Fig. 2).

The aim of a risk management model expanded to include sustainability is to maintain a balance in sustainable business, referred to below as ESG risk management (ESGRM). In this context, the newly integrated risk types cover a wide range of sustainability challenges, such as climate change, human rights, or working

[7] https://ec.europa.eu/info/sites/info/files/180131-sustainable-finance-final-report_en.pdf

[8] https://www.unpri.org

[9] https://www.unpri.org/pri/about-the-pri

[10] D. Bogataj and M.Bogataj 2007. Measuring the supply chain risk and vulnerability in frequency space. International Journal Production Economics 108 (2007) 291–301.

[11] It counteracts the case that, for example, in the case of a report on poor working conditions in a company's n-tier supply chain, it is necessary to work out whether the n-tier supplier is actually the supplier of the company at risk.

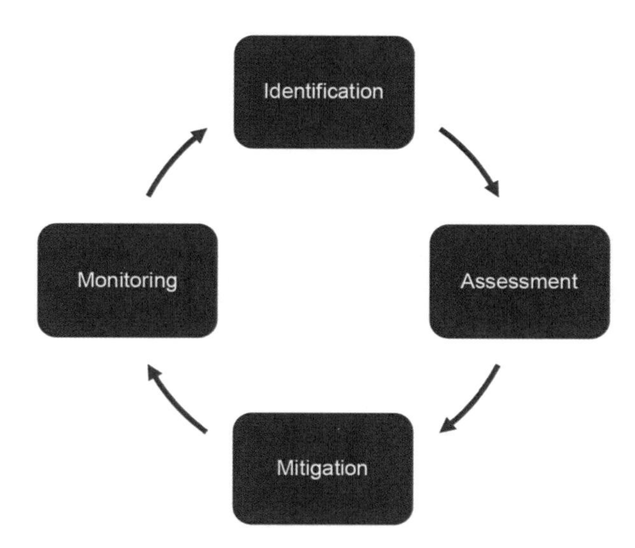

Fig. 2 Risk management process

conditions, among others. In particular, identifying risks in the diverse sustainability risk fields is challenging and time-consuming. This is because the risks are often not available as a structured data source but have to be collected from various data such as annual reports, studies, financial data, news reports, or blog and forum contributions.

In order to make the flood of internal and external risk information usable for decision-makers in organizations, it needs to be condensed and summarized. The information provided should be relevant, of high quality, up to date, and meaningful. The basis for this is provided by the concept of information systems. They are understood as systems that process information, i.e., capture, enrich, transform, store, and provide it. Operational information systems are understood as socio-technical systems that include human and machine components as task carriers.[12] Particularly in modern information systems, further task scopes are solved by the machine through AI. AI is understood in a broader sense as a task performed by a computer on a human level.

With regard to ESG risk management, a large number of different data and rating products[13] as well as software solutions already exist. However, even if they are supposed to map the same functionalities, the measurements have different, structural logics and not uniform calculation rules.[14] On the contrary, their underlying logics are continuously extended and adapted by commercialization efforts and the

[12] http://www.wirtschaftslexikon24.com/d/informationssystem/informationssystem.htm

[13] Hill, J. Environmental, social, and governance (ESG) investing. In A Balanced Analysis of the Theory and Practice of a Sustainable Portfolio; Academic Press: Kidlington, UK, 2020.

[14] Chatterji, A.K.; Durand, R.; Levine, D.I.; Touboul, S. Do ratings of firms converge? Implications for managers, investors and strategy researchers. Strateg. manage. J. 2016, 37, 1597–1614.

accompanying customer requirements,[15] with this leading to low applicability in the context of ESGRM processes and low suitability for risk management across process phases, counteracting a quasi-standard that supports the achievement of the SDGs. Academia and industry focus on researching or commercializing functional properties of their data products, rather than designing a socio-technical system to manage operational as well as strategic opportunities and risks. This is the guiding research question: What does a typology look like that makes it possible to classify relevant ESG risks?

3 Concept for the Development of a Taxonomy for the Classification of ESG-Relevant Opportunities and Risks

In order to answer questions about contemporary problems, qualitative studies, such as expert interviews, are often used to explore properties and requirements.[16,17] However, since various data products already exist, but there is a lack of a suitable typology to characterize them, a different procedure is chosen in this study. In the context of information system research, taxonomies are suitable to classify relevant objects according to their relationships and characteristic properties following a structured approach.[18] To consider ESGRM as a contemporary phenomenon in its real-world context, a case study-based approach is suitable.[19] This allows to generalize findings and to develop a deep understanding. To this end, the study is divided into two phases. In phase 1, ESG controversies are considered, compared, and analyzed. In the second phase, a taxonomy is developed based on the findings from the case study, and this is then developed and evaluated empirically and theoretically in the course of three iterations.

Figure 3 shows the substeps. The procedure allows to combine conceptual (k) and empirical (e) findings. The procedure was based on the process model developed by Nickerson et al. (2013) for the design and evaluation of taxonomies in the field of information system research.[20]

---

[15] Lozano, R. (2015). A holistic perspective on corporate sustainability drivers. Corporate Social Responsibility and Environmental Management,22(1), 32–44.

[16] For example, Coqueret, G. (2020). ESG Equity Investing: A Short Survey.

[17] For example, Zaccone, M. C.; Pedrini, M. (2020). ESG Factor Integration into Private Equity.

[18] Nickerson, R.C., Varshney, U., Muntermann, J., 2013. A method for taxonomy development and its application in information systems. Eur. J. Inf. Syst. 22 (3), 336–359.

[19] Yin, R.K., 2014. case study research: design and methods, 5 ed. Sage publications, Thousand Oaks, CA.

[20] Nickerson, R.C., Varshney, U., Muntermann, J., 2013. A method for taxonomy development and its application in information systems. Eur. J. Inf. Syst. 22 (3), 336–359.

Appraoch	
1. Determination of meta characteristics	
2. Definition of termination criterion	
3e. from empiricism to concept	3k. from concept to empiricism
4e. Identification of (new) object subsets	4k. Conception of (new) object features and dimensions
5e. Identifizierung von gemeinsamen Eigenschaften sowie Objektgruppen	5k. Analyze objects for the above features and dimensions.
6e. Grouping properties into dimensions to create a (revised) taxonomy	6k. Preparation of the (revised) taxonomy
7. While determination criteria are not met, repeat step 3.	

Fig. 3 Procedure model for the systematic development of a taxonomy

3.1 Structure of the Case Base (Empirical Data Basis)

Scientific data sources[21] as well as practice reports[22] are used to build the ESG case base. The selection criteria used are:

- Freely available information must be available on the ESG case.
- Sufficient information related to ESGRM must be available for the ESG case.
- In the case of incompletely documented ESG cases, features are available that, in the sense of data triangulation, allow the context of the case to be reconstructed via background research.

The multilingual case base includes 48,334 records[23] related to ESG controversies worldwide. The ESG controversies are stored as raw data and are fully available in a business intelligence infrastructure. For taxonomy development, 33 case studies are selected from the case base,[24] enriched by manual research if necessary. After data triangulation, all case studies show references to both ESG and risk management.

[21] Databases: ScienceDirect, IEEE Explore and the Web of Science, Google Search, etc.

[22] Practice reports: MSCI, EcoVadis, Arabesque, intuitive.AI, McKinsey, BCG, Accenture, and KPMG, among others.

[23] Source: intuitive.AI GmbH.

[24] Data triangulation through, among others, news articles, statements and press releases, statistical data, websites of companies or NGOs, and reports of rating agencies.

Type of ending condition	Termination criterion	Description / Question
Subjective	Concise	Does the number of dimensions lead to overloading?
Objective	Concise	In the last iteration, no object was merged with a similar object or split into multiple.
Subjective	Robust	Do the dimensions and characteristics provide for differentiation among objects sufficient to be of interest? Given the characteristics of sample objects, what can we say about the objects?
Subjective	Comprehensive	Can all objects or a (random) sample of objects within the domain of interest be classified? Are all dimensions of the objects of interest identified?
Objective	Comprehensive	All objects or a representative sample of objects have been examined
Subjective	Extendible	Can a new dimension or a new characteristic of an existing dimension be easily added?
Subjective	Explanatory	What do the dimensions and characteristic explain about an object?

Fig. 4 List of discontinuation criteria (Adapted from Nickerson et al. 2013)

3.2 Analysis and Evaluation

- Step 1: In the first step, meta characteristics are identified. The underlying premise here is that the characteristics have a high degree of distinctiveness. First, a distinction is made between *risk-specific* and *data-specific* meta dimensions. It is important that all dimensions and characteristics of the taxonomy can be assigned to the meta dimensions. This ensures that the taxonomy is focused on relevant ones.[25] In the application of the taxonomy, the extension of the *data-specific* meta dimension by the data *source-specific* dimension takes place in the second iteration.
- Step 2: In the second step, two objective and five subjective termination criteria are defined. Figure 4 presents the dropout criteria, assigns them to a category, and provides either an explanation or a question detailing the dropout criterion.

3.3 Iteration 1: Conceptual Development (from Concept to Empiricism)

In the first iteration, steps 3 k to 6 k are performed (Fig. 5). Conceptual-empirical dimensions and characteristics are derived from ESGRM literature and practice reports. Partial findings from this phase are taken up and presented from Sect. 2, Introduction ESG Risk Management and Information Systems.

[25] Remané, G., Nickerson, R.C., Hanelt, A., Tesch, J.F., Kolbe, L.M., 2016. A taxonomy of carsharing business models. In: Paper Presented at the Thirty Seventh International Conference on Information Systems (ICIS), Dublin, Ireland.

Meta dimensions	ESG risk dimension	Question	Characteristics				
Risk-specific	Type	What type of risk is present?	Environmental	Social	Governance		
	Relationship	What does the risk refer to, an organization or location?	Organization	Location			
	Reference	What exactly does the risk refer to?	Organization Group	Organization site	Geographical area	Geo-coordinates	
	Risk management phase	In which ECRM phase can the risk be classified?	Identification	Assessment	Mitigation	Monitoring	
	Assignment	Can the risk be assigned to a product or industry?	Product class	Industrial class			
Data-specific	Risk-related	How can the risk reference be established?	Implicit	Explicit	Standardized		
	Entity Reference	How can the reference to an organization or location be established?	Implicit	Explicit	Standardized		
	Temporal projection	Does the risk refer to a date in the future, present or past?	Past	Present	Future		
	Completeness	Is data triangulation necessary for one of the ECRM phases to be triggered?	Incomplete	Complete			
	Data type	What type of data is it?	Image	Voice	Textual	Nominal	Numeric
Source-specific	Data velocity	At what frequency are risks updated?	Real time	Near real time	Batch		
	Data access type	On which medium can the risk data be found?	Website	Web Service	Data feed	File transfer	
	Accessibility	Does the data source provide free access to its risks?	Open	Closed			
	Rights of use	May the content be used freely?	Unrestricted	Restricted			
	Data structure	What is the structure of the risk data offered by the source of information	Structured	Semi-Structured	Unstructured		
	Data format	Is it a proprietary data format?	Proprietary	Non-proprietary			

Fig. 5 ESG risk taxonomy

- **Input**: ESG and CSR literature and practical reports
- **Output**: Initial taxonomy

3.4 Iteration 2: Empirical Development (from Empiricism to Concept)

In the second iteration, the initial taxonomy is extended empirically and conceptually. For this purpose, it is applied to 30 randomly selected case studies, and a qualitatively structured data analysis is performed.[26] The empirical derivation of features is performed according to Yin (2014) until the termination criteria are met.[27] To do this, each case study is classified using the taxonomy and checked to see if new dimensions or features need to be added to characteristically describe each case study. In this iteration, the further differentiation of the meta dimensions also takes place.

- **Input**:
  - Initial taxonomy
  - Thirty case studies
- **Output**: Revised taxonomy

3.5 Iteration 3: Empirical Evaluation (from Empirical to Conceptual)

In the context of the empirical-conceptual evaluation, the applicability of the taxonomy is demonstrated by means of three case studies. Thereby, missing information is searched for, and an extended case context is established by data triangulation, which allows an argumentative-deductive analysis.[28] In this course, the risk management process phase is adopted as the final dimension. The characteristics of the dimension are derived from the literature.

[26] Miles, M.B., Huberman, A.M., Saldana, J., 2013. qualitative data analysis: A Methods. Sourcebook, vol. 3. Sage Publications, Los Angeles.

[27] Yin, R.K., 2014. case study research: design and methods, 5 ed. Sage publications, Thousand Oaks, CA.

[28] Wilde, T., & Hess, T. (2006). Method spectrum of business informatics: Overview and portfolio formation (No. 2/2006). Arbeitsbericht, Institut für Wirtschaftsinformatik und Neue Medien, Fakultät für Betriebswirtschaft, Ludwig-Maximilians-Universität.

- **Input**:
  - Revised taxonomy
  - Three case studies
- **Output**: Evaluated taxonomy with three ESG fingerprints

The evaluated ESG taxonomy shows 3 meta dimensions with a total of 15 ESG risk dimensions. All dimensions are further specified by a superordinate question. The total of 47 characteristic attributes allows 14,929,920 consistent combinations to characterize ESG risk objects.

4 Application of the Concept to Develop an ESG Fingerprint for AI-Based Information Systems

Selected case studies such as air and water pollution (E), child labor (S), and corruption (G) will be used to demonstrate the applicability of the ESG risk fingerprint. Based on the ESG risk database, three case studies from 2019 are used. The case studies are briefly explained below:

4.1 Case Study 1: Air and Water Pollution (E)

On 26.09.2019, the Neue Zürcher Zeitung reports on a fire in a chemical factory in the French city of Rouen. Despite unclear cause, parts of the factory building of the company Lubrizol exploded, in which more than 5200 tons of chemicals have been set on fire and after a short time caused miles of smoke. Residents within a radius of 500 meters were asked to stay at home and keep doors and windows closed. The extent of the damage caused by air and water pollution to the inhabitants, agriculture, and the Seine River region was very serious at the time,[29] but the extent of the damage could be assessed only after a delay.

4.2 Case Study 2: Child Labor in the Supply Chain (S)

On 19.09.2019, BBC report cites child labor in Turkey. Despite attempts by companies like Ferrero to make supply chains 100% transparent, child labor violations continue to be made public. In this case, Turkey is found to have a complex supply

[29] https://www.nzz.ch/panorama/grossbrand-in-chemiefabrik-in-nordfrankreich-schulen-geschlossen-ld.1511454

ESG Risk Dimension	Characteristic		
ESG Risk Event	Fire in chemical factory	Child labour in the supply chain	Suspected corruption
Source	nzz.ch	20min.ch	handelsblatt.com
Type	Environmental	Social	Governance
Relationship	Organization	Organization	Organization
Reference	Organization Location	Geographical area	Geographical area
Risk management phase	Identification	Identification	Identification
Assignment	Industrial class	Product class	Industrial class
Risk Reference	Implicit	Implicit	Implicit
Entity Reference	Implicit	Implicit	Implicit
Temporal projection	Present	Past	Future
Data Completeness	Complet	Complet	Complet
Data Velocity	Batch	Batch	Batch
Data access type	Website	Website	Website
Accessibility	Open access	Open access	Open access
Rights of use	Restricted	Restricted	Restricted
Data structure	Unstructured	Unstructured	Unstructured
Data format	Non-proprietary	Non-proprietary	Non-proprietary
Data type	Textual	Textual	Textual

Fig. 6 ESG fingerprint on selected ESG case studies

chain that is often difficult to penetrate. The only way to make the supply chain transparent is to make the path traceable directly to the plantation.[30]

4.3 *Case Study 3: Corruption (C)*

On 21.10.2019, the Handelsblatt once again reported on the suspicion of corruption of the Dax group Fresenius Medical Care (FMC) in various African countries. Despite multi-million dollar settlements in the USA, the Frankfurt public prosecutor's office was now also investigating at this time. The investigation is based on numerous bribery incidents in the years 2007–2016. In total, several millions flowed to various clinics, managers, and doctors in numerous countries.[31]

[30] https://www.20min.ch/story/stammen-die-nuesse-fuer-nutella-aus-kinderarbeit-731658317548

[31] https://www.handelsblatt.com/unternehmen/industrie/medizintechnikhersteller-gegen-fmc-mitarbeiter-wird-auch-in-deutschland-wegen-korruption-ermittelt/25137374.html?ticket=ST-2717123-cgbxdPdlIP6WrqGzRcKb-ap2

4.4 Application of the Taxonomy to Case Studies for ESG Fingerprint Development

The application of the risk information from the case studies is intended to demonstrate the general applicability of the taxonomy. The goal is to extract patterns, so-called ESG fingerprints, from the case base. First results are presented in the form of ESG fingerprints in Fig. 6.

Case study 1 has the ESG risk type "Environmental" because the explosion caused air and water pollution. Likewise, a direct relationship to the company Lubrizol and the reference to the site in Rouen can be established. Due to named materials, the superordinate industrial class can also be inferred. Due to explicitly named entities (here: Lubrizol) and explicitly named risks (here: air and water pollution), the risk information can be classified as complete. The risk information is accessed batchwise via the website of a publicly accessible news provider. The data are available in textual form and are thus considered unstructured. The risk information is up to date with respect to the current risk situation in Rouen.

Case studies 2 and 3 differ only slightly from case study 1. For example, case examples 2 and 3 do not specify an exact location, but a geographical area. Compared to case example 1, case example 2 mentions specific product classes and products such as Nutella. Case example 2 refers, among other things, to risk reports that come from the past, whereas in case example 3, the suspicion of corruption is investigated and may consequently represent a future risk event.

4.5 Potentials for the Use of Big Data and Artificial Intelligence

The evaluation of the taxonomy shows how patterns, so-called fingerprints, are recognized within a set of ESG risk objects. The three fingerprints generated demonstrate that a possible focus within the risk identification phase has high potential in machine learning. The risk information is unstructured and contains tacit knowledge for ESGRM in the form of text. In the work of Hassan (2019) *Enhancing Supply Chain Risk Management by Applying Machine Learning to Identify Risks,* a framework is introduced that builds on a structurally similar framework in the context of supply chain risk management.[32] The framework is transferable and applicable to the present case of ESG risks due to the high similarity of the fingerprint to the experiment conducted there. The research question of how a conceptual design of an artifact has to be shaped in order to be able to classify supply

[32] Hassan (2019): Enhancing Supply Chain Risk Management by Applying Machine Learning to Identify Risks, Business Information Systems, 191–205.

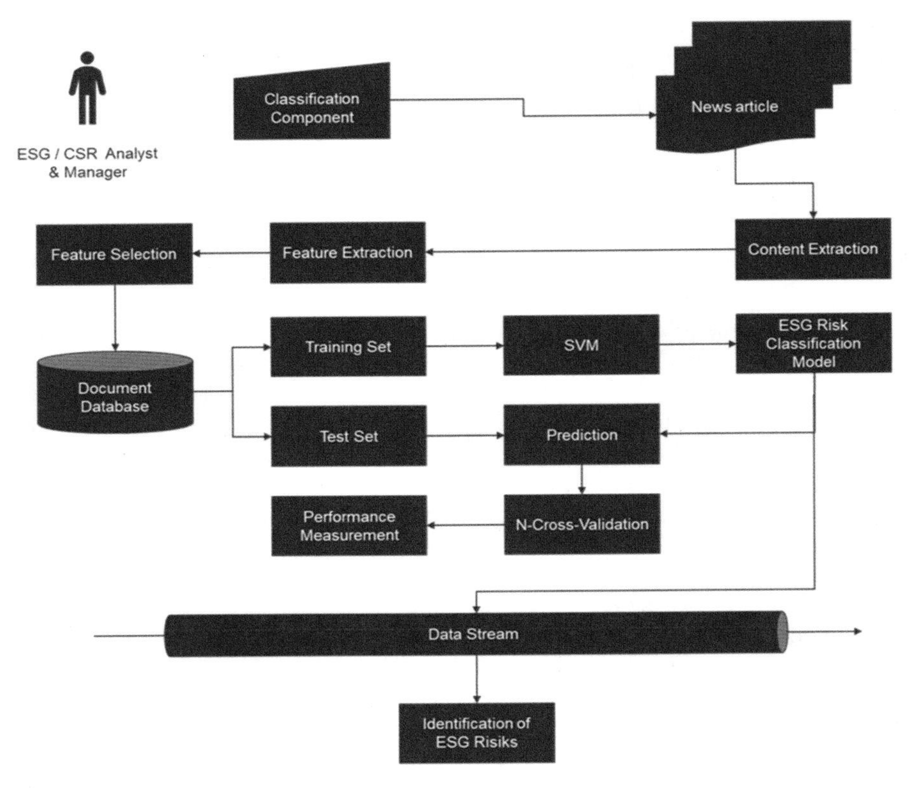

Fig. 7 Concept of artificial intelligence-based identification of ESG risks using support-vector machine (SVM)

chain risks in unstructured data forms the basis. The model is adapted to the present case and tested for theoretical applicability (Fig. 7).

The ESG analyst or manager is the expert for ESG risks. Based on the experience and expertise, he classifies news stories, according to the ESG fingerprint. The annotated data set of ESG risk case studies is processed in a second step by content and word extractions and written into a document database. The database is considered as the basis for machine learning techniques for automated recognition of ESG fingerprint information from unstructured news stories.

Through real-world experiments, Hassan (2019) demonstrated the relevance of the model. The use of technologies such as big data and AI can increase the speed to identify supply chain risks with high accuracy. Due to the high structural similarity of ESG and supply chain risks, as in this case, of risk information in unstructured news reports, as well as the high similarity of taxonomies, a general applicability of the model to ESG risks can be assumed. Consequently, ESG risks can be identified automatically with the help of AI. Areas of application for the method are therefore partially automated risk management, corporate and ESG due diligence processes, and novel ESG rating procedures.

5 Summary and Outlook

ESG shows great potential for integrating more sustainability into decision-making processes in organizations and companies and has been shown to lead to higher customer retention and acquisition and improved prospects for collaborations with other entities and has a positive impact on corporate reputation.[33,34] However, it is becoming apparent that it is not enough to develop new technologies and information systems in isolation from organizational processes. To be successful, an information system must be designed, created, and implemented as a socio-technical system starting from domain-specific processes. In this context, the development of the taxonomy is a necessary contribution to reduce complexity and to design ESGRM systems across process phases. Besides, an integrated view on ESGRM-relevant controversies is shown and empirically evaluated with the demonstrated ESG fingerprint.

The development of the ESG taxonomy as well as the resulting ESG fingerprint is suitable for characterizing ESGRM-related objects. Further insights can be expected once the evaluated taxonomy is applied to the entire case base and extensive fingerprints or patterns can be identified from the evaluation. These patterns will serve as a basis both for designing a holistic ESGRM system that enables integrative risk management across process phases and for deriving requirements for AI-based methods.

Since all three evaluation examples from the third iteration are classified in the risk identification phase, the authors consider it useful to draw a large sample from the entire case base and examine it in a comprehensive study. This investigation will allow further potential for the ESGRM phases to be identified.

Pajam Hassan is Managing Director of intuitive.AI GmbH and an expert in big data analytics in the field of CSR risks and supply chain management. He was awarded for his AI-based research in the automotive industry and successfully designed and implement one of the first AI-based early warning systems within a multinational corporation with a purchasing volume greater than € 100 bn. As a project portfolio manager, he combines data science practice and research in the field of big data analytics to shape the digital transformation in a most synergetic fashion.

[33] Ajina, A. S., Japutra, A., Nguyen, B., Alwi, S. F. S., & Al-Hajla, A. H. (2019). The importance of CSR initiatives in building customer support and loyalty. *Asia Pacific Journal of Marketing and Logistics.*

[34] Parcha, J. M. (2017). How much should a corporation communicate about corporate social responsibility? Reputation and amount of information effects on stakeholders' CSR-induced attributions. *Communication Research Reports*, *34*(3), 275–285.

Dr. Frank Passing is Managing Director of intuitive.AI GmbH and an expert in the field of natural language processing for strategic early intelligence of companies, technologies, innovations, and markets. He was awarded for his AI-based research in the automotive industry and the development of an innovative digital business model. He is the author of numerous scientific publications, studied at INSEAD, and received his PhD from the University of Bremen, where he is still a lecturer for "AI in Business."

Prof. Dr. Jorge Marx Goméz is Full Professor and Chair of Business Information Systems/Very Large Business Applications (VLBA) at the Carl von Ossietzky University of Oldenburg. His research interests include business information systems, federated ERP systems, business intelligence, and environmental management information systems. He is a member in specific departments of the "Gesellschaft für Informatik e.V.," the German Association of University Professors and Lectures and the German Forum for Interoperability. Furthermore, he is involved in industrial-driven consortia, such as the German Oracle users Group e.V. (DOAG) or the "SAP Arbeitskreis."

It's Only a Bot! How Adversarial Chatbots can be a Vehicle to Teach Responsible AI

Astrid Weiss, Rafael Vrecar, Joanna Zamiechowska, and Peter Purgathofer

Abstract We are currently witnessing an ever-growing entanglement of intelligent technology with people in their everyday lives, creating intersections with ethics, trust, and responsibility. Understanding, implementing, and designing human interactions with these technologies is central to many advanced uses of intelligent and distributed systems and is related to contested concepts, such as various forms of agency, shared decision-making, and situational awareness. Numerous guidelines have been proposed to outline points of concern when building ethically acceptable artificial intelligence (AI) systems. However, these guidelines are usually presented as general policies, and how we can teach computer science students the needed critical and reflective thinking on the social implications of future intelligent technologies is not obvious. This chapter presents how we used adversarial chatbots to expose computer science students to the importance of ethics and responsible design of AI technologies. We focus on the pedagogical goals, strategy, and course layout and reflect how this can serve as a blueprint for other educators in broader responsible innovation contexts, e.g., nonchat AI technologies, robotics, and other human-computer interaction (HCI) themes.

1 Introduction

In 2018, the Austrian Council on Robotics and Artificial Intelligence published a white paper entitled "Shaping the Future of Austria with Robotics and Artificial Intelligence." Several of these types of papers and reports have been published on national and international levels, and most suggest guidelines and recommendations

A. Weiss (✉) · R. Vrecar · J. Zamiechowska · P. Purgathofer
TU Wien, Faculty of Informatics, Institute of Visual Computing and Human-Centered Technology, Human-Computer Interaction (HCI) Group, Vienna, Austria
e-mail: astrid.weiss@tuwien.ac.at; rafael.vrecar@tuwien.ac.at; joanna.zamiechowska@tuwien.ac.at; peter.purgathofer@tuwien.ac.at

R. Schmidpeter, R. Altenburger (eds.), *Responsible Artificial Intelligence*, CSR, Sustainability, Ethics & Governance, https://doi.org/10.1007/978-3-031-09245-9_12

for developers regarding how to produce ethical artificial intelligence (AI) and create responsible innovation.

It is an appealing goal to ensure that AI-aided technologies are ethical; however, while it is currently popular to highlight this as a target, there is no agreed-upon route to achieve this implementation-wise. Additionally, there is a lack of concepts on teaching responsible AI to computer science (CS) students who will become the future developers of these technologies.

In this chapter, we present lessons learned and reflections on the master course Exploring Disruptive Technologies that we taught at TU Wien, Austria. Typically, courses on AI are offered to students of technical studies, such as CS, informatics, engineering, and others. In contrast, subjects that develop skills in recognizing and understanding ethical issues and responsible innovation are usually offered to students in humanities and social sciences. Therefore, one can reasonably expect that the CS student body is unlikely to have expertise in relevant humanities foundations. Because of this, we considered *adversarial chatbots* to be an opportunity to expose students to a broad range of new views and new ways of thinking about their work. In particular, chatbots can serve as a useful tool for training students in primary education in CS to be aware of how related fields, such as sociology, psychology, philosophy, and ethics, deal with AI-aided technology and how these perspectives can be useful for practitioners in designing, implementing, and evaluating technologies. This broad perspective should help students become critical and reflective future scholars and technology designers or developers.

This book chapter is structured as follows. Section 2 provides an overview of the state of education concerning exposing CS students to the ethical dimensions of technology and teaching resources for responsible AI. Section 3 outlines the course we developed at TU Wien, outlining its pedagogical goals, the course format, and the assignments. Next, in Section 4, we present the outcome of the course, presenting the *student projects* and *guidelines for ethical chatbot design* that students derived at the end of the semester. We close the chapter in Sect. 5 with a reflection from the students (two of whom are co-authors) and from the lecturers' perspective and conclude in Sect. 6 with an outlook on how this master course format could be leveraged into a plan for a bachelor-level course on responsible AI. This chapter presents subjective experiences from one instance of a master course attended by only six students. However, given that there is little AI ethics and responsible innovation education and substantial pressure to create it, we consider this chapter a relevant and hopefully inspiring contribution for fellow researchers.

2 Background

Artificial intelligence ethics originally emerged as a subfield of AI research in the past 15 years, and in parallel, the research stream of responsible innovation emerged from science and technology studies. The awareness of the need to include societal influence and ethics in CS education has existed for approximately the same period.

Twenty years ago, these topics were considered so relevant that they were added to the ACM/IEEE curriculum. Some governments are currently suggesting including AI education in primary schools (Chan, 2019). Western Universities, in comparison, are said not to be taking the topic seriously enough in their education (O'Neil, 2017). The policy framework "Responsible Research and Innovation" circumscribes four dimensions of societally responsible research and innovation processes: ongoing reflection on innovation processes, anticipation of societal implications, deliberation on ethical and value-related aspects, and responsiveness (i.e., the flexibility to adapt to unforeseen problems and potential). Therefore, responsible AI should go beyond conventional ethical review and approval, enabling researchers, developers, and society as a whole to consciously handle technology-related risks and challenges (Owen et al., 2012).

However, how can we teach the necessary skills and mindset for it? One reason for the lack of courses addressing these topics might be that neither AI ethics nor responsible innovations are sufficiently established research fields to offer an indisputable curriculum (or even a textbook) for teaching it. We must find methods to teach the future developers of technology not only to follow the national and international guidelines on AI and ethics, but to be critical thinkers aware of their influence when it comes to responsible innovation.

2.1 Exposing CS Students to AI Ethics and Responsible Innovation

Fiesler and colleagues conducted a qualitative analysis of 115 syllabi from a total of 202 identified AI ethics courses in university technology curricula (Fiesler et al., 2020). The main topics covered in the explored "tech ethics" courses included (listed by frequency) law and policies, privacy and surveillance, philosophy, inequality, justice and human rights, AI and algorithms, social and environmental impact, civic responsibility and misinformation, AI and robots, business and economics, professional ethics, work and labor, design, cybersecurity, research ethics, and medical/health. The goals and learning outcomes of the studied courses included critiquing, identifying issues, making arguments, improving communication, understanding multiple perspectives, creating solutions, considering consequences, and applying rules.

Fiesler and colleagues also explored those who typically teach courses on "tech ethics" (Fiesler et al., 2020). Arguments have supported two positions, that philosophers and social scientists trained in ethics should teach technology ethics and that CS lecturers should teach it "to emphasize to students that social impact issues are a fundamental part of computer science, not some tangential topic that they take somewhere else" (Johnson, 1994). Interestingly, their survey revealed that most classes are taught within CS departments, the home department, whereas their disciplinary background is more often in philosophy or information science rather than CS.

2.2 *Teaching Resources for Responsible AI*

The research field of "responsible AI is concerned with ensuring that the forum has sufficient power over the actor in the algorithmic accountability relationship" (Slavkovik, 2020, p.3). The field has numerous subinterests, including establishing a professional code of ethical conduct for AI research for developers and practitioners or developing strategies for assessing the ethical influence and value alignment of an AI application. Responsible AI also addresses overseeing the development of AI guidelines and ensuring that they are meaningful and actionable. However, it is not easy to include ethical and responsible AI topics in CS courses without being detached from the technology development process. Project-based teaching of human-robot interaction has proven to be a useful vehicle to expose technologist students to social aspects and implications of technology (Young, 2017).

Another idea regarding teaching and communicating responsible robotics to students and researchers of various disciplines was an approach based on a board game developed within the EU project REELER (2020). The primary goal of this board game (also available in a digital version) is to raise awareness of the increasing entanglement of robotic technology with people in society. A similar approach is the "IMAGINE Responsible Robotics" card-based engagement method (Sigl et al., 2021). Both approaches aim to develop common ground in inter- and transdisciplinary projects to develop novel robotic solutions. Therefore, both are suggested to be used in the initial stages of idea generation. Such playful approaches are helpful to create sensitivity for responsible AI in research approaches; however, as critically mentioned by the developers of the card game, such tools "can easily be misused as one-point interventions with little effect on actual research practices." (Sigl et al., 2021, p.1) Consequently, we must design teaching approaches that create lasting meaning for students.

3 Exploring Disruptive Technologies Course

The *Exploring Disruptive Technologies* master course at TU Wien aims to teach students what a particular technology can disrupt. It is now certain that AI applications are socially disruptive technologies. They change aspects of society, and it is in our interests to know how and why, whereas AI applications directly affect our job profiles and are changing the range of required skills that make one employable (Slavkovik, 2020). The course was tailored to the areas of expertise of the two lecturers (both authors of this chapter, with CS and sociology backgrounds) and their specific personal active research interests: privacy and security and human-centered human-robot interaction.

For the development of this course for the winter term 2020/2021, we started thinking of what might be a disruptive AI-aided technology in the personal lives of CS students. First, we thought of 3D printing a robot and following the example of a human-robot interaction project as the core of the course. However, because of the COVID-19 pandemic, we needed to find another nonembodied technology that serves the same purpose, but can be used for virtual distance learning. We finally produced the idea to focus the course on chatbots and a project in which students should aim to create "the most unethical/adversarial chatbot".[1]

3.1 *Pedagogical Goals*

The main goal of the course was to critically examine adversarial chatbots in theory and practice. From that example, students should learn the necessary skills to extrapolate possible futures and learn to observe the difference between the hyperbole usually surrounding new technologies and the social implications. Chatbots became prominent in 2016, "the year of the bot," when they were described as the new apps by Microsoft CEO Satya Nadella. Since then, we saw a rapid growth with more than 300,000 active bots with 8 billion messages exchanged every month only on Facebook. This growth is partially driven by commercial interests and the fact that simple chatbots are a fast and always-on service solution (Ruane et al., 2019). Therefore, we considered adversarial chatbots to be especially suitable for that course, as this is an AI most students have likely already encountered in one way or another. This type of AI-aided technology can serve as a starting point for discussions on responsible AI, which requires a shift in how students approach, discuss, and work through the challenges and ideas in the project work. Therefore, we aimed not to provide "concrete truths" about what a chatbot is by building small systems with extensive trial-and-error testing (which is a more "traditional" CS approach) but to present techniques for engaging with the technology and its implications from a human-centered perspective. Consequently, we emphasized oral discourse and peer feedback instead of programming and implementing working chatbots.

[1] A shorter description of the course was published as a short paper at INTERACT 2021 (the 18th International Conference promoted by the IFIP Technical Committee 13 on Human-Computer Interaction): Weiss, A., Vrecar, R., Zamiechowska, J., & Purgathofer, P. (2021, August). Using the Design of Adversarial Chatbots as a Means to Expose Computer Science Students to the Importance of Ethics and Responsible Design of AI Technologies. In IFIP Conference on Human-Computer Interaction (pp. 331–339). Springer, Cham.

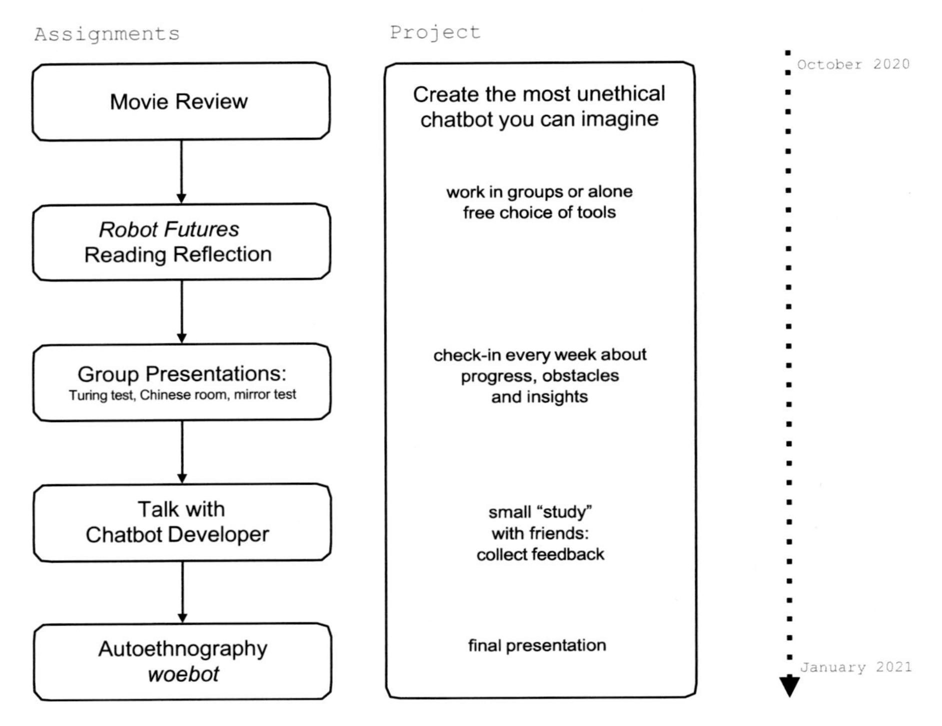

Fig. 1 Flow of the course

3.2 Course Format

The course format was based on weekly 90-minute time slots via video chat (Zoom). It was a small seminar-like course with six participating students. An additional Slack channel was created for asynchronous communication and discussion, and CryptPad was used to document the sessions and write assignment descriptions. Each Zoom class can be broken down into three components: (1) reflection on prepared materials (assignments) from the previous week, (2) input from the instructors, and (3) discussion of the status of the student project. The overall course structure, as held from October 2020 to February 2021, is presented in Fig. 1.

In the following section, we briefly describe the inputs and assignments that were part of the course.

3.3 Inputs and Assignments

At the beginning of the semester, the students had to conduct a movie review. We asked them to watch a movie of their choice in which robots or AI play a significant role. They were to pay close attention to how the film depicts the interaction between humans and intelligent technology; we recommended that they take notes as they watched. The review should not be a review of the movie; instead, all of the comments should address how humans and robots interact and communicate with each other. We proposed the following questions to guide the students: (1) What channels or modalities do people use to communicate with the robots, and how does their communication evolve? (2) What modes of expression do robots use to communicate with people? What about with each other? (3) What roles do the robots have in society? What are their effects? How do people react to the robots—positively, negatively, in what way? (4) Comment on what you think are the hard/easy social and technical problems involved with developing human-robot interaction of the sort shown in the movie and include potential ethical issues. Some examples of movies we suggested are *Robot and Frank*, *Ex Machina*, *Big Hero 6*, *Wall-e*, *Moon*, *The Iron Giant*, *Star Wars*, *Silent Running*, *Short Circuit, 2001*, *Hitchhiker's Guide to the Galaxy*, *AI*, *I Robot*, *Metropolis*, *Ghost in the Shell*, and *Astro Boy*; however, students could freely choose their movie.

The second task was a reading reflection of two chapters of *Robot Futures* (Nourbakhsh, 2013). This book starts every chapter with short dystopian stories and then discusses the current state of the art in related research and why such scenarios are partially likely. We considered this reading reflection to be an ideal starting point for debate in the course. The first chapter entitled "New Mediocracy" addresses the potential future of personalized advertisements, in which robotic technology can bridge between the digital and real world. It insightfully raises questions on privacy when human behavior is systematically tracked and used to manipulate desire. Therefore, the chapter served as a basis to discuss the interplay and dependency of economics and technology as powerful forces. The second chapter entitled "Dehumanizing Robots" tells a story of future challenges of distinguishing between humans and chatbots and aspects of morality in human-robot relations. This chapter served as a starting point to discuss the differences between artificial and human intelligence.

Students had to read the chapters in preparation for the Zoom class and write a short reflection (two to three paragraphs, not more than 500 words). Students were told that the reflections should serve as a basis for instructors and students to assess where the discussion is heading before the Zoom class. Moreover, they were told that their reflections should not just summarize the chapters, but should also reflect their understanding and thoughtful discussion of the material. We suggested that they focus on specific quotations in the reading that they think are particularly relevant and explain why they chose them. Alternatively, they could describe the most critical insights or pose the most important questions the reading raised for them and explain why. They could also include points that they planned to raise about the reading in class. The points of focus for these reflections were diverse. Some students were more interested in the scientific questions behind the portrayed futures, whereas others considered the ethical implications.

As the next assignment, students had to conduct group research on the topics of the Turing test (Saygin et al., 2000), Chinese room (Searle, 2006), and the mirror test

(Haikonen, 2007). The Turing test should deepen the reflection on the differences between human intelligence and AI. The Chinese room argument reflects whether machines can ever have a consciousness, and the mirror test questions whether machines can be self-aware. The presentations pointed out that the Turing test was already a well-known example for most students. However, the Chinese room argument and mirror test offered many new insights in the contexts of agency, human-likeness, and deception. Based on these insights, the similarities of human and machine intelligence and the crucial aspects in human-agent interaction [embodiment, personality simulation, the uncanny valley phenomenon (Mori et al., 2012), etc.] were discussed.

In a session shortly afterwards, we invited chatbot developer Dr. Barbara Ondrisek (self-proclaimed "Bot Mother" and "enthusiastic software developer") for a talk to share her experiences and provide students with advice regarding designing valuable chatbots for users. She presented different projects she was involved in and stated what makes a chatbot "good" from her viewpoint, which can be considered a business perspective. The talk was relevant input for the student projects, which was the second main pillar of the course. The two main takeaway messages for students were that (1) one should always think about the "personality"/ "character" and work with it in the interaction development to increase user satisfaction and that (2) there should always be a carefully considered way to "exit a conversation" with the bot to deal responsibly with customer requests. All students used this advice in their projects.

Last, the participants had to conduct an autoethnography of Woebot (https:// woebothealth.com, accessed September 27, 2021), a chatbot that helps users as a digital therapist. In advance, the students were provided with information and tips regarding the concept of autoethnography (Ellis et al., 2011) and received additional literature on the method. They were asked to install the Woebot app on their smartphone and use it extensively for several days. Afterwards, they were invited to discuss their experiences with other students. However, if one did not want to work with Woebot for privacy reasons, literature research on the technology was also sufficient. Some students experienced these interactions as "uncomfortable and weird." Moreover, we discussed the intersection of "AI and health care" and "persuasive technology" afterwards. Furthermore, the students tried to break the conversation flow with Woebot, which resulted in surprising and often funny conversations, which was also insightful for the students and their projects.

3.4 Student Project

The second main pillar of the course was a student project in which students were tasked to create a chatbot that was as unethical as they could imagine. Whether they focused on implementing an actual testable object with functional interaction or focused on conceptualizing a design for the bot was entirely in their hands. As we later observed with the different student projects, the focus was also completely different.

The students were encouraged to use story-boarding, machine learning frameworks, sketches, and, most importantly, project discussion methods with others and gather feedback during Zoom sessions as they provided updates on their projects. Reflecting on the progress and "unethical aspects" was incredibly important during the project. The students were generally allowed to work in groups, although all projects except one were done individually. Two students conducted the mentioned project and therefore also had a more extensive scope than the rest. More importantly, the students were to reflect on the fact that many different, but essential aspects besides machine learning must be considered when developing a chatbot, especially ethics and responsible design. Our firm belief is that this perspective is crucial to ensure that students gain knowledge about the effort, which is necessary to develop innovative technologies responsibly. After completing this course, they should appreciate the value that responsible development provides and understand why this extra effort is necessary.

A minimalist user evaluation should foster these newly developed values and motivate them to consider ethical aspects concerning the users when developing and offering these technologies. The results and a working demonstration were presented at the end of the course.

4 Outcome

We aimed for two major course outcomes: the student projects and student-developed guidelines reflecting what they learned regarding designing ethical chatbots. We describe both in the following sections.

4.1 Student Projects

Inspirational Quotes Bot A group of two students implemented an inspirational quotes bot that aimed to support conversations with "deep" quotes from the internet. This goal was achieved using machine learning and a database of sample conversations. The unethical aspect that the team of students wanted to demonstrate was that "suitable quotes" identified through a machine learning algorithm might be problematic. They demonstrated how their bot answered test messages such as "Don't you want to end your life sometimes, too?" with "Sometimes, one has to move on and let go of things one does not like anymore. Be brave to make your plans a reality."

Fitness Guru The fitness guru bot aimed to help overweight people lose weight. However, as it was modeled character-wise after a famous sports guru, it insulted them if they provided data that implied an unhealthy diet or something similar. Therefore, the unethical aspect was part of the character choice and, subsequently, the dialogue flow.

Phishing Attack Bot As there are numerous phishing or scare attacks on the internet, often initiated by spam email, another student thought of a way to misuse a chatbot to gain the victim's trust. However, this was from a distinct perspective than the others, as the chatbot was not designed to be unethical per se; instead, a specific unethical and criminal (if conducted in real life) use case was imagined. Instead of visiting a web page to steal user data, it used a "virus hoax" to scare the user, and the user had to interact with the chatbot to "remove" the virus. The chatbot then pretended to help remove the virus by directing the user to download a sophisticated antivirus tool that required a credit card transaction. While the downloaded software could infect the computer, the credit card data could also be stolen.

Customer Service Agent This project proposed a custom-built customer service agent in the form of a chatbot. Although this bot could be used on a commercial website, such as a webshop, to solve problems when using the page, it had the hidden agenda to pressure the user to positively review a specific product to generate a free advertisement for the product from their users. The unethical aspect of this project was deception: the bot pretended to be helpful, but deceived users into providing positive reviews.

Building Rapport This project proposed a chatbot that passes the Turing test and pretends that it is an actual human and not a bot. The bot emulates a human-like conversation using humor, slang, and wrongly spelled words, which are not expected from a bot. The unethical aspect of this bot is that it deceives the user into an erroneous belief. Furthermore, social media could be a possible use case, as this bot could post comments that appear to be from an actual human to foster a particular opinion or belief. The implementation was done using Google Dialogflow and Google Assistant technologies.

4.2 Guidelines

The last classroom activity was a retrospective examination of the project, other assignments, and discussion results. In this session, the students and instructors derived the following guidelines on ethical chatbot design:

- Be transparent: Openly show that your chatbot is a chatbot.
- Use a label or explain: Be open about the intentions.
- Do not try emulating human behavior: For example, avoid human names and emotions.
- Avoid assigning gender to the chatbot.
- Provide a way to chat with a human and offer a way to end the conversation.

- Try to detect urgency and relay the chat to a human.
- Be aware of vulnerable users or groups.
- Be aware of people with cognitive limitations or disabilities.
- Always question the values encoded in the chatbot when offering options: Try to be comprehensive and offer a "none of the above" option.
- Aspire to be sensitive to context: For example, in terms of culture, do not assume everyone celebrates on December 24.
- Inform users about how their data and input are handled, as mandated by General Data Protection Regulation (GDPR), and offer people a chance to opt out.
- Assume the worst about the chatbot and design for it.

5 Reflection

Overall, the course format worked well with some management, even in the distance learning format, and did not encounter any evident issues. In the beginning, students struggled with the critical discussion of the provided materials and assignments, and instructors felt that not all students were convinced from the learning method. However, this improved over time, particularly after the invited talk by Dr. Ondrisek as starting point for the more practical work. As soon as the work on the student projects started, students could relate to prior materials and discussions and realize how well it all interconnected.

Concerning the covered topics, students reported that they found the invited talk very inspiring, making it well suited as a starting point for their project work. Two students struggled with the Woebot task. One student openly shared actual therapy needs in his family and considered this app irresponsible, and another student decided to research the app, because they did not feel comfortable sending messages on her mood to the bot.

Finally, the short evaluation of the student projects through friends appeared to be highly effective in demonstrating the relevance of responsible innovation. Students proactively asked for help setting up their evaluation studies to not "harm" their participants. They became interested in research ethics and briefing and debriefing participants, and we provided them with additional literature on that topic (Rea et al., 2017; Geiskkovitch et al., 2016).

Comparing the guidelines developed by the students with the social and ethical considerations of conversational AIs (CAIs) (Ruane et al., 2019) indicates the successes of the course and future areas for improvement. The authors of these social and ethical considerations explain the importance of trust and transparency and how users can make informed choices in their interactions with a CAI. They suggest making the CAI status nonhuman and making its motivations and capabilities explicit, which aligns with the student guidelines.

Similarly, Ruane and colleagues explained that users expect CAIs to be neutral and unbiased and that their data are secure. However, this is often not the case. The authors describe the topic of user privacy as "paramount" and increasingly important to address on a societal and legislative level due to the encroachment of these technologies into more aspects of our lives. Users are mostly unaware of the amount and scope of data collected on them and how these data are used. These aspects are also partly addressed in the student guidelines on data handling. Similar to their suggestion to follow GDPR, Ruane and colleagues favored legislative and legal compliance concerning data privacy protection. The authors also aligned with the student guidelines on designing CAIs to be gender-neutral or gender-fluid. This character design avoids reinforcing gender stereotypes that are purposefully or unintentionally programmed into the CAI and avoids influencing the user's interaction with the agent. Dehumanizing and not anthropomorphizing the chatbot also helps preventing a user from subconsciously placing undue trust in it. Finally, Ruane and colleagues discuss the dangers of unsupervised learning, where the CAI could learn profanity, abusive language, or personal data from users and incorporate this into future conversations with different users. The importance of controlling the learning data and chatbot responses was demonstrated and extensively discussed in the context of the inspirational quote chatbot.

This comparison demonstrates that the course managed to convey the relevance that a developer or designer must always consider the unintended ethical consequences of their work to mitigate possible harm. As we observed from the student projects, even well-meaning chatbots can become problematic. As the projects progressed, the ethical considerations became increasingly nuanced and complex. Emerging themes of transparency, privacy, user rights, protecting people, and the larger social influence of technology became central discussion points. As such, the principles of digital humanism and placing the person's well-being at the center of design considerations underscore the guidelines constructed during the course.

5.1 Student Perspective

Our main goal was to present reflective techniques for engaging with technology and its implications to the students through the provocative task of thinking about unethical chatbots for the student project. We aimed to foster reflective stances and helped students link their project work with the literature, state-of-the-art research, and public discourse on responsible AI.

As a voluntary final submission, we asked the students for a written reflection on their learning experience, which four out of six students completed. The following quotes suggest that we succeeded:

> I honestly can say that the lecture substantially changed my view on chatbots. First and foremost because I did not see how many negative implications can be caused by malfunctioning chatbots as people probably can be encouraged by the bot to do things to harm themselves or others.

Following the constructivist learning theory, we believe that what one learns is determined greatly by the diverse and holistic ways one can think about a subject matter. Therefore, our meta-goal was that students learn skills necessary to extrapolate possible futures from the chatbot example. Again, student feedback suggested that this was indeed the case:

> Furthermore, I underestimated the ethical implications that come with designing these. For me, chatbots before were just small gadgets which regularly totally mess up and annoy me when changing details on my, e.g., phone contract, as they are often used to replace human assistance in my experiences [...]. The lecture was insightful from many perspectives and also thinking about malicious use cases broadened my horizon in a way that we always have to think twice [...] when elaborating if our intended design can have negative implications [...].

Another student stated that the course was "adventurous":

> We didn't have such strict tasks with rigid deadlines like in other courses, but together we explored unethical chatbots, which I find really adventurous. [...] Frankly, I never had in my technological education any focus on ethical aspect or social consequences of what I as an engineer create. Very big advantage of this course was opportunity to train creative thinking, as we weren't just ordered to perform very concrete strict tasks but had an opportunity to think about possible usage of our chat bot.

5.2 *Teacher Perspective*

The experience with the master course Exploring Disruptive Technologies using chatbots for teaching responsible AI made us assess how to transfer that to a bachelor course in a comparable manner. Integrating ethics into purely technical program courses involves several challenges. Among others, we must combat the "I am just an engineer" mindset and that ethics should be someone else's job (Slavkovik, 2020). The student projects in the master course that should focus on implementing an adversarial chatbot helped sufficiently challenge this mindset and determine the effect that the technology and their work can have on an ethical level. However, a project-centered approach is infeasible in an introductory class on responsible AI of around 300 bachelor students.

Instead, one could think about more invited talks by developers and reflect on their experiences with the research literature. Movie reviews, reading reflections, and autoethnographies of existing AI-aided technologies can also be reasonably managed in large classes. Discussion panels could stimulate discourse, and students could volunteer to obtain extra points. Students can prepare for the panel based on reading reflection and could be asked to represent a specific opinion or point of view. They would not have to share their own opinions but could discuss the matter as a type of roleplay. Finally, one could use playful methods regarding responsible robotics (REELER, 2020; Sigl et al., 2021) as a starting point to familiarize students with the topic.

6 Conclusion

How do we achieve responsible AI in practice? The idea that it can be achieved *downstream* "with the deployment of technology" (i.e., when the technology is already developed and ready for use) is unconvincing. If we want to shape the ethical, legal, and social character of AI technologies, we need to start at least *midstream* in the research development and design process or, even better, *upstream*, for example, through public discourse or in educating future developers (Fisher et al., 2006).

It is challenging to make the value-laden aspects in technology development projects visible and tangible for students. Studies have indicated that researchers sometimes orient themselves toward what is fundable or publishable during the academic socialization processes and do not pay sufficient attention to the broader societal issues in the progression of their academic careers (Fochler, 2016). Therefore, we consider it even more crucial and relevant to start early in bachelor's degree studies with project-based courses on responsible AI, later consolidating this in a master course.

This chapter demonstrated how adversarial chatbots could be useful for exposing CS students to a broader, socially embedded view on AI-aided technology. We consider the two-pillar approach of reflective assignments linked to a student project to be the most promising. However, the project focus must be on the ethical dimension and not on developing a working bot.

References

Chan, D. (2019). *Primary students to be taught AI in Guangzhou schools government has decided to give priority to the development of AI, information and biopharmaceutical industries*. Accessed September 27, 2021, from https://asiatimes.com/2019/07/primary-students-to-be-taught-ai-in-guangzhou-schools/.

Ellis, C., Adams, T. E., & Bochner, A. P. (2011). Autoethnography: An overview. *Historical Social Research/Historische sozialforschung, 36*(4), 273–290.

Fiesler, C., Garrett, N., & Beard, N. (2020). What do we teach when we teach tech ethics? A syllabi analysis. In *Proceedings of the 51st ACM Technical Symposium on Computer Science Education*, pp. 289–295.

Fisher, E., Mahajan, R. L., & Mitcham, C. (2006). Midstream modulation of technology: Governance from within. *Bulletin of Science, Technology & Society, 26*(6), 485–496.

Fochler, M. (2016). Beyond and between academia and business: How Austrian biotechnology researchers describe high-tech startup companies as spaces of knowledge production. *Social Studies of Science, 46*(2), 259–281.

Geiskkovitch, D. Y., Cormier, D., Seo, S. H., & Young, J. E. (2016). Please continue, we need more data: An exploration of obedience to robots. *Journal Human Robot Interaction, 5*(1), 82–99.

Haikonen, P. O. (2007). Reflections of consciousness: The mirror test. *AAAI Fall Symposium: AI and Consciousness*, pp. 67–71.

Johnson, D. (1994). Who should teach computer ethics and computers & society? *ACM SIGCAS Computers and Society, 24*(2), 6–13.

Mori, M., MacDorman, K. F., & Kageki, N. (2012). The uncanny valley [from the field]. *IEEE Robotics & Automation Magazine, 19*(2), 98–100.
Nourbakhsh, I. R. (2013). *Robot futures*. MIT Press.
O'Neil, C. (2017). *The ivory tower can't keep ignoring tech*. Accessed September 27, 2021, from https://www.nytimes.com/2017/11/14/opinion/academia-tech-algorithms.html.
Owen, R., Macnaghten, P., & Stilgoe, J. (2012). Responsible research and innovation: From science in society to science for society, with society. *Science and Public Policy, 39*(6), 751–760.
Rea, D. J., Geiskkovitch, D., & Young, J. E. (2017). Wizard of awwws: Exploring psychological impact on the researchers in social HRI experiments. In *Proceedings of the Companion of the 2017 ACM/IEEE International Conference on Human-Robot Interaction, HRI '17* (pp. 21–29). Association for Computing Machinery.
REELER. (2020). *BuildBot*. Accessed September 27, 2021, from https://reelertoolbox.ab-acus.com/buildbot/.
Ruane, E., Birhane, A., & Ventresque, A. (2019). Conversational AI: Social and ethical considerations. In *AICS*, pp. 104–115.
Saygin, A. P., Cicekli, I., & Akman, V. (2000). Turing test: 50 years later. *Minds and machines, 10*(4), 463–518.
Searle, J. (2006). *Chinese room argument, the Encyclopedia of Cognitive Science*.
Sigl, L., de Pagter, J., & Papagni, G. (2021). *"Imagine responsible robotics" - a card-based engagement method*.
Slavkovik, M. (2020). Teaching AI ethics: Observations and challenges. In *Norsk IKT-konferanse for forskning og utdanning, no. 4*.
Young, J. E. (2017). An HRI graduate course for exposing technologists to the importance of considering social aspects of technology. *Journal of Human-Robot Interaction, 6*(2), 27–47.

Astrid Weiss is an Assistant Professor on Human-Computer Interaction at the Institute of Visual Computing and Human-Centered Technology at TU Wien (Austria). She received her doctorate degree in Social Sciences in 2010 and her habilitation in Human-Computer Interaction in 2022, both at the University of Salzburg (Austria). She is one of Austria's key figures in the interdisciplinary research field of human-robot interaction (HRI), as evidenced by numerous articles, lectures, and conference organizations. In 2018, she was elected as member of the Young Academy of the Austrian Academy of Sciences. Astrid Weiss was one of the two lecturers of the Exploring Disruptive Technologies seminar at TU Wien in 2020.

Rafael Vrecar is currently pursuing his PhD studies in Human-Computer Interaction at TU Wien (Austria). He graduated in BSc Software and Information Engineering in the fall of 2020 and finished his additional Bachelor with Honors Program shortly afterward (both TU Wien). He wrote his master thesis in the field of human-robot interaction and graduated in MSc Media and Human-Centered Computing in May 2022 (TU Wien). His other research interests are rooted in the fields of usability and IT security. He attended the Exploring Disruptive Technologies seminar at TU Wien in 2020.

Joanna Zamiechowska finished her undergraduate degree in Electrical Engineering at the University of Illinois at Chicago, after which she worked as an energy specialist and then in performance assurance at Siemens Building Technologies. She is currently pursuing her master's degree in Data Science at TU Wien (Austria) with a special interest in blockchain technologies and natural language processing. She attended the Exploring Disruptive Technologies seminar at TU Wien in 2020.

Peter Purgathofer is an Associate Professor on Interactive Systems at the Institute of Visual Computing and Human-Centered Technology at the TU Wien (Austria). His habilitation discusses history, theory, and practice of design methodology, focused on the design of interactive software (2005). Since 2007, he is coordinator for the "media informatics" bachelor and master programs at TU Wien (Austria). Peter Purgathofer was one of the two lecturers of the Exploring Disruptive Technologies seminar at TU Wien in 2020.

Concerted Actions to Integrate Corporate Social Responsibility with AI in Business: Two Recommendations on Leadership and Public Policy

Francesca Mazzi

Abstract Businesses are increasingly adopting AI solutions. Governments, investors and consumers increasingly focus on their accountability for the environmental and social impact of their activities. To address this challenge, corporate social responsibility should be integrated with AI in business by design and by default. This chapter attempts to contribute to this goal providing two recommendations addressing leadership and public policy. Firstly, leaders can adopt a three-level mindset framework. Such framework embeds ethical considerations and the Sustainable Development Goals as a benchmark for impact assessments in the whole lifecycle of AI. Secondly, AI regulation and policy harmonisation can facilitate the adoption of such framework by businesses and consequently the maximisation of positive externalities of AI in business. The two recommendations are contextualised with insights from a dialogue with four projects in Latin America using AI for the Sustainable Development Goals.

Keywords AI regulation · Harmonised AI framework · Sustainable development goals · AI leadership · AI for the Sustainable Development Goals

1 Introduction

Most private and public organisations around the world are employing or about to employ artificial intelligence (AI).

AI can be defined as "a machine that can behave in ways that would be called intelligent if a human were so behaving" (Quotation from the 2006 re-issue in McCarthy et al. 2006). However, as argued by Floridi (2019), it does not mean that the machine is intelligent or able to think. It behaves according to the instructions received through training and data, without any autonomous consideration on

F. Mazzi (✉)
Saïd Business School, University of Oxford, Oxford, UK
e-mail: Francesca.Mazzi@sbs.ox.ac.uk

R. Schmidpeter, R. Altenburger (eds.), *Responsible Artificial Intelligence*, CSR, Sustainability, Ethics & Governance, https://doi.org/10.1007/978-3-031-09245-9_13

the impact of its activity. Therefore, it is not possible to say that AI behaves in a "good" or a "bad" way per se (Sabater-Mir et al., 2019; Moore, 2019).

The types of AI used in the public sector concur to the performance of public tasks, which are presumably in the public interest (Chen & Zhou, 2019). For the purpose of the present chapter, we will assume that acting in the public interest means to act in consideration of the greater good of a community. Thus, we could say that the AI used by public sector is supposed to "behave" in a "good" way, as it serves a greater good. In contrast, the private sector employs AI for objectives that depend on the purpose of their organisations, and consequently without any requirement to consider the public interest, apart from the boundaries of public order and morality eventually imposed by the applicable laws. Thus, the AI used in the private sector could "behave" in a good or bad way, depending on the impact on the community (Patelli, 2019).

However, there is a growing concern regarding the impact of private use of AI on the population, with scholars arguing that it could amount to a situation of public interest, highlighting, for example, the risks it might pose to democracy (Manheim & Kaplan, 2019). Moreover, draft regulations of AI in Europe indicate that private organisations will be required to justify the decisions incorporated in the design of AI for accountability purpose, which seems likely to lead to ethical auditing (Koene et al., 2019; Mökander et al., 2021). Also, the growing focus on the field of business ethics, the relevance of corporate social responsibility (hereinafter "CSR") for brands' reputation (Mahmood & Bashir, 2020) and the increasing attention to ESG parameters in finance[1] indicate that the values and the purpose embedded in AI design will have repercussion in the financial performance of private organisations. In a nutshell, private actors (especially big corporations) are increasingly required to report to both consumers and investors on the impact of their actions on communities.

Thus, leaders in the era of AI must address several challenges, such as the integration of human intelligence with artificial intelligence, the changing skills required for jobs, the environmental impact of technologies and others (Antonescu, 2018). This chapter focuses on one of them: the integration of CSR with the adoption and deployment of AI.

Integrating CSR with the adoption and deployment of AI is proposed as mindset that includes considerations on AI as a technology, the business model, the long-term business strategy and the external impact.

[1] Biermann, F, Kanie, N, Kim, RE (2017) Global governance by goal-setting: The novel approach of the UN Sustainable Development Goals. Current Opinion in Environmental Sustainability 26: 26–31.; Kanie, N, Biermann, F (eds) (2017) Governing Through Goals: Sustainable Development Goals as Governance Innovation. MIT Press, London, UK.; Stevens, C, Kanie, N (2016) The transformative potential of the sustainable development goals (SDGs). International Environmental Agreements 16: 393–396. Bouteligier, S (2011) Exploring the agency of global environmental consultancy firms in earth system governance. International Environmental Agreements: Politics, Law and Economics 11(1): 43–61.

Specifically, we argue that AI ethics (Floridi & Cowls, 2021) and AI for social good (specifically, AI for the Sustainable Development Goals) (Taddeo & Floridi, 2018) are two essential elements in the process of integrating CSR with AI in business and that both business leaders and policymakers should facilitate the integration process.

The chapter adopts a programmatic approach: it provides two recommendations to facilitate such integration activity based on the assumption that concerted actions from both private and public actors are desirable to achieve a maximisation of AI externalities of the private sector. The first recommendation interests business leadership (what leaders need to do), and the second public policy (what leaders need). As a case study, the chapter uses insights obtained through a dialogue with four AI4SDGs projects from Latin America. The peculiarities of such case study are both the fragmented normative framework and the centrality of the socially good purpose of AI in the projects. Because of these characteristics, the case study is not intended to generalise nor to verify the hypothesis of the framework with statistical significance. It allows for contextualisation of the two recommendations for descriptive and exemplificative purposes.

In terms of business leadership, the chapter proposes what can be defined as a "mindset framework", designed based on the attributes that leaders should have in the era of AI according to Heukamp (2020). It is a three-level framework for leaders to envision the integration of CSR with the use of AI: through the adoption of AI (new AI), in the tasks that the AI perform (applied AI) and in the scale and scope of its application (potential AI). The chapter contextualises the framework with the experience of the four AI4SDGs projects; it discusses the application of the framework in different sectors and its limitations.

In terms of public policy, the chapter argues in favour of (1) AI regulation that incentivises AI for social good and (2) regional harmonisation that maximises AI externalities for social good, building on the positive effects of harmonised regulation for businesses in the EU. The chapter provides a contextualisation of the recommendation with the experience of the four AI4SDGs projects in the fragmented normative frameworks of Latin America, and it explains how regulatory harmonisation in the region could optimise the AI for social good externalities.

The chapter is structured as follows: Sect. 2 sets the scene, describing the interconnections between CSR, ethics and the UN Sustainable Development Goals and defining the scope of the discussion; Sect. 3 presents and discusses the mindset framework as a method to integrate CSR with AI in business; and Sect. 4 explains the relevance of the policy framework to facilitate the integration process and the maximisation of AI externalities.

2 Setting the Scene: CSR, Ethics and SDGs

The aim of the chapter is to provide two recommendations to facilitate the integration of CSR with AI in business. We shall define CSR: we consider the work of Sheehy (2015), which defined it as a form of private business self-regulation focused on the environmental and social impacts of business; and of Benedict Sheehy and Federica Farneti (n.d.), which highlighted how CSR includes a host of individual and collective rights and ethical considerations. From these elements, we provide a working categorisation of the interests at stake and parties involved to visualise CSR as a system. We classify the business as the provider of CSR, the environmental and social impact as the object of CSR, ethics as the set of guiding principles of CSR and individuals and collectivity as recipients of CSR. The proposed categorisation reflects the five dimensions described in the work of (Dahlsrud, 2008), which analysed 37 definitions of CSR: however, the economic dimension of CSR, which prescribes preserving profitability and would lay in the business as provider of CSR, is outside of the scope of the present chapter.

From our definition, it is self-explanatory that CSR, ethics and environmental and social sustainability are interconnected.[2] Since the focus of the chapter is the integration of CSR with the adoption of AI, we shall now introduce two elements that are fundamental for our purpose: the first one is AI ethics. Ethics has been defined as the set of guiding principles of CSR: Dahlsrud (2008) showed that the terms “values”, “ethical values”, “ethical behaviours” and “ethical expectations” are frequent in CSR definitions, and Sheehy (2015) reported that CSR has been viewed as a manifestation of business ethics. AI ethics is therefore an essential element of integrating CSR with AI. The second one concerns the United Nation’s Sustainable Development Goals (hereinafter “SDGs”). As we have seen, a positive environmental and social impact represents the ideal objective of CSR. Businesses as providers of CSR are part of Earth System Governance, and they can and shall contribute to the SDGs (Dahlmann et al., 2019; Nylund et al., 2021). Therefore, the SDGs can be used as a benchmark to evaluate the impact of the business use of AI.

These two elements are interconnected. Indeed, as shown by Josh Cowls et al. (n.d.), the respect of ethical principles is essential to foster the development of AI for SDGs.

In this chapter, we will provide insights from a dialogue with four projects that have AI for SDGs as core business (hereafter “AI4SDGs”) as they meet the five criteria set by Cowls et al. (2021). As shown by Cowls et al. (2021), AI4SDGs projects offer unprecedented opportunities across many domains, considering the

[2] See Benedict Sheehy and Federica Farneti (n.d.) for the relationship between sustainability and CSR and Bansal, P, Song, HC (2017) Similar but not the same: Differentiating corporate sustainability from corporate responsibility. Academy of Management Annals 11(1): 105–149. Mitchell, RK, Agle, BR, Wood, DJ (1997) Toward a theory of stakeholder identification and salience: Defining the principle of who and what really counts. Academy of Management Review 22(4): 853–886.

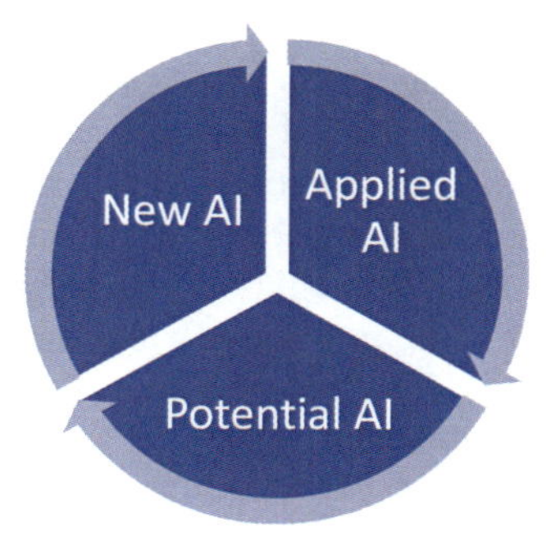

Fig. 1 The three levels of the mindset framework

global nature of the SDGs. However, AI4SDGs projects as a category are still poorly understood: research on the potential of AI for social good is flourishing (Tomašev et al., 2020; Khamis et al., 2019; Vinuesa et al., 2020), but an overview of best practices and lessons learned deriving from existing AI4SDGs projects attracted no research efforts yet. The Oxford Initiative on AI4SDGs aims, amongst others, to fill this gap, developing best practices and lessons learned from existing AI4SDGs projects. In the context of such macro-area of research, a channel of communication with the projects was established, and this chapter reports few preliminary findings obtained through a dialogue with four AI4SDGs projects from Latin America.[3]

3 A Recommendation on Business Leadership: Adopting a Three-Level Mindset Framework

We have defined the scope of the discussion by specifying why it is desirable to achieve a goal, i.e. the integration of CSR with AI, and what is the meaning of such goal, i.e. AI ethics and SDGs as constituent parts (guiding principles and objective) of CSR, businesses as providers of CSR and individuals and communities as recipients. We shall now move to how to achieve such goal. In this section, a tentative three-level mindset framework for leaders is presented, based on the key "be" leadership attributes in the era of AI according to Heukamp (2020): ethical, unbiased, humility, adaptability, vision – purpose, engagement, trust and privacy.

The three levels are *new AI*, *applied AI* and *potential AI*. "New AI" concerns the process of designing and training the AI and the reason why it is employed by the business. "Applied AI" concerns what AI does, i.e. what is the function that it performs and what is the impact of its application. "Potential AI" concerns what AI can do, as of what it can do in other geographical areas, in other sectors, for other purposes (Fig. 1).

[3] An initial survey was launched from April to June 2021.

According to the framework, leaders should adopt an ethical framework for AI and the SDGs as a benchmark in each of the three levels.

In new AI, the ethical framework leads to inter alia verification of algorithmic fairness and transparency. In applied AI, it leads to an evaluation of the negative and positive impacts of the use of AI and, in general, of the business activity. In potential AI, it helps in prioritising the next steps as either to address the negative impacts of the AI or to implement other uses of AI to provide further positive contributions.

This allows for the integration of the "ethical, unbiased and privacy" attributes (Heukamp, 2020).

The SDGs benchmark is used in new AI to incorporate the business vision (intended as purpose, mission statement), and it integrates the attribute "vision – purpose". In applied AI, it is employed to measure the impact of the business model, integrating "humility" attribute (Heukamp, 2020). In potential AI, it is used to evaluate opportunities to further contribute through the improvement of both internal and external deficiencies and through expansions or collaborations. This potential can be then integrated with the business vision and new AI.

The adoption of such mindset framework aims at favouring a cyclical, constant monitoring and revision of the business impact and the business vision, which allows to integrate the "adaptability, engagement and trust" attributes (Heukamp, 2020).

As an example, in the new AI, phase leaders can question why do they adopt AI (how does it contribute to the business vision) and whether it is designed ethically; in the applied AI, leaders can evaluate what is the impact of AI and of the business model in terms of ethics and contribution to SDGs; in the potential AI phase, they can question how to address eventual negative impacts identified through applied AI and how to use AI to further contribute to the SDGs (Fig. 2).

The framework is designed to encourage the integration of CSR with AI in business by design and by default. Moreover, it allows to think about AI dynamically, to continuously reconsider its application within the business and in relation to the SDGs and to exploit its potential by reducing negative externalities and maximising positive externalities. It promotes proactive leadership in addressing societal goals and favours transparency and accountability.

3.1 Contextualising the Framework: A Case Study of Four AI4SDGs Projects in Latin America

In this section, the mindset framework is contextualised with the experience of four AI4SDGs projects. Specifically, we intend to highlight why adopting an AI4SDGs framework would build on the strengths of the business vision, and it would allow to address the weaknesses.

Three projects out of four have expanded their scope since their first real-world application, two of them by expanding the project in another geographical area and one collaborating with another entity in another geographical area. All the projects

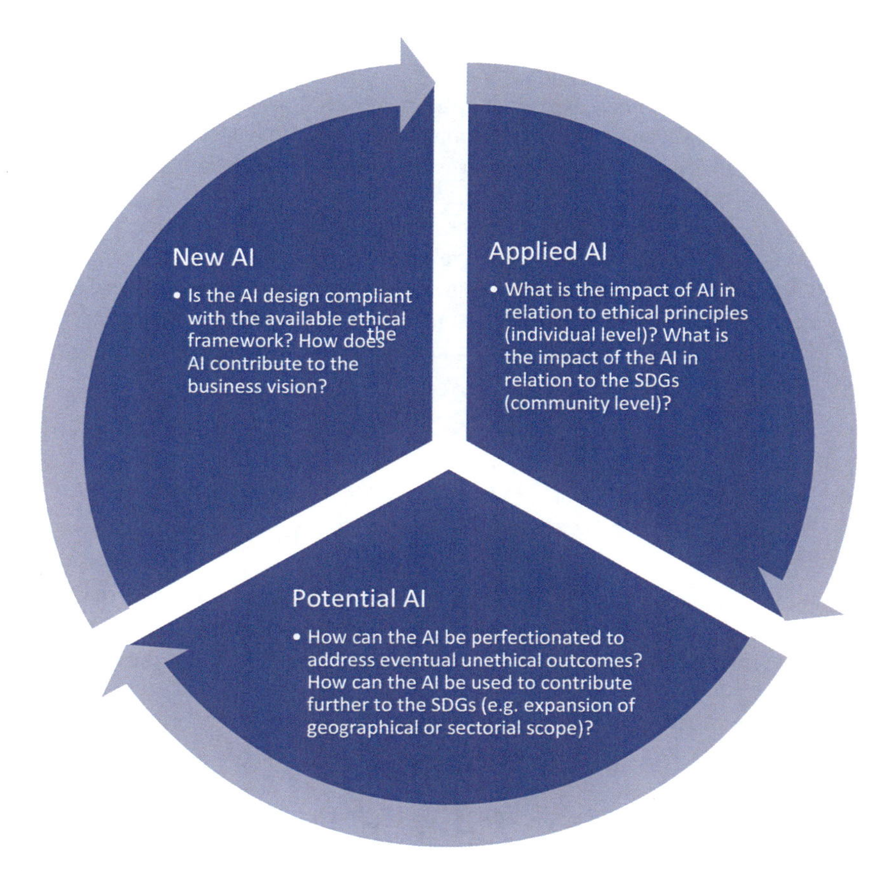

Fig. 2 The three-level mindset framework as a cycle

would be willing to expand by amplifying the scope of the project/technology, amplifying the targeted geographical area, collaborating with other projects and adding new functionalities. The implemented uptake strategies include scale the project to other groups/persons and geography, agile methodology and lean start-up, partnerships and alliances. All projects consider their structure and model to be replicable to address goal(s) elsewhere in the world and would be willing to create partnership with other AI4SDGs projects. Feedback on what could be done better include scaling the project to a higher number of participants.

When asked to identify three elements that went well and proved to be key success factors of their projects, the answers included the following: project's scalability, partnerships with NGOs that are used to work on the ground, interactions with the open data community, visibility with development banks and identification of key collaborators. When asked to identify elements that could have gone better and would have made a difference in the project, the answers included interactions with governments at different levels and interaction with organisations that meet the same SDGs but with a different approach.

We have observed challenges that appear to be common to the projects: lack of resources (financially and in terms of human resources, both specialised such as data analysist and of assistance), uneasy data access and lack of governmental schemes to support initiatives.

It is evident from the insights of the four AI4SDGs projects that the mindset framework would facilitate scaling up, expansion, collaboration and partnerships. Moreover, the cyclical process would allow to identify challenges and find methods to overcome them, as suggested in most of the AI ethical frameworks proposed by different governments (OECD, 2021) and organisations ('Understanding Artificial Intelligence Ethics and Safety', n.d.).

3.2 *The Application of the Three-Level Mindset Framework in Different Sectors and Its Limitations*

Having contextualised how the proposed mindset framework aligns with the purpose of projects addressing AI4SDGs, this section describes how it can be applied in different sectors and by businesses whose purpose does not focus on the contribution to SDGs.

Firstly, the adoption of the proposed mindset framework requires an evaluation of both the business model and the business vision/purpose that exceed the AI dimension.

The interest of companies towards SDGs is growing (Cordova & Celone, 2019). In most sectors, it is possible to link the service/product provided to the consumer to the SDGs (Urlings, 2020). The examples range from food (Djekic et al., 2021) to manufacturing (Martín et al., 2021).

In terms of business model, the adoption of the proposed mindset requires a complex measurement of negative and positive impacts (Martín et al., 2021). Metrics to evaluate ESG parameters represent available tools to perform these measurements, and they show the complex scale and magnitude of such assessment. For example, Sustainability Accounting Standards Board (SASB) has developed a complete set of 77 industry standards, and although they have been used to calculate firms' contributions to SDGs (Consolandi et al., 2020), there is no consensus yet on the parameters and on the hierarchy of such parameters (Serafeim & Yoon, 2021).

However, the idea of performing such assessment using the SDGs as a benchmark is not new, and some companies have already started publishing reports of a similar kind.[4] It is intuitive that big companies are the precursors, since such

[4]CNBC/Schneider Electric sustainability report 2021 available at https://www.se.com/ww/en/about-us/sustainability/sustainability-reports/?gclid=Cj0KCQiAnaeNBhCUARIsABEee8Wq0NBOAZFfh45al0qawnoOZ0o4uWK2l46YTM6GoAB_xfp-30v195QaAoLyEALw_wcB&

assessments are time-consuming and require resources, and consequently they are more affordable for them than for SMEs or start-ups. However, the adoption of these reports is desirable as big companies can act as role models, on the one hand, and gain the know-how and expertise on how to develop such assessment, so that such methodology can become increasingly accessible and potentially standardised. Despite the regulatory uncertainty in this area, the lack of a top-down approach allows business to develop bottom-up strategies to anticipate and invite policy action.

The proposed framework has several limitations at this level of abstraction and subsequent limitations at more granular levels of abstractions. To name a couple, it does not address questions concerning the hierarchy of ethical principles nor the complexity of identifying the constituent elements of CSR for multinational firms operating in different countries. There are multiple questions left unanswered; however, these are areas where further research is desirable.

4 A Recommendation on Public Policy: AI Regulation and Policy Harmonisation

The previous section discussed what leaders need to do to integrate CSR with AI in business; this section focuses on what leaders might need to be facilitated in this process. We argue that public policy plays a crucial role in aiding the adoption of the proposed mindset framework, as regulations and policies can create a sandbox in which sustainable business can flourish. Integrating CSR with AI aims to maximise the positive externalities of AI, inter alia by contributing to the achievement of the Sustainable Development Goals, which is in the public interest (Jackson, 2020). Therefore, policymakers should also be interested in building a framework for the purpose of maximising positive externalities of business activities investing in AI. In this section, we argue that (1) a regulatory framework for AI and (2) harmonised policy at a regional level help in achieving such purpose. The impact of AI regulation on firms' behaviour is debated (Lee et al., 2019). However, we argue that regulating AI creates legal certainty and safety boundaries and can direct investments towards sustainable and ethical AI (Smuha, 2021). Regional harmonisation favours partnership (Chipofya et al., 2009) between business of different countries and scalability of sustainable AI solutions (Moşteanu, 2020).

A best example is the European Union. In its attempt to regulate the digital, which has a quintessentially extra-territorial dimension, the EU has provided a harmonised framework concerning data protection, and it is working to regulate the use of

---

gclsrc=aw.ds#xtor=SEC-477-GOO-[Sustainable_Development_Report_BMM]-[491027166120]-S-[%2Bschneider%20%2Bsustainability%20%2Breport]&utm_source = google&utm_purpose = marketo&utm_campaign = UK_202101_SEM_GlobalSustainableDevelopmentGoals_Global_BRTextEN_2210183&utm_term = %2Bschneider%20%2Bsustainability%20%2Breport

AI. The foreseeable positive consequences of regional harmonisation and regulation in the field of AI include the scalability of AI projects at regional level between different countries, the prioritisation of EU citizens' rights and security, the definition of digital sovereignty's boundaries, the collaboration between countries to regulate the digital space, the definition of a common vision for the region, resilience towards external influences and others.[5]

4.1 The Experience of Four AI4SDGs Projects in Latin America: Regional Fragmentation of AI Policies and Regulations

In this section, we contextualise the recommendation with the experience of the four AI4SDGs projects in Latin America. When asked whether they performed an ethical impact assessment of the AI, only one project gave a positive answer. Moreover, the projects lamented the absence of governmental support in the development phase, not only financially but also in terms of direction. This shows that without regulation, there is a lack of legal certainty and of guidance in terms of AI ethics, which leaves harm prevention and risk assessments to business discretion. A desirable regulation ideally would include the two constituent elements of CSR discussed in Sect. 2, i.e. AI ethics, which is already part of most proposed AI regulations, and SDGs considerations, which could be implemented in the forms of incentives or legal requirements to ensure an higher minimum standard of environmental and social impact. Moreover, in Sect. 3.1, the insights from the four projects revealed that although three projects already expanded their scope, all the surveyed projects would be willing to further expand and/or collaborate with other AI4SDGs projects. Thus, it is worth investigating how policymakers can facilitate the "scaling up" of existing solutions to apply them more broadly in Latin America (and potentially beyond, but considerations on broader levels of harmonisation are outside of the scope of the present chapter). From a public policy perspective, the replication of a successful AI4SDGs project elsewhere reduces the costs of duplication, as well as the environmental impact (Cowls et al., 2021), and it helps in tackling pressing social challenges, such as health and education. The creation of a coherent regional framework for AI in terms of policies and regulations could be an opportunity to maximise of AI4SDGs development and, consequently, benefits.

However, AI policies and regulations are different and fragmented in the region. Telecommunications Management Group released a report in February 2020 that stressed the need for Latin American countries to develop appropriate AI

[5] See European Commission, 2021, Proposal for a REGULATION OF THE EUROPEAN PARLIAMENT AND OF THE COUNCIL LAYING DOWN HARMONISED RULES ON ARTIFICIAL INTELLIGENCE (ARTIFICIAL INTELLIGENCE ACT) AND AMENDING CERTAIN UNION LEGISLATIVE ACTS. COM/2021/206 final.

frameworks in specifically for what concerns data protection, liability and ethics and for governments to revise current legal frameworks or release new ones at an early stage of AI development, to avoid inhibition of AI innovation due to legal uncertainty and to ensure accountability and redress mechanisms available in light of the increasing involvement of AI in decision-making processes.[6]

The work of the Inter-American Development Bank "Artificial Intelligence for Social Good in Latin America and The Caribbean" released in July 2020 showed that the progress made by the governments of 12 selected countries to incentivise AI for social good differs substantially from country to country.[7] Since then, there were new national strategies[8] and other initiatives to foster collaboration[9], but regional harmonisation is still not on the table. Although Latin American businesses are adopting AI at scale (almost 80% of large Latin American businesses are using AI), the lack of regional cohesion and political stability are holding the ecosystem back.[10]

---

[6] https://www.tmgtelecom.com/wp-content/uploads/2020/03/TMG-Report-on-Overview-of-AI-Policies-and-Developments-in-Latin-America.pdf

[7] 'Artificial Intelligence for Social Good in Latin America and the Caribbean: The Regional Landscape and 12 Country Snapshots | Publications' (n.d.-a, n.d.-b)

[8] Chile and Peru made further steps, for example ('National Laboratory for Artificial Intelligence Policy Initiative', n.d.) ('Peru's National AI Strategy (first Draft) Policy Initiative', n.d.).

[9] The database of projects fAIr Lac ('Home | FAIrLAC' n.d.), supported by the Inter-American Development Bank, represents a great opportunity to facilitate cooperation between existing projects in the region, and it facilitates a dialogue between public and private entities for a responsible and ethical use of AI.

[10] ('The Global AI Agenda: Latin America', n.d.) Latin America's AI ecosystem would benefit from greater policy continuity and regional collaboration as it would incentivise the use of AI to help governments, policymakers and organisations tackling critical issues in the region, such as corruption, violence, weak institutions and challenging socioeconomic conditions, that are part of the global 2030 agenda (zero hunger, no poverty, reduced inequalities, sustainable cities and communities). A research concerning foreign AI and cloud investments in the region argues that the wave of digitalisation may exacerbate patterns of unequal and combined development in Latin America, absent targeted policies (Seoane & Facundo, 2021). Therefore, national and most of all regional AI policies are urgently needed to increase the beneficiaries of the digitalisation of Latin America. Moreover, a research shows that, absent policies targeting employment, 55% of all formally employed workers in Brazil is in jobs with high or very high risk of automation (Albuquerque et al., 2019). Thus, the role of AI in society should be used to support the planning of economic and social interventions, to anticipate transformation of the labour market. Indeed, for example, policies concerning requalification of existing workers and training of new ones would result in equal opportunities so that everyone can access the benefits brought by AI. A study by Microsoft in relation to Argentina shows that job requalification would bring a number of highly qualified jobs that would represent more than half of the jobs (56%). This would result in an increase in qualified jobs by 25% between 2020 and 2030, and in significant gains in requalification for most industries, which would translate into a general improvement in wages ('Futuro del trabajo: en los próximos diez años, Argentina podría tener un 56% de empleo calificado si maximizara la adopción de inteligencia artificial' 2019).

4.2 *Identification of a Forum for Policy Harmonisation and Limitations*

We have argued that AI regulation and regional harmonisation can foster the integration of CSR with AI in business. We have shown that slowly, AI regulation is forthcoming, with the examples of the EU and of individual countries in Latin America. However, policy harmonisation is still uncertain in most regions of the world. In this section, we argue that a first step to initiate the public debate is the identification of a forum.

Development banks can play a role in identifying potential fora and facilitating the dialogue. The Inter-American Development Bank, for example, suggested that forums devoted to AI at the international level involving Latin American countries, individually or as a bloc, should be used by Latin American governments to align their positions on AI issues in general. The forums include the following: the United Nations (the Group of Friends on Digital Technologies), Digital 9, the Pacific Alliance and the GEALC Network.[11]

Indeed, international trade organisations could facilitate such dialogue. The need to align AI policies could represent the opportunity to resume the convergence between Mercosur (Argentina, Brazil, Paraguay and Uruguay) and the Pacific Alliance (Chile, Colombia, Peru and Mexico) that was a result of the Chilean initiative named "Convergence in Diversity", launched in 2014 under the Michelle Bachelet's administration and her Foreign Affairs Ministry, Heraldo Muñoz. In 2018, the presidents of Mercosur and the Pacific Alliance established the Plan of Action, which consolidated the agenda of rapprochement with topics that included trade facilitation, regulatory cooperation and a digital agenda (Itamaraty, 2018). These topics represent a perfect overarching framework for the development of a regional AI policy.

Alternatively, interregional organisations between countries that share values can promote such policy alignment. For example, Prosur (Forum for the Progress and Development of South America) could be a relevant forum for such discussion. Prosur was created to be coordination mechanism supporting public policies, democratic values, powers' independence, the market economy, the social agenda and sustainability.[12]

Another potential forum could be created in the context of – or based on the model of – other networks created to address other digital challenges. An example is the Ibero-American Data Protection Network (RIPD, for its initials in Spanish). The RIPD is an organisation comprised of the data protection authorities of Andorra, Argentina, Chile, Colombia, Costa Rica, Mexico, Peru, Portugal, Spain and

[11]('Artificial Intelligence for Social Good in Latin America and the Caribbean: The Regional Landscape and 12 Country Snapshots | Publications' n.d.-a, n.d.-b)

[12]Texto de la declaración del Presidente Duque sobre Prosur en entrevista con 'Oye Cali' available at https://id.presidencia.gov.co/Paginas/prensa/2019/190114-Texto-de-la-declaracion-del-Presidente-Duque-sobre-Prosur-en-entrevista-con-Oye-Cali.aspx

Uruguay, with additional participation from data protection entities in other Latin American countries as well as Europe and Africa.[13] In January 2019, RIPD approved "Recommendations for the Processing of Personal Data by Artificial Intelligence".[14] The objective of the recommendations was to advise developers on how to incorporate regulatory requirements on personal data processing into their AI products. Such structure of independent authorities could be used to develop a regional AI framework.

This section identified potential fora to discuss harmonisation at regional level in Latin America. It shall be acknowledged that many factors influence political stability and cohesion in the region, and they are outside the scope of this chapter. Moreover, the focus was on Latin America for consistency reasons. Identifying a forum to harmonise AI policies and frameworks can be useful both for other regions (based on geographical proximity) and for groups of countries (based on shared values) to incentivise the integration of CSR with AI in business and maximise the positive externalities.

5 Conclusion

AI represents both a challenge and an opportunity for businesses and governments. The pace of technological progress is unprecedented: it requires concerted actions to lead the Fourth Industrial Revolution in the direction of social and environmental sustainability. This chapter aimed at providing examples of how public and private actors can and should work towards the same goal, arguing that business leaders should integrate CSR with AI by default and by design and that policymakers should incentivise such integration, inter alia with AI regulation and policy harmonisation to maximise positive externalities. Embedding ethical considerations in business processes and assessing the business impact with the SDGs as a benchmark are not easy tasks. Most businesses, irrespective of size and sector, would have to unveil uncomfortable practices. Many of them are likely to fall short on proving that positive externalities outweigh negative externalities. However, adopting the suggested three-level mindset allows to incorporate the improvement of both internal and external negative impacts as part of the business vision. If green and ethics washing is avoided, such mindset and consequent business practices can ultimately transform the contribution to the SDGs into products and add economic value to business. Moreover, public policy has both the duty and the right to play a major role in supporting such a sustainability transition, for example, with systems of incentives. Research is most needed in these areas, as it will contribute to shape future societies.

[13] Supra, n. 6, p. 9.

[14] Available at https://www.redipd.org/sites/default/files/2020-02/guide-general-recommendations-processing-personal-data-ai.pdf

References

'Artificial Intelligence for Social Good in Latin America and the Caribbean: The Regional Landscape and 12 Country Snapshots | Publications'. (n.d.-a). Accessed September 20, 2021a, from https://publications.iadb.org/publications/english/document/Artificial-Intelligence-for-Social-Good-in-Latin-America-and-the-Caribbean-The-Regional-Landscape-and-12-Country-Snapshots.pdf.

'Artificial Intelligence for Social Good in Latin America and the Caribbean: The Regional Landscape and 12 Country Snapshots | Publications'. (n.d.-b) Accessed September 20, 2021b, from https://publications.iadb.org/publications/english/document/Artificial-Intelligence-for-Social-Good-in-Latin-America-and-the-Caribbean-The-Regional-Landscape-and-12-Country-Snapshots.pdf.

'Futuro del trabajo: en los próximos diez años, Argentina podría tener un 56% de empleo calificado si maximizara la adopción de inteligencia artificial'. (2019). News Center Latinoamérica. 3 December 2019. https://news.microsoft.com/es-xl/futuro-del-trabajo-en-los-proximos-diez-anos-argentina-podria-tener-un-56-de-empleo-calificado-si-maximizara-la-adopcion-de-inteligencia-artificial/.

Albuquerque, P. H., Melo, C. A., Saavedra, P. B., de Morais, R. L., & Peng, Y. (2019). The robot from Ipanema goes working: Estimating the probability of jobs automation in Brazil. *Latin American Business Review, 20*(3), 227–248. https://doi.org/10.1080/10978526.2019.1633238

Antonescu, M. (2018). Are business leaders prepared to handle the upcoming revolution in business artificial intelligence? *Calitatea: Acces La Success, 19*(S3), 15–19.

Chen, F., & Zhou, J. (2019). AI in the public interest. In Closer to the machine: Technical, social, and legal aspects of AI. https://opus.lib.uts.edu.au/bitstream/10453/140958/2/AI%20Book%20-%20Chapter%204%20revised%20clean%20version.pdf.

Chipofya, V., Kainja, S., & Bota, S. (2009). Policy harmonisation and collaboration amongst institutions – A strategy towards sustainable development, management and utilisation of water resources: Case of Malawi. *Desalination, 248*(1), 678–683. https://doi.org/10.1016/j.desal.2008.05.119

Consolandi, C., Phadke, H., Hawley, J., & Eccles, R. G. (2020). Material ESG outcomes and SDG externalities: Evaluating the health care Sector's contribution to the SDGs. *Organization & Environment, 33*(4), 511–533. https://doi.org/10.1177/1086026619899795

Cordova, M. F., & Celone, A. (2019). SDGs and innovation in the business context literature review. *Sustainability, 11*(24), 7043. https://doi.org/10.3390/su11247043

Cowls, J., Tsamados, A., Taddeo, M., & Floridi, L. (2021). A definition, benchmark and database of AI for social good initiatives. *Nature Machine Intelligence, 3*(2), 111–115. https://doi.org/10.1038/s42256-021-00296-0

Cowls, J., et al. (n.d.). *How to design AI for social good: Seven essential factors*. SpringerLink. Access October 14, 2021, from. https://link.springer.com/article/10.1007/s11948-020-00213-5.

Dahlmann, F., Stubbs, W., Griggs, D., & Morrell, K. (2019). Corporate actors, the UN sustainable development goals and earth system governance: A research agenda. *The Anthropocene Review, 6*(1–2), 167–176. https://doi.org/10.1177/2053019619848217

Dahlsrud, A. (2008). How corporate social responsibility is defined: An analysis of 37 definitions. *Corporate Social Responsibility and Environmental Management, 15*(1), 1–13. https://doi.org/10.1002/csr.132

Djekic, I., Batlle-Bayer, L., Bala, A., Fullana-i-Palmer, P., & Jambrak, A. R. (2021). Role of the food supply chain stakeholders in achieving UN SDGs. *Sustainability, 13*(16), 9095. https://doi.org/10.3390/su13169095

Floridi, L. (2019). What the near future of artificial intelligence could be. *Philosophy & Technology, 32*(1), 1–15. https://doi.org/10.1007/s13347-019-00345-y

Floridi, L., & Cowls, J. (2021). A unified framework of five principles for AI in society. In L. Floridi (Ed.), *Ethics, governance, and policies in artificial intelligence* (Philosophical Studies Series) (pp. 5–17). Springer International Publishing. https://doi.org/10.1007/978-3-030-81907-1_2

Heukamp, F. (2020). AI and the leadership development of the future. In J. Canals & F. Heukamp (Eds.), *The future of management in an AI world: Redefining purpose and strategy in the fourth industrial revolution* (IESE business collection) (pp. 137–148). Springer International Publishing. https://doi.org/10.1007/978-3-030-20680-2_7

Jackson, E. A. (2020). *Importance of the Public Service in Achieving the UN SDGs*. MPRA Paper. 18 March 2020. https://mpra.ub.uni-muenchen.de/101806/.

Khamis, A., Li, H., Prestes, E., & Haidegger, T. (2019). AI: A key enabler of sustainable development goals, part 1 [industry activities]. *IEEE Robotics Automation Magazine, 26*(3), 95–102. https://doi.org/10.1109/MRA.2019.2928738

Koene, A., Clifton, C., Hatada, Y., Webb, H., & Richardson, R. (2019, April). *A governance framework for algorithmic accountability and transparency*. https://doi.org/10.2861/59990.

Lee, Y. S., Larsen, B., Webb, M., & Cuéllar, M.-F. (2019, November). *'How would AI regulation change firms' behavior?* Evidence from thousands of managers'. https://research.cbs.dk/en/publications/how-would-ai-regulation-change-firms-behavior-evidence-from-thous.

Mahmood, A., & Bashir, J. (2020). How does corporate social responsibility transform brand reputation into brand equity? Economic and noneconomic perspectives of CSR. *International Journal of Engineering Business Management, 12*(January), 1847979020927547. https://doi.org/10.1177/1847979020927547

Manheim, K., & Kaplan, L. (2019). 'Artificial intelligence: Risks to privacy and democracy' 106. https://yjolt.org/sites/default/files/21_yale_j.l._tech._106_0.pdf.

Martín, G., Mercedes, A. P., Álvarez, J. O.-M., Villalba-Díez, J., & Morales-Alonso, G. (2021). New business models from prescriptive maintenance strategies aligned with sustainable development goals. *Sustainability, 13*(1), 216. https://doi.org/10.3390/su13010216

Mökander, J., Axente, M., Casolari, F., & Floridi, L. (2021, November). Conformity assessments and post-market monitoring: A guide to the role of auditing in the proposed European AI regulation. *Minds and Machines, 32*(2), 241–268. https://doi.org/10.1007/s11023-021-09577-4

Moore, J. (2019). AI for not bad. *Frontiers in Big Data, 2*, 32. https://doi.org/10.3389/fdata.2019.00032

Moşteanu, Narcisa Roxana. 2020. 'Green sustainable regional development and digital era'. *Green buildings and renewable energy: Med green forum 2019 - part of world renewable energy congress and network*, by Ali Sayigh, 181–197. Innovative renewable energy : Springer International Publishing. https://doi.org/10.1007/978-3-030-30841-4_13.

National Laboratory for artificial intelligence policy initiative. (n.d.). Access October 14, 2021, from https://oecd.ai/en/dashboards/policy-initiatives/http://aipo.oecd.org/2021-data-policyInitiatives-27150.

Nylund, P. A., Brem, A., & Agarwal, N. (2021). Innovation ecosystems for meeting sustainable development goals: The evolving roles of multinational enterprises. *Journal of Cleaner Production, 281*(January), 125329. https://doi.org/10.1016/j.jclepro.2020.125329

OECD. (2021). *State of implementation of the OECD AI principles: Insights from national AI policies*. OECD. https://doi.org/10.1787/1cd40c44-en

Patelli, L. (2019). Ai Isn't neutral. *Strategic Finance, 101*(6), 11–12.

Peru's national AI strategy (1st draft) policy initiative. (n.d.). Access October 14, 2021, from https://oecd.ai/en/dashboards/policy-initiatives/http://aipo.oecd.org/2021-data-policyInitiatives-27146.

Sabater-Mir, J., Torra, V., & Aguiló, I. (2019). *Artificial Intelligence Research and Development: Proceedings of the 22nd International Conference of the Catalan Association for Artificial Intelligence*. IOS Press.

Seoane, V., & Facundo, M. (2021). Chinese and U.S. AI and cloud multinational corporations in Latin America. In T. Keskin & R. D. Kiggins (Eds.), *Towards an international political economy of artificial intelligence* (International political economy series) (pp. 85–111). Springer International Publishing. https://doi.org/10.1007/978-3-030-74420-5_5

Serafeim, G., & Yoon, A. (2021). *'Stock Price reactions to ESG news: The role of ESG ratings and disagreement'. SSRN scholarly paper ID 3765217*. Social Science Research Network. https://doi.org/10.2139/ssrn.3765217

Sheehy, B. (2015). Defining CSR: Problems and solutions. *Journal of Business Ethics, 131*(3), 625–648. https://doi.org/10.1007/s10551-014-2281-x

Sheehy, B., & Farneti, F. (n.d.). Corporate social responsibility, sustainability, sustainable development and corporate sustainability: What is the difference, and does it matter? *Sustainability*. Accessed December 2, 2021, from https://doi.org/10.3390/su13115965.

Smuha, N. A. (2021). From a "race to AI" to a "race to AI regulation": Regulatory competition for artificial intelligence. *Law, Innovation and Technology, 13*(1), 57–84. https://doi.org/10.1080/17579961.2021.1898300

Taddeo, M., & Floridi, L. (2018). How AI can be a force for good. *Science, 361*(6404), 751–752. https://www.science.org/doi/abs/10.1126/science.aat5991

The global AI agenda: Latin America. (n.d.). *MIT technology review*. Accessed 20 September 2021. https://www.technologyreview.com/2020/06/08/1002864/the-global-ai-agenda-latin-america/.

Tomašev, N., Cornebise, J., Hutter, F., Mohamed, S., Picciariello, A., Connelly, B., Belgrave, D. C. M., et al. (2020). AI for social good: Unlocking the opportunity for positive impact. *Nature Communications, 11*(1), 2468. https://doi.org/10.1038/s41467-020-15871-z

Understanding Artificial Intelligence Ethics and Safety. (n.d.). *The Alan Turing Institute*. Accessed December 2, 2021, from https://www.turing.ac.uk/research/publications/understanding-artificial-intelligence-ethics-and-safety.

Urlings, L. (2020). Benchmarking for a better world: Assessing corporate performance on the SDGs. In *Assessment of responsible innovation*. Routledge.

Vinuesa, R., Azizpour, H., Leite, I., Balaam, M., Dignum, V., Domisch, S., Felländer, A., Langhans, S. D., Tegmark, M., & Nerini, F. F. (2020). The role of artificial intelligence in achieving the sustainable development goals. *Nature Communications, 11*(1), 233. https://doi.org/10.1038/s41467-019-14108-y

Francesca Mazzi is a postdoctoral research fellow in AI and Sustainable Development at Saïd Business School, University of Oxford.

AI and Leadership: Automation and the Change of Management Tasks and Processes

Isabell Claus and Matthias Szupories

Abstract Until now, executives have mainly focused on the management of "human intelligence" (HI versus AI) in the company. However, executives increasingly need to shift towards automating processes and routines using AI. A step-by-step approach includes the identification of challenges, effectiveness and efficiency potentials in the company and the search for solutions, the introduction of AI solutions with a sustainable impact on the daily work of the employees and the flow of business processes and the establishment and further development of the employees and the organisation with the progressive use of AI within the organisation. These extensive and often far-reaching changes request for new management skills and knowhow, both professionally and personally. A case study in the area of continuous environment analysis for companies illustrates a concrete application that shows that AI and human intelligence complement each other very well and open up new possibilities with regard to the effectiveness and efficiency of leadership, which ultimately should always be geared towards maintaining and creating competitive advantages and future viability.

1 The Combination of Artificial and Human Intelligence

Management, both in theory and in practice, has so far essentially focused on managing "human intelligence" (HI versus AI). Internal interdependencies, environmental relationships and actions are coordinated and proactively influenced. The goal is the same across all sectors: business success, whether it is to maximise sales, margins or market share.

I. Claus (✉)
thinkers GmbH, Vienna, Austria
e-mail: isabell.claus@thinkers.ai

M. Szupories
DEUTZ AG, Cologne, Germany
e-mail: matthias.szupories@deutz.com

R. Schmidpeter, R. Altenburger (eds.), *Responsible Artificial Intelligence*, CSR, Sustainability, Ethics & Governance, https://doi.org/10.1007/978-3-031-09245-9_14

The challenge for human intelligence is to be able to abstract all relevant processes in a company and of its environmental relationships with stakeholders (including customers, investors, shareholders, suppliers and the public) as well as their interdependencies and degrees of effectiveness. Added to this is the difficulty that all companies compete to achieve efficiency and effectiveness advantages through strategic and operational moves. Thus, it is a permanent change process in an ever-changing environment and with changing stakeholders. This also means that the speed on the one hand and the flood of data from internal and external communication channels, management systems and personal conversations on the other hand overtax a human being and his or her absorption capacity. This is where AI can come in to support executives. It should offer building blocks that are easy to use and reduce complexity to a level that is manageable for humans.

This is how AI is used for this purpose: The focus is on the ability to evaluate and process mountains of data in a very short time. The focus is on an intelligent reduction of mass. The strengths of human intelligence are then built on the result: It should recommend and execute actions—always based on current information and analytics. This includes standardised, recurring or ad hoc actions. In the operative business, the "right" lever should be selected to achieve the company's goals.

In this framework, AI does not replace human intelligence, but rather helps it to better follow the environment, which is currently already too complex for the brain. AI helps to follow the rapid and comprehensive development and to act and react more quickly.

On the other hand, human intelligence provides the basis for AI: AI is not "intelligent" by itself. If it is wrongly selected or trained, an automatism will not evaluate and present much usable material. Instead, processes (analytics, recommendations for action and routines) still need to be defined by human intelligence who will, for example, take the necessary background for the business field into account. The necessary data quality must be generated and the ongoing operation and effectiveness checked and ensured.

In short, AI can better search, process and make large data sets "manageable". However, the decision-making process in strategic and operational management stays with and is enforced and, if necessary, adapted by executives. AI prepares a much better analysis and recommendation basis for human intelligence compared to the current situation in which executives often lack data-driven decisions due to the missing ability and intelligent tools for data processing and analytics. At the end of the day, competitive advantages in a market, industry or environment are achieved by the tandem of human and machines.

2 Leadership with AI: Why There Is No Alternative

Against the background of the aforementioned, executives will need to increasingly focus on automating processes and routines by means of AI in the coming years. Consequently, current management practice which (almost) exclusively targets human intelligence must be looked at more closely and decided in which areas AI can support and positively influence. The translation into systems is then the second step.

Time is not an insignificant factor in this context. As mentioned, the goal of management is usually to achieve (future) competitiveness and competitive advantages in order to grow and expand. In this competitive "race", the speed of the market competitors versus one's own speed is decisive. The aim should therefore be to achieve a high level of effectiveness and efficiency through automation. In this context, the skills of the executives and the entire team—not just the AI used—will determine success.

The data basis, the depth and breadth of it, is better the more it is tailored to the individual organisation and its executives. At the moment of a decision, it should provide the most ideal information base possible and moreover elicit more confidence in decisions, such as large, strategic budget decisions. Achieving this new, much more comprehensive market transparency—nationally, internationally or globally—is not achievable without AI, because human resources are not able to process this amount of data. The use of AI thus offers essential effectiveness components for corporate management on the one hand, but also completely new types of efficiency components that can represent either a saving of resources or a time advantage in the market.

In addition to the factors of being able to move faster, more encompassing and more situation-specific, AI improves problem-solving competence and better stores learnings from previous experiences. Automatisms in prosperous or recessionary times and markets repeat themselves, so AI can help to reduce error frequencies or, in other words, warn sufficiently not to always repeat mistakes. Through more transparency in existing data and information, crises and early warning indicators as well as market opportunities should be recognised better and earlier. Here, the field of AI is developing very quickly and is already providing solutions in practice.

In addition to automating recurring, reliable or foreseeable processes, good corporate management also needs a successive further development of the scope for action taken into account in automation and newly emerging possibilities for action. Currently, systems mostly only fulfil highly standardised tasks. However, there are numerous initiatives that attempt to incorporate creativity into the self-learning process. Because advanced companies and providers agree: In the long run, simple booking rhythms and production routines will not provide the decisive competitive advantage. Combining AI and human intelligence means to combine data, experience and routines with inspiration and risk-taking and will constitute another important stage in the development of AI.

The conclusion, then, is that the tasks and ways of working of corporate management are on a clearly recognisable path towards greater incorporation of automation possibilities. This offers new management advantages but also the competitive situation may even force executives to act in this field. Therefore, there is no way around dealing with AI and finding the right individual path within the framework of the new possibilities.

3 The Optimum and Pace of Development

What still prevents us from automating complex "brain tasks" today is our "ignorance" of the way the brain works as a whole in order to "imitate" this way of working through technology. Moreover, perfect information transparency is an ideal of economics that we will probably never fully achieve. Risk-free decisions are thus a long way off or unattainable. The core of human intelligence thus also contains the possibility of failing or being wrong. Nevertheless, we can evaluate data better, faster and more specifically and show options for action digitally. How much is then invested in decision-making and implementation, however, will remain—for the foreseeable future—in the hands of executives and thus human intelligence.

If we imagine the ideal state in which we know how our brain works as a whole, we probably would not need so long to teach machines to execute similar processes. De facto, however, this is exactly what remains hidden from us. In countless studies and investigations, individual characteristics, abilities and processes are being researched with the help of various approaches and methods and in several disciplines. In this way, we come to new insights piece by piece. At the same time, however, there is no holistic overview of all the knowledge that humanity has already generated about the brain and how these "pieces of the knowledge puzzle" might fit together. AI and, in particular, one of its special abilities take us a step further in this problem: the processing and evaluation of mass data using mathematical and statistical methods. If mass data processing is linked, for example, in the field of brain research, with human intelligence, the greater is the likelihood of transparency in what happens in our heads and what actually constitutes intelligence in order to reproduce it artificially. This is where the potential of human and artificial intelligence working together becomes apparent: massive progress.

Status quo, however: We have to live with the knowledge we have, develop it further and turn it into success factors in the business environment in the best possible way. The urge to automate even tasks with intelligence requirements varies among executives. The factors that influence the will to progress with automation include:

- Competitive pressure
- Pressure for efficiency
- Access to knowledge about solutions and their potentials and modes of operation
- The resources available for purchasing and experimenting with relatively new technologies
- Fears with regard to change
- The time pressure in decisions
- The pace of relevant events

Last but not least, the "tone from the top" and the "enablement" of the workforce are important.

If these factors are prevailing, the ideal situation described at the beginning of Chapter "Artificial Intelligence: Companion to a New Human "Measure"? A Brief Outlook" is not achievable, but a significant improvement takes place.

4 Leadership Encompasses Implementation Strength

If an executive is dealing with the use of AI, this work and its progress can be divided into three basic steps, whereby these processes are constantly repeated.

1. The recognition of (current and future) challenges, effectiveness and efficiency potentials in companies and the search for solutions for their systematisation, digitalisation and automation
2. The introduction of AI solutions with sustainable impact on the work design of employees and the flow of business processes
3. Further development of the employees and the organisation with the use of AI within the company

In summary, the last step is the broad institutionalisation and strategic transformation of working methods and processes in the organisation as a whole, taking into account the pace of implementation and the values of the organisation and its stakeholders.

How can these steps now be positively influenced by AI?

4.1 Recognising AI Potential and Finding Solutions

The use of AI does not differ from the use of human intelligence in terms of its objective: Ultimately, it is about competitiveness and advantages. Analyses should be carried out and options for action identified so that the best option for a team, a department and ultimately the entire company is selected, implemented and pursued, taking into account various scenarios.

The development of intelligence thus goes back to the basic genes of business management theory since the 1990s, in that the awareness and endeavour matured that companies are not successful by chance, but can actively influence their economic success and their dependence on their environment, i.e. the ecological, social and economic framework conditions.

AI should therefore achieve partial goals and later an overall goal that is as comparable to the results achieved by means of human intelligence. However, this should be done with stringent consideration of significantly greater combination possibilities, a greater breadth and depth of data, almost infinite repetition routines and, ideally, the coverage of essential business processes in (almost) real time.

However, AI cannot do this without human intelligence as a necessary basis: AI learns from human intelligence—usually continuously, not once. This means that human intelligence must be able to describe business processes and systems clearly and in the best possible way and select promising applications. Only when business processes and systematics are optimised, standardised and systematised can they be resiliently designed with AI.

The basis for an executive is therefore not only a deep strategic and operational knowledge of the systematics of his or her own business field but also a technical understanding or, alternatively, the ability to abstract or imagine, in order to be able to make use of technology or—alternatively—to make use of the knowledge and skills of technology experts, to be able to describe tasks and goals sufficiently and to outline and systematise the business field accordingly with the goal of successfully implementing processes or analyses in intelligent systems. Only then systematisations can be translated into AI-based systems, tools, programmes and routines that offer advantages over human intelligence. In other words, AI can become the accelerator to put the core of strategic thinking and action into practice, whether in a local or global context.

The global context in particular gives AI an additional boost. For example, various programmes, languages, systems and content are developed in Asia, America and Europe. Matching and allocating the abundance of data and identifying trends is too much for an individual in terms of quantity, information search and speed of solution. Here, AI can create enormous time and cost advantages that cannot be substituted in the battle for efficiency and effectiveness.

However, as will become apparent, AI is not always a ready-made software that is downloaded, installed and immediately usable. Even if convenience is a central requirement, today it usually still requires continuous engagement with AI applications and potentials. In this respect, the principle of what is required of an executive is not much different from what is required of a team's development: developing potentials and orchestrating and perfecting the interaction of people and technology.

As in other areas, the multitude of AI solutions available on the market offers an excellent selection and comparison possibilities for the technology part. On the other hand, it also demands ongoing attention and constant learning from an executive in order to make the right choice for a system and its building blocks. Not significantly different from the continuous development of one's own team, the focus on AI needs and its further development potentials is required for full development and the greatest possible return on investment (ROI) for an organisation.

4.2 *Success Factors for the Implementation of AI Systems*

Once the selection is made, the next step is implementation. Apart from the technical implementation, a comprehensive AI introduction means the potential—and also the goal—to bring about noticeable organisational changes. While this will be forward-looking for the organisation as a whole, it can mean major upheavals—not always in a positive sense—for individual employees or entire teams. Involvement, communication, training and active change management steered by the executive are necessary to make an AI introduction a success. Acceptance by employees is an important factor for this success: Just as dissonance can arise between employees that impedes the progress of the team, rejection or fear of AI systems can also reduce success or, in the worst case, jeopardise it completely.

It is not only positive feedback from employees that needs to be taken into account. The executive's own attitude towards AI integration is also important. It often determines the speed of achieving results and the support of the employees: Authentic interaction and the will to promote progress by an executive allow the necessary resources to successfully establish AI applications. In addition, there will always be a negative resonance in dealing with AI, because where information collection, analysis and decision-making processes are optimised and automated, less manual work is required. AI is therefore often accompanied by an automation of business processes, which, similar to automation in production, has the potential to reduce staff. For the executive, this means an area of tension: on the one hand, planning costs for automation which save personnel costs and on the other hand, taking away the fear of AI from the employees involved and showing the potential for success as a field of growth.

Thus, the project planning, introduction and implementation of AI is a very human challenge and requires emphatic, strong and fear-free curiosity and leadership of the employees in an exciting, fast and knowledgeable world as well as change management processes that could affect either the work as such or individuals.

4.3 *Institutionalising and Holisting Implementation*

AI will by no means replace the use of human intelligence in the foreseeable future. Rather, it is necessary to regularly question the distribution of work between human intelligence and AI, to reappraise it and to find new systematics and combinations in order to generate competitive advantages. This "endurance run" is a demanding requirement for executives.

Understanding an AI application as purely a software tool often undermines their potential. Even if the intelligence aspect has so far consisted of mathematical formulas and methods, parts for which we use, human intelligence today, especially repetitive and at the same time unstandardised tasks, can be automated—a novelty compared to traditional software application. This also means that either human intelligence builds on the results of AI, parts of it are replaced or entirely new possibilities for human intelligence arise through AI.

Executives are in the lead to bring business and work processes to a new level instead of adapting them marginally or not at all. If only the data width, depth and time factor of knowledge acquisition are optimised, but the following decision-making and implementation processes are not touched, AI exploits little of its actual potential or achieves only little development of the organisation. Therefore, the thirst for knowledge and the will to implement are essential to positively influence and develop a modern organisation.

5 Case Study: AI for Continuous Monitoring of a Company's Business Environment

In management, intelligent dashboards are increasingly replacing manual reporting and selective human analyses and evaluations. The advantage is, among other things, that data on turnover, market share, market size, product and customer mix and yield can be mapped with all detailed master trees, aggregated and scaled in real time and at any time and these can additionally be provided with predictions for future developments. Gone are the days when PDF or Excel formats show sales without details of underlying customers, products or services. Only analytics of various levels and perspectives of large data sets enable promising interpretations and actions.

In addition to the conventional "internal reporting view", external environmental factors are also decisive for business success. No company operates in a vacuum, and no organisation operates in a closed space but is confronted with a multitude of globally operating competitors, substitutes from new market entrants or changing customer behaviour. Lack of foresight, lack of early warning signs or a lack of timely responses to changing conditions can be devastating.

It is crucial to adapt to, anticipate or even positively stimulate changing customer requirements and stakeholder environments. This is an ideal field for AI to detect and connect relevant data in internal systems, industry-wide knowledge bases to social networks and any website worldwide.

The term "Connect", for example, has for some years already made clear how business areas can be developed: Whether "CarConnect", which indicates "Advanced Service", warns of tyre bursting at an early stage or even maps SOS routines in the event of an accident. Or the refrigerator, which controls the shopping list via missing milk and eggs, which would otherwise have cost time, effort and leisure or simply would have caused the user to go without if forgotten. And, finally, the washing machine, which can be loaded in the morning and activated remotely when the user arrives. These fields are examples of routines whose automation makes daily life easier and more pleasant. These examples can be supplemented by numerous routines in companies—including research activities, pick-up services or repetitive communication measures.

One can see from these examples not only how numerous tasks with a repetitive character and time-consuming nature are changed but also how data can quickly lead to new services that enhance products and companies and achieve success.

In addition to a company's products and services, data about economic and social development (beside other data) is also very relevant for companies. There are countless barometers and indices. But which of these are relevant for anticipating economic developments? How do I create a general recognition or an early warning system of these factors on a global basis? Also, new topics or behavioural patterns arise which need to be taken into account.

How can such complex, numerous and diverse topics be identified and managed? Human resources are not feasible on this scale. The conventional approaches, for example, those of on- or offline target group surveys, are costly and yet lead to moderately satisfactory results. The multitude of information provided by credit agencies, regulation or reporting obligations creates a very selective and isolated transparency, which only when sufficiently bundled and repeated in large numbers brings transparency and insight into consumer behaviour, user structures and their ongoing changes. Even if all ethical questions of data protection and other regulations have to be taken into account here, this spectrum can provide more far-reaching and, above all, much more resilient databases than questionnaires and surveys can depict today.

For companies and their executives, the search for and extraction of knowledge and data becomes increasingly complex the broader the portfolio, customer base or geographic spread. AI, on the other hand, is designed to process and incorporate large masses of data, and so it is irrelevant to them whether a local, regional or international business environment is to be analysed. The abundance of information and the speed of change are more of an advantage than a disadvantage, because they lead to systems learning even more and ever more precisely from a large amount of data.

Whereas today individual surveys still serve as the basis for statistical trends or manual web searches seem expedient for gaining knowledge, in many cases, AI can evaluate and aggregate decisive and often even publicly accessible information in a more goal-oriented and transparent way. Announcements, press releases or other contributions and publications by competitors or (potential) customer companies are examples of this. AI leads, among other things, to executives being informed much faster and better by a substitute of a customer or a competitor's offer than would be transmitted by normal hearsay. Today, on the other hand, in the vast majority of cases, neither executives nor employees take note of much of the actually relevant information and decide on investments, budgets and strategies without considering important data and thus knowledge. The probability of success of these investments and plans is questionable, as it contains a considerable proportion of random and uncertain components. AI can therefore not only be faster here but also significantly increase the focus in order to prevent or avoid wrong decisions.

In this use case, the capabilities of AI make a significant value contribution that is not achievable through human labour. AI goes through massive amounts of data; checks it for relevance to the business area, a team or a company; separates relevant from irrelevant data; and processes and therewith prepares relevant knowledge for executives and their teams. The result is a new level of transparency with regard to all relevant events and processes in the company's environment and thus a noticeably better data basis for decisions and ongoing company management. It also enables significant time savings with regard to repetitive tasks such as searching for information. Studies have shown that office workers spend an average of 2 hours a day searching for information (The Social Economy—Unlocking Value and Productivity Through Social Technologies, p. 47, McKinsey Global Institute, 2012). This means that an employee spends an entire business day per week on searching. However, this immense time investment does not in itself create value.

Only relevant information that is subsequently analysed by human intelligence and becomes a basis for action creates value for an organisation.

AI takes over the repetitive search and deliver search results that are very well pre-qualified in terms of content. Employees then use their time to evaluate the information and act accordingly based on the data. This is the approach that thinkers. ai is taking for companies and public institutions with regard to continuous environment analyses based on data from the web. It enables a new dimension of data-based corporate management. The time-consuming work of searching is reduced to a minimum, employees work with the search results and analyses, and companies achieve competitive advantages by being able to decide, invest and plan more quickly and data-driven in all business-relevant topics. AI solves a cross-industry problem: The massive flood of publicly accessible data and information on the web, which grows every second, is efficiently collected, filtered and processed in the best possible way to proceed with human intelligence.

Consequently, executives will increasingly distant from the day-to-day management of repetitive, not directly value-creating tasks of a team. Tasks that require analytical strength, competence and curiosity are increasing. In addition, cognitive abilities for a comprehensive understanding of corporate contexts, systems and environmental factors are in focus. These must be present and strengthened not only in the executives themselves but also in all employees who work with a system like thinkers.ai. In this way, such a system fulfils its purpose of being an intelligent source of knowledge and a basis for sustainably successful and efficiency-increasing management. It is a cornerstone of the ever-advancing interaction of people, business and IT in the evolving knowledge society and a companion on the way to re-designing products and services ("Connect-X") as a reaction to recognised trends.

Against this background, a basic understanding of what AI can generally do or an ability to abstract from it is very helpful. However, systems like the one presented here are not IT systems for specialists: They are designed for users and can be used accordingly without expert IT understanding or training. The requirement for an executive is therefore explicitly not to hire and build up only more IT specialists in the future. Rather, the focus is on understanding of a company's business field and environment—the ability that AI needs in order to achieve results that can be converted into competitive advantages.

6 Conclusion

AI opens up new opportunities for executives, but at the same time also demands new abilities from them. Factors from the internal and external environment of the company determine how quickly an organisation must or should change in order to remain fit for the future or to expand competitive advantages. A well-considered and well-structured approach taken by the executive with the goal of a sustainable effect of automation and also dealing with the social factors with regard to the team or the workforce is recommended.

Dr Isabell Claus is Managing Director and co-founder of thinkers.ai, a provider of AI-based search engine technology for businesses. As a serial entrepreneur, she was previously in the co-founders management team of a cybersecurity company that was among the top 100 fastest-growing companies in the EMEA region for five consecutive years. In 2020, she was awarded "Entrepreneur of the Year" in Austria. She studied Economics (Business and Law) and International Business Administration at the Vienna University of Economics and Business Administration, earned her doctorate there and completed several stays abroad at Harvard University, London School of Economics, Singapore Management University and Dubai University College.

Dr Matthias Szupories has worked for many years in the automotive industry for brands such as Peugeot, Sixt, MAN and Volkswagen Financial Services and most recently at DEUTZ in specialist and management positions. Currently, as Head of Central Sales and Marketing and Regional Manager of the second largest sales region, he was responsible for increasing customer satisfaction, developing new business models and thereby increasing turnover and EBIT. He also holds several supervisory and board positions. As an advocate of dual education, he completed his diploma part-time at the FHDW in Bergisch Gladbach and later earned his doctorate in industrial economics at the Pan-European University in Vienna and Bratislava via advanced courses. Since 2013, he has worked as a lecturer at the FHDW and Prof. Stefan Bratzel's Automotive Center and as a guest lecturer at other universities.

Achieving CSR with Artificially Intelligent Nudging

Dirk Nicolas Wagner

Abstract No longer limited to the factory hall, automation and digitization increasingly change, complement, and replace the human workplace also in the sphere of knowledge work. Technology offers the possibility of creating economically rational, autonomously acting software—the machina economica. This complements human beings who are far from being a rational homo economicus and whose behavior is biased and prone to errors. This includes behaviors that lack responsibility and sustainability. Insights from behavioral economics suggest that in the modern workplace, humans who team up with a variety of digital assistants can improve their decision-making to achieve more corporate social responsibility. Equipped with artificial intelligence (AI), machina economica can nudge human behavior to arrive at more desirable outcomes. Following the idea of augmented human-centered management (AHCM), this chapter outlines underlying mechanisms, opportunities, and threats of AI-based digital nudging.

1 Introduction

It is widely accepted that digitalization and artificial intelligence request new proposals from management science (Chui et al., 2018; Jarrahi, 2018; Davenport & Ronanki, 2018). Scholars and practitioners are challenged to derive suitable approaches to human resource management and development in a world with increasing automation and digitalization (Stone et al., 2018). Ever since the sudden stop of the insightful Hawthorne experiments, management science just like managerial practice has struggled to balance the drive for automation and replacement of relatively expensive human labor with an adequate development of human resources. A typical reaction is to do no more but to pursue more and better computer skills (Crick, 2017; Peyton Jones, 2011).

D. N. Wagner (✉)
Karlshochschule International University, Karlsruhe, Germany
e-mail: dwagner@karlshochschule.org

R. Schmidpeter, R. Altenburger (eds.), *Responsible Artificial Intelligence*, CSR, Sustainability, Ethics & Governance, https://doi.org/10.1007/978-3-031-09245-9_15

With the rise of cognitive computing and artificial intelligence, a paradigm shift is overdue: a shift from the era of rationalization and automation to a new era of collaboration between man and machine (Malone, 2018; Wilson & Daugherty, 2018). Proposals have been made to first and foremost develop human skills in the tradition of human factors science (Salvendy, 2012; Kanki et al., 2010) and in line with the requirements of so-called human-agent collectives (Jennings et al., 2014) which, due to technological progress, currently emerge across most if not all industries. This has been called augmented human-centered management (Wagner, 2021). As predecessors of such an approach in technology-oriented high-risk industries like aviation have shown, a process of augmentation requires a long-range approach to training and development in order to achieve desired results (Kanki et al., 2010). This involves time and investment which in many industries may not easily be available.

This chapter suggests that technology can help to achieve augmented human-centered management by nudging (Thaler & Sunstein, 2008) employees and managers to make better decisions. Evidently, such an endowment would be particularly relevant for corporate social responsibility (CSR). An example of digital nudges that enhance CSR is when Microsoft Outlook asks the sender of a corporate email late at night if she wants to consider waiting and automatically sending the email next morning during the usual working time of the recipients.

The chapter is structured as follows: First, the socioeconomic context of human-agent collectives and the dilemma of the contemporary homo economicus are introduced. Then, the concept of augmented human-centered management (AHCM) is outlined before the possibilities of digital nudging as a means to achieve AHCM are explored.

2 The Emergence of Human-Agent Collectives

Ongoing and increasingly powerful digitalization means that machines are in the process of becoming actors in their own right. They do not only compete more often with human labor, but, increasingly, they also influence human action and as such sometimes enhance options and on other occasions limit options available to humans (Carr, 2014). Or, in other words, they become part of the decision-making architecture within which human individuals find themselves. This is well illustrated by the developments in game of chess: Following the defeat of the human grandmaster in chess by the supercomputer Deep Blue in 1996, Garry Kasparov was among the initiators of a re-definition of the game of chess who let humans cooperate with chess computers to compete with other man-machine teams in so-called freestyle chess or advanced chess tournaments. Here, the human player is still the ultimate decision-maker, but the computer offers options and proposals and prompts more reflective decisions. It was found that strong human players or supercomputers were not competitive against relatively weak human players using standard chess computers when these organized their team effectively by implementing superior processes (Kasparov, 2008; Cowen, 2014).

The game of chess is only one of the many domains where people team up with artificial agents to achieve goals. More generally, Varian (2014) observes that today computers are in the middle of virtually every transaction and traces this back to dramatic cost decreases in computers and communication. Jennings et al. (2014) identify socio-technical systems in which humans and smart software (agents) interact as human-agent collectives (HAC). Just like a computer that is already in the middle of a transaction today, HAC now emerging in many industries are likely to step by step and with increasing influence shape the work and social environment for humans. Examples from different industries are the crew on the flightdeck of a contemporary airliner that is assisted by software that relies on tens of thousands of sensors distributed across the plane (Yedavalli & Belapurkar, 2011), the farmer who is guided by precision agriculture technology (Kitouni et al., 2018), the product manager who uses conversational commerce approaches and who deploys software agents to interact with customers, the psychotherapist who works with embodied conversational agents to provide internet-based cognitive behavior therapy in preventative mental health care (Suganuma et al., 2018), or smart logistics management software that directs human labor in warehouses (Mahroof, 2019). Technological change considerably influences the working environment for humans. In HAC, the roles of humans and agents co-evolve.

According to Jennings et al. (2014), the era of issuing instructions to passive machines is over, and humans start to work in tandem with highly interconnected artificially intelligent agents that act autonomously. These environments are considered to be open and characterized by flexible social interactions. Here, "sometimes the humans take the lead, sometimes the computer does, and this relationship can vary dynamically" (Jennings et al., 2014, p. 80). The notions "flexible autonomy" and "agile teaming" (Jennings et al., 2014, p. 82) describe a short-lived nature of teams with a varying degree of human involvement and with authority relations that are not considered to be fixed but context-dependent. The pro-active involvement of machines in information gathering and filtering, analytical, and decision-making processes raises questions of social accountability and responsibility. Since software often operates "behind the scenes" (Jennings et al., 2014, p. 85), its rationale and actions are regularly not readily available to the involved humans.

The open nature of HAC means that "control and information is widely dispersed among a large number of potentially self-interested people and agents with different aims and objectives. [. . .]. The real-world context means uncertainty, ambiguity, and bias are endemic and so the agents need to handle information of varying quality, trustworthiness, and provenance" (Jennings et al., 2014, p. 82).

3 Homo Economicus and Machina Economica

Across most if not all industries, managers persistently pursue productivity increases for their organizations (Drucker, 2001; Malik, 2010). This drive for efficiency has led to a focus on automation (Frey & Osborne, 2017). For decades, managerial practices have been informed by management theory taught at business schools to undergraduate as well as graduate students and on executive courses and programs. Management theory in turn substantially draws from economic theories like neo-classical microeconomics, the theory of the firm or transaction cost economics (e.g., Douma & Schreuder, 2013). Over the decades, this has led to a situation where the economic conception of the self-interested human being as boundedly rational utility maximizer is today widely reflected in the way how companies are structured and organized. Managers perceive their organizations to be populated by the so-called homo economicus. And stakeholders of these organizations perceive the managers to even more closely correspond with this model of man. These developments have been reviewed and criticized as far as management education and development is concerned (Ghoshal, 2005). And, for many years, high-profile issues like the cases of Enron, Tyco, BP's Deepwater Horizon, the Bangladesh factory disaster, the VW Dieselgate scandal, and even global climate change as a whole stand for problematic practical implications of a drive for productivity based on contemporary managerial and economic concepts.

It does not come as a surprise that steps toward automation in business follow in the footsteps of the impetus economics has on strategy and organization. Where technologically and economically feasible, automation becomes a preferred choice on the route to higher levels of productivity. And a closer look quickly reveals why: "machina economica" promises to be the better "actor economicus." Algorithms can be programmed to follow the calculus of rational choice and utility maximization under constraints, and this has for a long time been well received by computer science (Huberman, 1988; Wagner, 2001). This substantial incentive for automation has led to the assumption that it is desirable to remove humans from the value chain whenever possible. However, such strategies repeatedly failed. While humans are regularly considered to be weakest link in the chain, they are still tasked with crucial roles and often expected to monitor automation and to intervene in case of problems. For such phenomena, Bainbridge (1983) coined the term "ironies of automation." A review by Baxter et al. (2012) showed that despite technological progress, the ironies of automation continue to persist. Ultimately, the dilemma of increasing responsibility in increasingly complex environments while being sidelined by automation does not sit well with how human actors are endowed by nature (Gazzaley & Rosen, 2017).

As an interim conclusion, it can be noted that machines may be the better "actor economicus" but the irony is that stubborn automation may push humans even more into an unnatural "homo economicus" role. In this sense, the managerial approaches taken at the retail and technology company Amazon.com Inc. appear to be symptomatic: "If you're a good Amazonian, you become an Amabot" (Kantor & Streitfeld, 2015).

4 A Different Way of Thinking Complements

Instead of narrowly following the paradigm of automation, it can be enlightening to consider other forms of man-machine cooperation. And it is the field of behavioral economics where a way out of the just portrayed dilemma can be found. Here, the point of departure for the understanding of human behavior is so-called dual-process theories which refer to two different modes of thinking (Evans & Stanovich, 2013). Unlike the homo economicus model of man, humans are assumed to rely on an implicit, intuitive, and automatic mode of thinking (system 1) as well as on an explicit, controlled, and reflective mode of thinking (system 2).

Being confronted with 20,000 or more decisions every day (Pöppel, 2008), humans have been shown to primarily rely on system 1 thinking. Relying on gut feeling, mental shortcuts, and heuristics that have proven to work in the past is faster, requires less mental resources, and is just much more practical in everyday life (Gigerenzer, 2007), which explains why this is the principal way of human thinking and decision-making. However, this also means that humans behave "predictably irrational" (Ariely & Jones, 2008). The predisposition to reduce effort and deploying the unconscious automatic system 1 makes humans susceptible to cognitive biases. Prominent examples of biases are excessive optimism which can lead to unrealistic goals, confirmation bias which results in ignorance, status quo bias which creates resistance to change, or loss aversion which attaches too high values to possessions. Research continuously identifies new biases that in the meantime have accumulated to a substantial catalogue (Centre for Evidence-Based Medicine, 2021). All of those can be a source of irrational behavior and sometimes error. In summary, biases mean that humans find it difficult to assess options, they show limited self-control and distorted time preferences, they act emotionally, but they also express social preferences like reciprocity, fairness, and trust.

But system 1 thinking can be and is counter-balanced with the reflective mode of system 2 thinking. Every human being is able to let both systems cooperate. System 2 thinking builds on rational analysis. It is logical and deliberate which in turn requires time and energy (Kahneman, 2011). And this is exactly where "machina economica" and the concept of human-agent collectives come back in. For where humans struggle, the computer is in its element. It acts rationally and with calculus, comes along with constantly growing computing power, and tirelessly evaluates exponentially growing volumes of big data. With the help of machine learning, artificial intelligence can observe human behavior to then report back individually and precisely to the individual decision-maker. This means that instead of solely relying on the natural capacity of system 2 thinking, humans can benefit from external system 1 type processing of information by artificial agents. There is room for augmentation.

5 Augmented Human-Centered Management

Augmented human-centered management (AHCM) is a conceptual framework which identifies and describes the competencies that humans need to develop to successfully participate in highly automated and digitalized business environments (Wagner, 2021). This serves to close currently perceived gaps in how human resource management can support digitalization strategies (Fenech et al., 2019; Parry & Strohmeier, 2014) while keeping a human-centered approach (Bissola & Imperatori, 2019).

AHCM is rooted in humanism as defined by Erich Fromm as "a system centered on Man, his integrity, his development, his dignity, his liberty. [It is based] on the principle that Man is not a means to reach this or that end but that he is himself the bearer of his own end. It not just based on his capacity for individual action, but also on his capacity for participation in history, and on the fact that each man bears within himself humanity as a whole" (Fromm et al., 1961, p. 147, cit. in Aktouf & Holford, 2009, p. 108). Such a view can be perceived to be compatible with the integration thesis around which stakeholder theory has developed and which claims that it makes no sense to talk about business without talking about ethics and it makes no sense to talk about ethics without talking about business. This results in the conclusion that "it makes no sense to talk about either business or ethics without talking about human beings" (Freeman, 2010, p. 7).

To get there, a broader perception of fundamental human qualities is useful. As already portrayed, the capacity for combining system 1 and system 2 thinking means that humans are inductively rational which means that we mix inductive pattern recognition with deductive logic (Beinhocker, 2007). A relevant aspect of this is the capacity for tacit knowing (Polanyi, 1967) which puts humans in the position to know more than they can tell. But there are other relevant human traits. Sennett (2008) explores human craftsmanship which springs from synergistically combined activities of body and mind. Arendt (1960) stresses the human ability to make new beginnings which can be seen as a nucleus for innovation. And Frank (1988), in the tradition of Adam Smith and others, explores the capacity for morality and passion.

If, in this sense, "man is the measure of all things" (Aktouf & Holford, 2009, p. 101), clarification with regard to man's relation to technology is required. An approach that can be brought in line with a humanistic perspective on management comes from Davenport and Kirby (2016). They propose a paradigm shift away from automation (human replacement) and toward augmentation (human enhancement) and recommend to humans five strategies when working alongside machines:

1. Stepping Up: To work a level above machines and make decisions about augmentation
2. Stepping Aside: To leave the current job to the machine and pursue a job that machines are not good at
3. Stepping In: To monitor and improve a computer's automated decisions

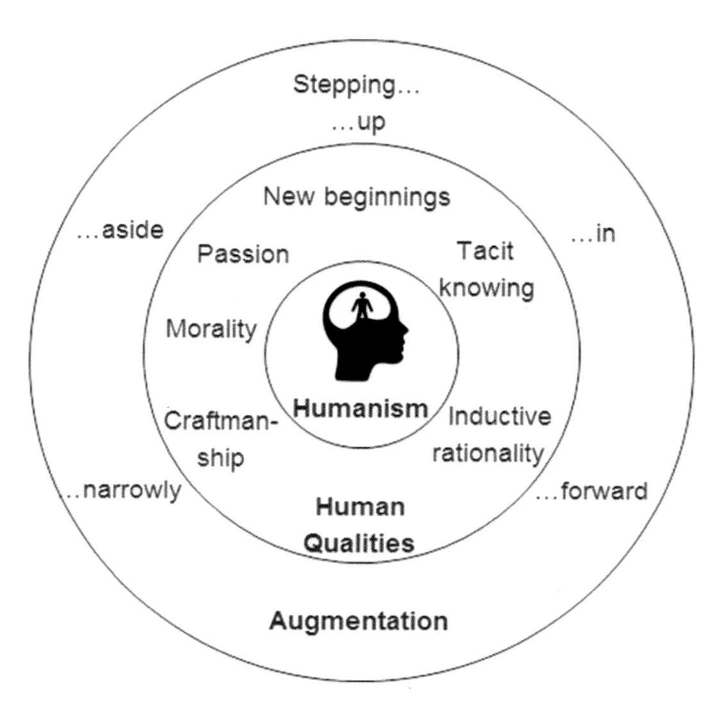

Fig. 1 Nucleus of augmented human-centered management. Source: Wagner (2021)

4. Stepping Narrowly: To find special area in one's profession that would not be economical to automate
5. Stepping Forward: To create future technology

The human qualities of inductive rationality, tacit knowing, craftsmanship, the ability to make new beginnings, as well as morality and passion represent enablers for these strategies as well as selection criteria for potential individual choices. Figure 1 summarizes the nucleus for AHCM as described above. AHCM aims to augment human competencies in line with the fundamental ideas of humanism as well as in accordance with the integration thesis of stakeholder theory in order to become more effective (doing the right things) and more efficient (doing things right) in the workplace. A specific form of application of stepping up (1), in (2), and forward (5) is to devise and work with artificially intelligent decision-making architectures that de-bias human decisions and help to make them better. As described below, this is achieved by allowing system 2-driven computers to nudge humans.

6 Augmentation with Digital Nudging

Knowledge about typically human decision-making can warn about potentially wrong decisions, especially in situations where there is too much information, the data available does not convey enough meaning, there is a need to act vast, or there is a need to consider past experience. Thus, knowledge about biases can be used to shape the context within which people make decisions and the information available to them. This may be done with the intention to influence behavior. Richard Thaler and Cass Sunstein were the ones who turned the insights from psychology and behavioral economics into a profession which ever since has been called "nudging" (Thaler & Sunstein, 2008).

A nudge is "an aspect of the choice architecture that alters people's behavior in a predictable way without forbidding any options or significantly changing their economic incentives" (Thaler & Sunstein, 2008, p. 6). Thaler and Sunstein initially focused on the public and political domain. Nudges from the domains of health, where opt-in versus opt-out rules show an effect on organ donation, or the domain of personal wealth, where automatic enrolment into pension plans can reduce old-age poverty, became popular examples. But not only governments quickly adopted the idea of designing decision-making architectures with the intention to influence behavior without direct intervention into people's choices. Organizations, aiming to nudge at the workplace, were quick to follow.

And the trend was accelerated by the digital transformation of the economy and society. This soon enabled digital nudging which Weinmann et al. (2016) define as "the use of user-interface design elements to guide people's behavior in digital choice environments." Today organizations of all sectors but in particular technology firms and corporations deploy digital nudging to pursue external goals in marketing and sales and to internally shape their operations management. Prominent examples include Google's autofill function, recommender systems by Amazon or Netflix, but also Uber or Deliveroo nudging their drivers and riders to work longer hours or to speed up their performance (Scheiber, 2017; O'Byrne, 2019).

It is already evident that digitalization has set up nudging for exponential dissemination. Recent and current technological advances are bound to make artificially intelligent nudging an omnipresent phenomenon in human-agent collectives. What used to be analogue, static, and the same for all of the audience can now be digital, dynamic, interactive, and micro-targeted at specific individuals. To achieve this, digital data is transformed into useable intelligence that is incorporated into human work to augment human capabilities (Demirkan et al., 2015; Mele et al., 2018).

Hansen and Jespersen (2013) developed a useful framework to categorize nudges. They propose to assess the transparency of a nudge and to distinguish which mode of thinking a nudge does engage. As shown in Fig. 2, this leads to four categories of nudges that manipulate choices, manipulate behaviors, prompt reflective choices, and influence behavior. Following this framework, Caraban et al. (2019) reviewed research papers that presented digital nudges. The majority of the reviewed examples were found to be in the category prompting reflective choice (52%), while nudges

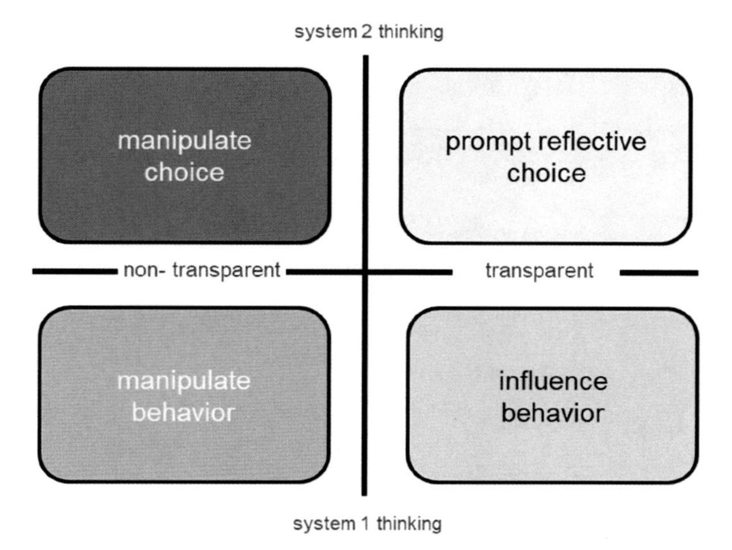

Fig. 2 Categorization of (digital) nudges. Source: own representation, adapted from Hansen and Jespersen (2013) and Caraban et al. (2019)

influencing behavior came in second (26%), which meant that nudges where people can transparently perceive the intentions of the decision-making architecture dominated the field. As will be explained in further detail below, these types of nudges are of specific interest when aiming to push for corporate social responsibility.

7 Nudges for CSR

Thaler and Sunstein also defined a nudge as "any factor that alters behaviors of humans, even though it would be ignored by homo economicus" (Thaler & Sunstein, 2008, p. 9). In turn, this means that nudging can help to rationalize human decisions and behavior. In this sense, the technique can be used to enhance corporate social responsibility by instigating corresponding behaviors among corporate citizens. While the approach is still in its infancy and certainly far from being systematically applied across industries, sufficient evidence is already available to review possible routes for implementation. The following selection of examples is primarily dedicated to serve as a source of inspiration for CSR practitioners:

CSR field of activity:	*Sustainability*
Example:	**Saving paper**
Description:	Intervention by the IT administrator to set "double-sided print" as the default option on office printers, which lead to a 15% reduction in paper consumption (Egebark & Ekstrom, 2016)

(continued)

Type of nudge (category):	Influence behavior
Type of nudge (specific):	Setting a default
Bias involved:	Status quo bias (leverage)
CSR field of activity:	*Sustainability*
Example:	**Saving jet fuel**
Description:	Flight captains being confronted with (a) feedback report on fuel consumption, (b) fuel-saving targets, and (c) charity donations upon target achievement. All three nudges led to reduced fuel consumption (Fetherston et al., 2017)
Type of nudge (category):	Prompt reflective choice
Type of nudge (specific):	Disclosure
Bias involved:	Spotlight effect (leverage)
CSR field of activity:	*Sustainability*
Example:	**Eco driving**
Description:	Drivers receive cockpit information or vibration alarms on steering wheel when deviating from fuel-efficient driving or showing aggressive driving behaviors (Ibragimova et al., 2015; Lee et al., 2011)
Type of nudge (category):	Influence behavior
Type of nudge (specific):	Reminder
Bias involved:	Spotlight effect (leverage)
CSR field of activity:	*Sustainability*
Example:	**Save energy in buildings**
Description:	A thermostat using sensors and machine learning to understand occupancy habits and properties of buildings to make people aware how to save energy (Mele et al., 2021)
Type of nudge (category):	Influence behavior
Type of nudge (specific):	Reminder
Bias involved:	Spotlight effect (leverage)
CSR field of activity:	*Sustainability*
Example:	**Energy monitoring**
Description:	The energy monitoring system called the "Never Hungry Caterpillar" uses an emotionally engaging animation by confronting the user with the digital representation of an animal that shows signs of suffering when the user deviates from ideal behaviors (Laschke et al., 2011)
Type of nudge (category):	Influencing behavior

(continued)

Type of nudge (specific):	Reinforcement by evoking empathy
Bias involved:	Affect heuristic (leverage)
CSR field of activity:	*Risk management/cyber security*
Example:	**Secure wireless networks**
Description:	Traveling and mobile employees can choose from wireless networks ranked and color-coded by level of security. This led to a significant increase in the rate of secure network selection (Turland et al., 2015)
Type of nudge (category):	Prompt reflective choice
Type of nudge (specific):	Positioning
Bias involved:	Status quo bias (leverage)
CSR field of activity:	*Risk management/cyber security*
Example:	**Secure passwords**
Description:	Deploying nudges to prompt people to increase the strength of passwords, including different images like a pair of watching eyes that activate social norms (Renaud & Zimmermann, 2019)
Type of nudge (category):	Prompt reflective choice
Type of nudge (specific):	Reminder
Bias involved:	Risk compensation
CSR field of activity:	*Employee wellbeing*
Example:	**Healthy snack choices**
Description:	Hiding unhealthy snacks and comparing healthy snacks to inferior alternatives on a snack ordering website (Lee et al., 2011; Thaler & Sunstein, 2008)
Type of nudge (category):	Manipulate behavior
Type of nudge (specific):	Framing by adding inferior alternative
Bias involved:	Decoy effect (leverage)
CSR field of activity:	*Employee wellbeing*
Example:	**Delaying emails**
Description:	Microsoft equipped its office software with digital nudges including a nudge to reduce after-office hours' impact on coworkers by asking users to delay late messages until the next morning (Raveendhran & Fast, 2021)
Type of nudge (category):	Prompt reflective choice
Type of nudge (specific):	Reminder
Bias involved:	Present bias
CSR field of activity:	*Employee wellbeing*

(continued)

Example:	**Ensuring good posture habits**
Description:	Smart wearable devices that encourages good postures by measuring the angle of the neck and alerting the user in case of poor posture angles (Mele et al., 2021)
Type of nudge (category):	Prompt reflective choice
Type of nudge (specific):	Warning
Bias involved:	Status quo bias
CSR field of activity:	*Employee wellbeing and stakeholder management*
Example:	**Avoiding unwanted social media disclosures**
Description:	Software plugin that reminds users of potential consequences of an intended social media post by making explicit that "....can see this" (Wang et al., 2014)
Type of nudge (category):	Prompt reflective choice
Type of nudge (specific):	Warning
Bias involved:	Availability heuristic (leverage)

At this stage, the nudges briefly portrayed above are isolated examples. For the specific context and decision-making situation, each case shows that human behavior can be augmented in order to support motives and goals typically as part of CSR goals of companies and organizations. The examples also illustrate the interplay between decentralized human behaviors across the corporation and centralized decisions on nudges made by so-called choice architects (Thaler & Sunstein, 2008, p. 3) and implemented with the help of digital technology in general and more and more often in the form of artificially intelligent agents in particular.

The strategic task for CSR executives and practitioners will be to derive and evolve a strategy for artificially intelligent nudging that systematically links decentralized human behaviors, centrally devised and digitally distributed as well as scaled nudges, and resulting changes in behaviors that contribute in measurable ways to the achievement of CSR standards and goals of the corporation. To get there and as surveyed above, detailed knowledge from behavioral economics and psychology on behavioral biases is available that can be used to design specific nudges. This is complemented by first recommendations and heuristics that serve to derive integrated nudging strategies. In this respect, for example, Fetherston et al. (2017) recommend incremental approaches that rely on empirical testing rather than on logic alone. They also caution that "good intentions don't automatically lead to positive behaviors" (ibid, p. 4). Caraban et al. (2020) propose "The Nudge Deck" as a support tool for technology-mediated nudging. Artificially intelligent, data-driven digital nudging can pair resulting choice architectures with social as well as personal data and thus allow for micro-targeting approaches to be implemented, techniques already well established in the field of marketing (André et al., 2018). Möhlmann (2021) recommends watching out for win-win situations between company and employee [and society], for example, by implementing personalized reward systems and by ensuring that the logic of algorithms is transparent and explained. This implies for nudging strategies to be positioned on the right-hand side of Fig. 2.

8 Conclusion

This chapter provided the reader with the relevant context as well as with the necessary building blocks to derive a strategy for artificially intelligent nudging that supports CSR. On this basis, a suitable framework for digital nudging to augment the behavior of corporate citizens can be derived and customized to the needs of the organization. Upon reflection of the content of this chapter, the reader is likely to conclude that the techniques of artificially intelligent nudging are not only available to serve the purposes of CSR but that they may be equally effective when being deployed to maximize profits in less responsible ways which includes but is not limited to exerting pressure or privacy violations (Möhlmann, 2021). Özdemir (2020) calls this "dark patterns."

While the very idea of nudging is not to limit the choices people have, it remains a serious intervention into their decision-making with the intention to influence behavior. Therefore, it is important to note that strategic corporate decisions about the goals of artificially intelligent nudging are fundamental steps that decide whether this approach will ultimately support or undermine CSR. The starting point for any strategy of any artificially intelligent nudging strategy will therefore be the degree of responsibility of the choice architects involved.

References

Aktouf, O., & Holford, D. (2009). The implications of humanism for business studies. In H. Spitzeck, M. Pirson, W. Amann, S. Khan, et al. (Eds.), *Humanism in business* (pp. 101–122). Cambridge University Press.

André, Q., Carmon, Z., Wertenbroch, K., Crum, A., Frank, D., Goldstein, W., et al. (2018). Consumer choice and autonomy in the age of artificial intelligence and big data. *Customer Needs and Solutions, 5*(1), 28–37.

Arendt, H. (1960). *The human condition*. Univ. of Chicago Press.

Ariely, D., & Jones, S. (2008). *Predictably irrational*. Harper Audio.

Bainbridge, L. (1983). Ironies of automation. In. *Automatica, 19*, 775–780.

Baxter, G., Rooksby, J., Wang, Y., & Khajeh-Hosseini, A. (2012). The ironies of automation... still going strong at 30? In *Proceedings of ECCE 2012 Conference*. Edinburgh, 29th-31st August.

Beinhocker, E. (2007). *The origin of wealth. Evolution, complexity, and the radical remaking of economics*. Harvard Business School Press.

Bissola, R., & Imperatori, B. (Eds.). (2019). *HRM 4.0 for human-centered organizations*. Emerald Publishing (Advanced series in management).

Caraban, A., Karapanos, E., Gonçalves, D., & Campos, P. (2019). 23 ways to nudge: A review of technology-mediated nudging in human-computer interaction. In *Proceedings of the 2019 CHI Conference on Human Factors in Computing Systems*, pp. 1–15.

Caraban, A., Konstantinou, L., & Karapanos, E. (2020). The nudge deck: A design support tool for technology-mediated nudging. In *Proceedings of the 2020 ACM Designing Interactive Systems Conference* (pp. 395–406). ACM.
Carr, N. (2014). *The glass cage. How our computers are changing us*. W. W. Norton & Company.
Centre for Evidence-Based Medicine - CEBM (2021). *Catalogue of bias*. Retrieved August 26, 2022, from https://catalogofbias.org/
Chui, M., Manyika, J., Miremadi, M., Henke, N., Chung, R., Nel, P. & Malhotra, S. (2018). Notes from the AI frontier. *Insights from hundreds of use cases*. Discussion Paper. McKinsey Global Institute.
Cowen, T. (2014). *Average is over - powering America beyond the age of the great stagnation*. Penguin Putnam.
Crick, T. (2017). Computing in education. In *An overview of research in the field*. The Royal Society.
Davenport, T., & Kirby, J. (2016). *Only humans need apply. Winners and losers in the age of smart machines*. Harper Business.
Davenport, T., & Ronanki, R. (2018). *Artificial intelligence for the real world*. Harvard Business.
Demirkan, H., Bess, C., Spohrer, J., Rayes, A., Allen, D., & Moghaddam, Y. (2015). Innovations with smart service systems: Analytics, big data, cognitive assistance, and the internet of everything. *Communications of the Association for Information Systems, 37*(1), 35.
Douma, S., & Schreuder, H. (2013). *Economic approaches to organizations* (5th ed.). Pearson.
Drucker, P. (2001). *The essential Drucker. The best of sixty years of Peter Drucker's essential writings on management*. HarperCollins.
Egebark, J., & Ekström, M. (2016). Can indifference make the world greener? *Journal of Environmental Economics and Management, 76*, 1–13.
Evans, J., & Stanovich, K. (2013). Dual-process theories of higher cognition: Advancing the debate. *Perspectives on Psychological Science, 8*(3), 223–241.
Fenech, R., Baguant, P., & Ivanov, D. (2019). The changing role of human resource management in an era of digital transformation. *International Journal of Entrepreneurship, 22*(2), 1–10.
Fetherston, J., Bailey, A., Mingardon, S., & Tankersley, J. (2017). *The persuasive power of the digital nudge. BCG perspectives*. The New Way of Working Series.
Frank, R. (1988). *Passions within reason. The strategic role of the emotions*. W. W. Norton & Company.
Freeman, E. (2010). *Stakeholder theory. The state of the art*. Cambridge Univ. Press.
Frey, C., & Osborne, M. (2017). The future of employment: How susceptible are jobs to computerisation? *Technological Forecasting and Social Change, 114*, 254–280.
Fromm, E., Marx, K., & Bottomore, T. (1961). *Marx's concept of man*. Frederick Ungar Publishing Co.
Gazzaley, A., & Rosen, L. (2017). *Distracted mind. Ancient brains in a high-tech world*. MIT Press.
Ghoshal, S. (2005). Bad management theories are destroying good management practices. *Academy of Management Learning & Education, 4*(1), 75–91.
Gigerenzer, G. (2007). *Gut feelings: The intelligence of the unconscious*. Penguin.
Hansen, P., & Jespersen, A. (2013). Nudge and the manipulation of choice: A framework for the responsible use of the nudge approach to behaviour change in public policy. *European Journal of Risk Regulation, 4*(1), 3–28.
Huberman, B. (Ed.). (1988). *The ecology of computation*. Elsevier.
Ibragimova, E., Mueller, N., Vermeeren, A., & Vink, P. (2015, April). The smart steering wheel cover: Motivating safe and efficient driving. In *Proceedings of the 33rd Annual ACM Conference Extended Abstracts on Human Factors in Computing Systems* (pp. 169–169). ACM.
Jarrahi, M. (2018). Artificial intelligence and the future of work: Human-AI symbiosis in organizational decision making. *Business Horizons, 61*(4), 577–586.
Jennings, N., Moreau, L., Nicholson, D., Ramchurn, S., Roberts, S., Rodden, T., & Rogers, A. (2014). Human-agent collectives. *Communications of the ACM, 57*(12), 80–88.
Kahneman, D. (2011). *Thinking, fast and slow*. Farrar, Straus and Giroux.

Kanki, B., Helmreich, R., & Anca, J. (Eds.). (2010). *Crew resource management*. Academic Press/ Elsevier.
Kantor, J., & Streitfeld, D. (2015). Inside Amazon: Wrestling big ideas in a bruising workplace. In *The New York Times*, 8/15/2015.
Kasparov, G. (2008). *How life imitates chess*. Arrow Books.
Kitouni, I., Benmerzoug, D., & Lezzar, F. (2018). Smart agricultural Enterprise system based on integration of internet of things and agent technology. *Journal of Organizational and End User Computing, 30*(4), 64–82.
Laschke, M., Hassenzahl, M., & Diefenbach, S. (2011). Things with attitude: Transformational products. In *Create11 Conference*, pp. 1–2.
Lee, M. K., Kiesler, S., & Forlizzi, J. (2011). Mining behavioral economics to design persuasive technology for healthy choices. In *Proceedings of the Sigchi Conference on Human Factors in Computing Systems*, pp. 325–334.
Lee, S. S., Lim, Y. K., & Lee, K. P. (2011). A long-term study of user experience towards interaction designs that support behavior change. In *CHI'11 Extended Abstracts on Human Factors in Computing Systems* (pp. 2065–2070). ACM.
Mahroof, K. (2019). A human-centric perspective exploring the readiness towards smart warehousing. The case of a large retail distribution warehouse. *International Journal of Information Management, 45*, 176–190.
Malik, F. (2010). *Management. The essence of the craft*. Campus.
Malone, T. (2018). *Superminds. The surprising power of people and computers thinking together*. Oneworld Publications.
Mele, C., Spena, T., & Peschiera, S. (2018). Value creation and cognitive technologies: Opportunities and challenges. *Journal of Creating Value, 4*(2), 182–195.
Mele, C., Spena, T. R., Kaartemo, V., & Marzullo, M. L. (2021). Smart nudging: How cognitive technologies enable choice architectures for value co-creation. *Journal of Business Research, 129*, 949–960. https://doi.org/10.1016/j.jbusres.2020.09.004
Möhlmann, M. (2021). Algorithmic nudges don't have to be unethical. In *Harvard Business Review*, April 2020.
O'Byrne, E. (2019). Slave to the algorithm? What it's really like to be a Deliveroo rider. *Irish Examiner,* 20.02.2019.
Özdemir, Ş. (2020). Digital nudges and dark patterns: The angels and the archfiends of digital communication. *Digital Scholarship in the Humanities, 35*(2), 417–428.
Parry, E., & Strohmeier, S. (2014). HRM in the digital age – Digital changes and challenges of the HR profession. *Employee Relations, 36*(4), 345–365.
Peyton Jones, S. (2011). *Computing at school*. BCS - The Chartered Institute for IT.
Polanyi, M. (1967). *The tacit dimension*. The University of Chicago Press.
Pöppel, E. (2008). *Zum Entscheiden geboren. Hirnforschung für Manager*. Hanser.
Raveendhran, R., & Fast, N. J. (2021). Humans judge, algorithms nudge: The psychology of behavior tracking acceptance. *Organizational Behavior and Human Decision Processes, 164*, 11–26.
Renaud, K., & Zimmermann, V. (2019). Nudging folks towards stronger password choices: Providing certainty is the key. *Behavioural Public Policy, 3*(2), 228–258.
Salvendy, G. (2012). *Handbook of human factors and ergonomics*. Wiley.
Scheiber, N. (2017). How Uber uses psychological tricks to push its drivers' buttons. *New York Times, 02*(04), 2017.
Sennett, R. (2008). *The craftsman*. Yale University Press.
Stone, C., Neely, A., & Lengnick-Hall, M. (2018). Human resource management in the digital age: Big data, HR analytics and artificial intelligence. In M. Pedro & M. Carolina (Eds.), *Management and technological challenges in the digital age* (pp. 13–42). CRC Press, Boca Raton (Manufacturing design and technology series).
Suganuma, S., Sakamoto, D., & Shimoyama, H. (2018). An embodied conversational agent for unguided internet-based cognitive behavior therapy in preventative mental health. Feasibility.

Thaler, R. H., & Sunstein, C. (2008). *Nudge: Improving decisions about health, wealth, and happiness*. Yale University Press.

Turland, J., Coventry, L., Jeske, D., Briggs, P., & van Moorsel, A. (2015). Nudging towards security: Developing an application for wireless network selection for android phones. In *Proceedings of the 2015 British HCI conference*, pp. 193–201.

Varian, H. (2014). Beyond Big Data. *Business Economics, 49*(1), 27–31.

Wagner, D. (2001). *Software-agents and Liberal order: An inquiry along the borderline between economics and computer science*. Universal Publishers.

Wagner, D. (2021). Augmented Human-Centered Management. Personalentwicklung für hochautomatisierte Geschäftsfelder. In R. Altenburger & R. Schmidpeter (Eds.), *CSR und künstliche Intelligenz*. Springer Gabler.

Wang, Y., Leon, P., Acquisti, A., Cranor, L., Forget, A., & Sadeh, N. (2014). A field trial of privacy nudges for facebook. In *Proceedings of the SIGCHI conference on human factors in computing systems*, pp. 2367–2376.

Weinmann, M., Schneider, C., & Vom Brocke, J. (2016). Digital nudging. *Business & Information Systems Engineering, 58*(6), 433–436.

Wilson, J., & Daugherty, P. (2018). Collaborative intelligence: Humans and AI are joining forces. *Harvard Business Review, 96*(4), 114–123.

Yedavalli, R., & Belapurkar, R. (2011). Application of wireless sensor networks to aircraft control and health management systems. *International Journal of Control Theory and Applications, 9*(1), 28–33.

Dirk Nicolas Wagner is an entrepreneur and Affiliate Professor of Strategic Management at Karlshochschule International University (GER). He studied economics and management at Université de Fribourg (CH) and Royal Holloway University of London (UK), graduating as a MBA in International Management and Dr.rer.pol. in the area of New Institutional Economics. Since the 1990s, he has been dealing with questions related to man and machine governance, and he regularly publishes with the Zukunftsinstitut in Frankfurt am Main. He is a Managing Partner of medium-sized business in the technical services industry in Europe.

Printed in the United States
by Baker & Taylor Publisher Services